Transportation Analysis

Managing Editors

Lucio Bianco, Rome

Agostino La Bella, Rome

Titles in this Series

L. Bianco · A. R. Odoni (Eds.)
Large Scale Computation and Information Processing
in Air Traffic Control

Nathan H. Gartner · Gennaro Improta (Eds.)

Urban Traffic Networks

Dynamic Flow Modeling and Control

With 98 Figures

Springer-Verlag

Berlin Heidelberg New York
London Paris Tokyo
Hong Kong Barcelona
Budapest

Professor Nathan H. Gartner
University of Massachusetts
Department of Civil Engineering
Lowell, MA 01854, USA

Professor Gennaro Improta
Università di Napoli "Federico II"
Via Claudio, 21
I-80125 Naples, Italy

ISBN-13:978-3-642-79643-2 e-ISBN-13:978-3-642-79641-8
DOI: 10.1007/978-3-642-79641-8

Softcover reprint of the hardcover 1st edition 1995

42/2202-543210 – Printed on acid-free paper

FOREWORD

The problems of urban traffic in the industrially developed countries have been at the top of the priority list for a long time. While making a critical contribution to the economic well being of those countries, transportation systems in general and highway traffic in particular, also have detrimental effects which are evident in excessive congestion, high rates of accidents and severe pollution problems.

Scientists from different disciplines have played an important role in the development and refinement of the tools needed for the planning, analysis, and control of urban traffic networks. In the past several years, there were particularly rapid advances in two areas that affect urban traffic:

1. Modeling of traffic flows in urban networks and the prediction of the resulting equilibrium conditions;

2. Technology for communication with the driver and the ability to guide him, by providing him with useful, relevant and updated information, to his desired destination.

These areas require quite distinct capabilities. The first area requires analytical capabilities such as those possessed by transportation engineers, operations researchers, applied mathematicians, or computer scientists: whereas the second area requires expertise in communication technologies, control systems, human factors, and vehicle technologies. There is a need for a close interaction and cooperation among these groups to produce the best possible research results.

To initiate and facilitate such an interaction, The First Capri Seminar was organized in June 21-23, 1989, The seminar was titled **Urban Traffic Networks: Dynamic Control and Flow Equilibrium**, and was jointly sponsered by the U.S. National Science Foundation and the Italian Consiglio Nazionale delle Ricerche, Participants from other countries were also invited. The aim of that seminar was to foster improved methodologies for traffic control and traffic flow modeling in urban areas through a fusion of recent technological developments and computational capabilities. A total of 26 papers were presented at that seminar and were included in a volume of preprints [Reference 1]. Revised and edited versions of 13 of those papers were published in two special issues of the international journal **Transportation Research** [References 2,3].

In view of the success of the first seminar, a second seminar was held in Capri, July 5-8. 1992. The seminar was titled the **Second International Seminar on Urban Traffic Networks**, in the hope and anticipation that it will turn into a regular forum for the presentation of the most recent research results and the exchange of ideas in this flourishing field. A total of 39 papers were accepted for presentation in the second seminar [Reference 4] covering a broad range of topics:

1 Modeling and control of dynamic traffic flow phenomena.

2 Interactions of ATMS (Advanced Traffic Management Systems) ADIS (Advanced Driver Information Systems) and network performance in the development of IVHS (Intellingent Vehicle Highway Systems).

3 Information requirements and communication needs for implementation of advanced strategies.

4 Computational techniques for large-scale network analysis.

5 New technologies for management and control of urban traffic networks.

This volume contains a selection of 14 of those papers that were reviewed, revised and edited for publication. The papers can be devided into two broad categories: (1) Information and Control Systems - Traffic Management Strategies, and (2) Dynamic Network equilibrium Models.

The first three papers in the first group are by Mahmassani and Peeta, by Bell et al. and by Kaysi et al. They all address problems of providing real-time information to guide suitably equipped motorists to their destinations. Route guidance systems are at the center of all advanced highway technology developments and these papers offer significant contributions to the understanding, modeling and design of such systems. The next paper, by Ben-Akiva at al, presents an on-line traffic prediction model for an inter-urban motorway network. The two papers by Cantarella and Sforza and by Al-Malik and Gartner develop models and techniques for the combination of signal control strategies with optimal route choices of drivers. The last paper in this group is by Bruno and Improta and deals with traffic control and assignment models taking into consideration the effects of emission constraints.

The second set of papers deals with dynamic traffic assignment models and the calculation of network equilibria under such conditions. The papers illustrate the fact that this is a very dynamic research area, which has not quite yet gelled into an established set of methodogies and techniques. Each paper addresses similar problems in a very distinct manner and provides a very specific view of the problem. Obviously, there is scope for further research until the most successful models will have established their validity and usefulness.

We hope that this volume will serve as a further contribution to the advancement of the science and pratice of urban traffic management and control.

Nathan H. Gartner **Gennaro Improta**

Acknowledgment

We wish to thank **Giuseppe Bruno** for his contribution to organizing and conducting the Second International Capri Seminar on Urban Traffic Networks.

We also gratefully acknowledge **Giuseppe Bruno** and **Marco Toro** for their assistance in the editorial work.

REFERENCES

(1) N.H. Gartner and G. Improta (Editors). Proceedings of the Italy-USA Joint Seminar or Urban Traffic Networks: Dynamic Control and Flow Equilibrium. June 1989.

(2) N.H. Gartner and G. Improta (Editors). Special Issue on Urban Traffic Networks: Dynamic Control and Flow Equilibrium. Transportation Research, Vol. 24B, No. 6, December 1990.

(3) N.H. Gartner and G. Improta (Editors). Special Issue on Dynamic Control of Urban Traffic Networks (Part II). Transportation Research; vol. 25A; No. 5, September 1991

(4) N.H. Gartner and G. Improta (Editors). Compendium of the Second International Seminar on Urban Traffic Networks. July 1992.

CONTENTS

Dynamic Network Equilibrium

Information and Control Systems

Traffic Management Strategies

SYSTEM OPTIMAL DYNAMIC ASSIGNMENT FOR ELECTRONIC ROUTE GUIDANCE IN A CONGESTED TRAFFIC NETWORK

Hani S. Mahmassani

Hayden Head Professor of Civil Engineering, The University of Texas at Austin, Department of Civil Engineering, Austin, Texas 78712, U.S.A.

Srinivas Peeta

Graduate Research Assistant, The University of Texas at Austin Department of Civil Engineering, Austin, Texas 78712, U.S.A.

This paper addresses the problem faced by a central controller seeking to optimize overall network performance through the provision of real-time routing information to suitably equipped motorists. Conceptual and mathematical formulations are presented for various scenarios that arise based on the amount of information available to the controller. Principal elements of a dynamic assignment formulation for electronic route guidance systems are discussed, and the associated difficulties for solution methodologies are illustrated. The ideal case of known time-dependent origin-destination flows over the whole planning horizon is formulated as a dynamic system-optimal assignment problem. Extensions and variants of the basic formulation are discussed for incomplete information availability to the central controller.

A solution approach is proposed for the ideal case of complete a priori information availability, in which the traffic flow in the network is explicitly modeled using a detailed traffic simulation model (DYNASMART). Traffic simulation circumvents all the major limitations of existing dynamic assignment formulations, particularly violations to the First-In, First-Out principle, and the "holding back" of traffic. In addition, the use of a simulation-assignment methodology enables correct dynamic flow modeling and convenient evaluation of an otherwise intractable objective function while accounting for time-dependent link-path incidence relationships.

1. Introduction

1.1 Problem definition

The system optimal (SO) dynamic traffic assignment problem arises in the context of Advanced Traveler Information Systems (ATIS), which provide tripmakers with real-time information about existing traffic conditions and/or instructions on route selection from their current location to their destinations. It is faced by a central controller with complete and/or partial, a priori and/or real-time information on time-dependent O-D trips and network characteristics, solving for paths on which to assign users in order to minimize the total travel cost in the system, subject to individual routing considerations.

Different formulations of the problem are obtained under various assumptions on information availability to the controller and the tripmakers. In this paper, we identify the principal elements that determine such formulations and the issues faced in their solution.

Several formulations are then given, corresponding to different controller information availability scenarios. A heuristic algorithm is described for the SO problem under complete controller information on the users' time-dependent O-D trip desires. The solution procedure employs a detailed traffic simulation model within the search framework, thereby circumventing some of the principal difficulties encountered in mathematical formulations of dynamic assignment problems. The implementation of several key elements of the procedure are also described.

1.2 Background review

Dynamic network assignment is under active development, for both the user equilibrium and SO problems. Existing formulations are not entirely satisfactory in terms of underlying assumptions and cannot be solved for realistic networks.

The bulk of contributions to the SO dynamic assignment problem have addressed the situation where known time-dependent flows are assigned from multiple origins to a *single* destination through the links of a network so as to minimize total system cost.

The first mathematical programming approach to this problem is due to Merchant and Nemhauser (1978). Their model was formulated as a discrete-time, non-linear, non-convex mathematical program and the corresponding algorithm solved a piecewise linear version of it. Congestion was treated explicitly using conventional link performance functions.

Carey (1986) reformulated the Merchant-Nemhauser problem as a well-behaved convex nonlinear program, with mathematical and algorithmic advantages over the original formulation. Extensions to multiple destinations or multiple commodities remain problematic because of non-convexity issues. Multiple destinations require the models to explicitly satisfy a "first-in, first-out" requirement that is essential from a traffic viewpoint. This requirement introduces additional constraints that complicate the formulation, and destroy many of its nice properties, (Carey, 1992).

A more recent line of work has considered constrained optimal control theory. The O-D trip rates are assumed to be known continuous functions of time, and the link flows are sought as continuous functions of time. Friesz et al. (1989) discuss optimal control formulations for both system optimal and user equilibrium problems. They propose a dynamic generalization of Beckmann's equivalent optimization problem for static user optimized traffic assignment in the form of an optimal control problem. Ran and Shimazaki (1989) used the optimal control approach to develop a general model of dynamic system optimal traffic assignment for an urban transportation network with many origins and many destinations. Ran and Boyce (1993) formulated a continuous dynamic user optimal traffic assignment model in which exit flows are treated as a set of control variables rather than as functions, so as to overcome difficulties posed by the non-linearity of the exit flow function for multiple origin-destination networks. Wie (1990) extended the model by Friesz et al. (1989) to include elastic time-varying travel demand, which leads to the implicit consideration of departure time choices. Wie also enumerates several limitations of this approach.

Boyce et al. (1991) used the optimal control theory approach to obtain a convex model for dynamic user equilibrium assignment by defining inflows and exit flows on links to be control variables. They discussed a methodology to solve the discretized version of the problem using the Frank-Wolfe algorithm and an expanded time-space network representation. However, the use of static link performance functions is a limitation of this model as such functions do not adequately model congested traffic behavior. The authors have not reported any implementation of this approach even on a test network.

Another direction of work with feedback regulation was introduced by Papageorgiou et al. (1990). A multivariable feedback regulator with integral parts and a simple bang-bang controller was developed and tested for a particular network traffic model.

However, the formulation does not establish the underlying mathematical properties and lacks a first-in, first-out requirement.

Ghali and Smith (1992) proposed a formulation for the SO dynamic traffic assignment problem for multiple origin-destination demands in which congestion arises exclusively at specified bottlenecks modeled as deterministic queues. A solution procedure is proposed by analogy with the static SO problem, using marginal link costs. Although the approach does not ensure system optimality, and has limitations due to certain assumptions on queuing, it addresses several of the troublesome traffic modeling issues that seriously limit the realism and validity of previous formulations. Smith (1991) proposed a dynamic user equilibrium model for peak period traffic flows on congested capacity-constrained urban road networks.

Motivated by the first-in first-out property, the model specifies a "no overtaking" condition and determines the relative priorities of vehicles at each node seeking to proceed along the various paths containing that node based on the past history of the vehicle (represented by a binary numbering scheme). Smith also proposed an algorithm for solving the model, though offered no proof of convergence.

In summary, the state of art is fragmented along several lines of work, none of which is entirely satisfactory in terms of the realism of the underlying assumptions. Key weaknesses remain in terms of representing dynamic traffic phenomena which are of the essence in congested networks. A comprehensive review and discussion of dynamic assignment and traffic simulation models for ATIS/ATMS are given in Mahmassani et al. (1992). In the next section, the principal elements of SO traffic assignment problem formulations for ATIS/ATMS applications are identified, along with the key issues faced in their solution.

2. System -optimal formulations - issues for ATIS context

This section discusses eight principal issues involved in formulating the SO dynamic traffic assignment problem for ATIS/ATMS applications.

2.1 Information available to controller

Different scenarios are possible based on the extent and type of information on O-D desires and network traffic conditions available to the controller. If complete information is available on the origin, destination and timing of all trips for the entire duration of interest, path assignment can be made for all time intervals in the beginning.

Partial information on O-D desires in the future can be modeled using a rolling horizon framework, possibly in connection with a stochastic formulation in which O-D trips are modeled as random variables. Solution of the complete information formulation is necessary to obtain a benchmark and a lower bound on system costs for other formulations.

2.2 Information available to travelers

Users with equipped vehicles are given information on network condition and/or instructions on the path to be taken to their destinations. In the SO problem, the controller provides users with routes. However, compliance influences system performance. Under descriptive (as opposed to normative) information supply strategies (see Mahmassani and Jayakrishnan, 1991, for a discussion of such strategies), system performance depends on user decisions in response to the specific information supplied. A related (normative) problem faced by the controller is to determine the optimal information supply strategy, namely what kind of predicted trip times should be supplied to which users in order for the resulting path choices to attain certain system-wide objectives.

2.3 Objective function evaluation

The SO dynamic assignment problem aims at optimizing some system-wide criterion like the total system travel time. The time-dependent nature of the assignment considerably complicates the computation of the objective function. For instance, paths followed by future O-D desires are likely to share common links with paths assigned to current trips (generally upstream), and thus influence the travel times experienced by current assignments. The path travel times experienced by vehicles are the net result of the complex spatial and temporal interactions taking place in the system over a period of time, virtually precluding the ability to evaluate the objective function analytically. Furthermore, analytic evaluation would entail correct representation of the dynamic traffic flow phenomena (queue formation and discharge, congestion build-up and dissipation), a task which is far from the capability of the state of the art in traffic flow modeling. For these reasons, simulation suggests itself as a plausible candidate for evaluating the objective function.

2.4 Traffic flow modeling

This point follows directly from the difficulty just mentioned of correctly representing the dynamics of traffic flow using the kind of analytic functions typically used in static equilibrium assignment models. Furthermore, the representation of flow as a continuum is not appropriate in the time-dependent case. Users entering the system at different times will experience different network conditions and will be assigned different paths. Continuous flow does not allow for distinction of vehicles based on whether they are equipped or not.

2.5 Path-based and link-based formulations

In virtually all existing traffic assignment models, static or dynamic, the link flows are the variables being solved for. However, for the ATIS/ATMS context, path-based assignments are called for because of the need to provide paths to the tripmakers. The problem with obtaining path flows from link flows using link-path incidence relationships is that uniqueness is not guaranteed. Furthermore, the solution of path-based formulations is likely to require partial enumeration of paths for each O-D pair, which is computationally burdensome.

Flows on arcs and paths are mathematically related through definitional identities known as the link-path incidence relationships.

While relatively straightforward in the static case, link-path incidence relationships are far from trivial in the time-dependent case. In the dynamic problem, unlike flows at steady-state, vehicles assigned to a path at a given time are not simultaneously present on all links forming that particular path. Therefore, link-path incidence relationships must recognize the time at which vehicles are actually present on a link.

2.6 Holding of traffic

In a network, it may often be advantageous, from a systemwide total delay standpoint, to favor certain traffic streams or movements over others (e.g. holding back traffic at the minor approach of an intersection in favor of the major approach). Unless otherwise specified, the solution of a SO assignment formulation may entail holding of traffic on one path in favor of traffic on other paths for some significant amount of time at points where the paths overlap or intersect. In other words, vehicles may be artificially delayed on a link for a time that exceeds what may be considered "fair" or "reasonable". Such a solution is probably not acceptable socially nor realistic operationally. When traffic simulation is used to model traffic for a given assignment, unintended holding is implicitly precluded, and no additional explicit constraints are needed.

2.7 First-in, First-out requirement

The physical behavior of traffic on a roadway link exhibits the so-called "first-in, first-out" (FIFO) property, creating a particularly vexing difficulty in the solution of mathematical programming formulations. The FIFO requirement states that traffic that enters a road at a particular time exits from the facility, *on average*, before traffic which enters in later periods. While individual vehicles may travel at different speeds and do pass each other, FIFO should not be violated when considering travel time, averaged over a reasonable number of vehicles entering the link in a given time interval. The problem does not arise in static assignment problems (single or multiple destinations) nor in dynamic assignment with a single destination. However, in dynamic assignment problems with multiple destinations, vehicles on different paths (from different O-D pairs) who share one or more common links may be moved across this arc in a manner that violates FIFO, for instance, if the downstream arc along one path is blocked but not along the other path(s). For SO problems, total travel costs could be lowered if some commodities (e.g. traffic between given O-D pair) could be temporarily held back on an arc, while allowing some other traffic types to proceed to downstream arcs. This form of holding back would violate FIFO, and is not generally physically possible, especially under congested conditions, as vehicles cannot make such "jumps" over traffic ahead of them.

FIFO is a serious liability from a mathematical programming standpoint. Carey (1992) proposed possible additional mathematical constraints to impose the FIFO rule. However, these constraints make the feasible set non-convex, destroying many of the computational and mathematical (analytic) advantages of the formulation. Carey suggests solving the problem without introducing an explicit FIFO restriction, then analyzing it for the seriousness of FIFO violations. However, no explicit procedure is proposed for this purpose. Smith's (1991) DUE model approximately bypasses the FIFO issue.

The above issues further highlight the relative advantages of a simulation-assignment strategy. Simulation moves vehicles based on their current location and speed, and FIFO is implicitly satisfied.

2.8 Temporal issues

The treatment of time in various aspects of the formulation and solution of dynamic assignment problems is an essential and subtle element of these problems. When time is discretized, the size of the time interval for assignment decisions, and its relation to the time step that may be used in the simulation of traffic movement, need to be determined.

The size of the assignment interval affects the size of the "packet" of O-D desires to be assigned jointly. If the time interval is relatively large, there will be several vehicles going from a particular origin to a particular destination in that time interval. However, their travel "experience" in the network may not be identical as vehicles at the beginning of the interval may experience different time-varying traffic conditions. A smaller assignment time interval implies more intensive computation, giving rise to the usual trade-offs between computational intensity and accuracy.

The above section has illustrated the difficulties involved in modeling the system-optimal assignment problem for the ATIS/ATMS context, and how a simulation-assignment strategy overcomes these problems.

3. Formulation of the system optimal dynamic assignment problem

3.1 Problem statement

Consider a traffic network represented by a directed graph G(N,A) where N is the set of nodes and A is the set of directed arcs.

A node can represent an origin and/or a destination and/or an intersection. We consider networks with multiple origins and destinations. Given a set of origin-destination trip desires, we wish to use an Advanced Traveler Information System (ATIS) during congested periods of traffic flow in the network. We consider a planning horizon (or peak period) of length T'. Under ATIS, some motorists with suitably equipped vehicles are provided with information on current network traffic conditions and/or advice on what route to follow to get to their destinations. We consider a normative route guidance information supply strategy, whereby a controlling agent (controller) assigns users to various routes in the network so as to satisfy some systemwide objectives. Three formulations are presented in this section, corresponding to three scenarios of the information available to the controller.

3.2 Information availability scenarios for the controller

For a normative assignment capability, the "ideal" scenario would be a fully informed system, where the controller has complete a priori information about every tripmaker in terms of origin, destination and timing of the trip, and uses this information to develop an integrated scheme that assigns to each user a path to the desired destination so as to achieve some system-wide objectives.

These functional capabilities are envisioned under the so-called "coordinated stage" of ATIS development (Mobility 2000, 1990a).

It is unlikely that the controller will have full information on O-D trip desires for the complete duration for which the assignment is to be made. A more probable scenario is one where information is available for a short duration into the future through detectors and advance information from drivers. A rolling horizon approach with forecasted future O-D desires is used to obtain models for system optimal assignment for this scenario.

An alternative scenario is that the controller has O-D trip desires only for the present period, and future O-D desires are treated as random variables with known probability distributions (based on historical data), giving rise to a stochastic programming formulation of the problem. Formulations resulting under each of the above information availability scenarios are presented next.

3.3 Definition of variables and notation

The following variables and notation are used in the various formulations:

i = subscript for origin node

j = subscript for destination node

n = node in the network, $n \in N$

a = arc (or link) in the network, $a \in A$

k = subscript for a path in the network

τ = subscript denoting the time interval in which assignment is made

t = subscript denoting current time interval

Δ = length of a time interval

T' = total duration (peak period) for which assignment is to be made

r^{τ}_{ij} = number of vehicles who wish to depart from i to j in period τ

r^{τ}_{ijk} = number of vehicles who wish to depart from i to j in period τ assigned to path k

$\delta^{\tau t a}_{ijk}$ = dynamic arc-path incidence indicator, equal to 1 if vehicles going to i and j assigned to path k at time τ are on link a in period t, i.e.

[$\delta^{\tau t a}_{ijk}$ = 1, if r^{τ}_{ijk} is on arc a during period t

= 0, if arc a does not belong to path k

= 0, if $\tau > t$

= 0, if $r^{\tau}{}_{ijk}$ is not on arc a during period t]

$T^{\tau}{}_{ijk}$ = path travel time for vehicles going from i to j assigned to path k at time τ

$x^{\tau ta}{}_{ijk}$ = number of vehicles (i to j) assigned to path k in period τ which are on link a at the beginning of period t

$d^{\tau ta}{}_{ijk}$ = number of vehicles (i to j) assigned to path k in period τ which enter arc a in period t

$m^{\tau ta}{}_{ijk}$ = number of vehicles (i to j) assigned to path k in period t which exit link a in period t

x^{ta} = total number of vehicles on link a at the beginning of period τ

d^{ta} = total number of vehicles which enter link a in period t

m^{ta} = total number of vehicles which exit link a in period t

C(n) = set of links directed towards node n

B(n) = set of links directed away from node n

3.4 Deterministic full information scenario

In this formulation, the controller has completed a priori information on O-D desires for the entire assignment duration. The controller assigns to each vehicle a path to its desired destination as the vehicle enters the network so as to optimize some indicator of system performance subject to the applicable constraints.

In the ATIS context, this formulation corresponds to the SO dynamic traffic assignment problem. The formulation incorporates dynamic link-path incidence variables that relate path flows to link flows. The fundamental difficulty in solving dynamic assignment problems (SO or otherwise) is that the dynamic incidence variables are themselves a function of the assignment, giving rise to a complicated fixed-point problem. Essentially, the resulting formulation, which involves nonlinearities in the objective function as well as in the constraints, yields generally undesirable mathematical properties that preclude the guarantee of global optimality. In addition, as explained in the previous section, the dynamic link-path incidence is not trivial. In the static case, flows assigned to a path exist on all the links along that path simultaneously, leading to constant known link-path incidence matrix. Such an assumption in the dynamic case would be clearly flawed, as vehicles starting along a path at a given time are not simultaneously present on all links of the path. The incidence matrix changes dynamically, considerably complicating vehicle conservation constraints and travel time calculations. The advantages of a rigorous formulation using dynamic link-path incidence relationships are offset by the difficulty and intricacies involved in attempting to solve it.

The formulation is detailed below :

Given:

r^{τ}_{ij}, $\forall$ i,j and $\tau = 1, \ldots\ldots\ldots\ldots T'$

Objective function:

Min. $\Sigma_\tau \Sigma_i \Sigma_j \Sigma_k (r^{\tau}_{ijk} . T^{\tau}_{ijk})$

or

Min. $[\ T(r^{\tau}_{ijk}), \forall\ i, j, k, \tau]$

Subject to:

1. $r^{\tau}_{ij} = \Sigma_k\ r^{\tau}_{ijk}$, $\quad \forall\ i, j, \tau$
2. $\Sigma_c m^{tc} = \Sigma_b d^{tb}$, $\quad \forall\ t, c \in C(n), b \in B(n), n \neq i \text{ or } j$
3. $x^{ta} = x^{t-1a} + d^{t-1a} - m^{t-1a}$, $\quad \forall\ t, a$
4. $x^{ta} = \Sigma_k \Sigma_\tau \Sigma_i \Sigma_j\ (r^{\tau}_{ijk} . \delta^{\tau ta}_{ijk})$, $\quad \forall\ t, a$
5. $T^{\tau}_{ijk} = \Sigma_t \Sigma_a [\delta^{\tau ta}_{ijk} \cdot \Delta]$, $\quad \forall\ i, j, k, \tau$
6. $\delta^{\tau ta}_{ijk} = f(r^{\tau}_{ijk})$, $\quad \forall\ i, j, k, \tau, t, a$
7. $d^{ta} = \Sigma_k \Sigma_\tau \Sigma_i \Sigma_j\ d^{\tau ta}_{ijk}$, $\quad \forall\ t, a$
8. $m^{ta} = \Sigma_k \Sigma_\tau \Sigma_i \Sigma_j\ m^{\tau ta}_{ijk}$, $\quad \forall\ t, a$
9. $\tau \leq t$
10. $\delta^{\tau ta}_{ijk} = 0 \text{ or } 1$
11. All variables (other than $\delta^{\tau ta}_{ijk}$) ≥ 0

There are two alternative forms for the objective function in the above formulation. The first states that the total travel time in the system is aggregate of the product of the number of vehicles assigned to a particular path (from a given origin to a given destination at a particular time) and the corresponding path travel time. This assumption is realistic when assignment intervals are reasonably small (in which case there are not more than two or three vehicles to a particular path from an origin to a destination). The nonlinearity of the objective function arises from the fact that the travel time on the path is itself a complicated non-explicit function of the number of vehicles assigned to the various paths of the network, via the dynamic link-path incidence.

The second form of the objective function simply states that the total travel time of all vehicles assigned to the various paths during the duration of ATIS application is some function of the assignment. This objective function can be evaluated by any available means. We do it through simulation.

Constraint (1) is a definitional constraint stating that O-D desires assigned to the various paths should sum up to the demand (conservation at the origin). Constraint (2) states that vehicles cannot be stored at intermediate nodes, that is, the number of vehicles exiting from all links incident on an intermediate node should equal the number of vehicles entering all links incident from that node at any given time. Constraint (3) represents the conservation of vehicles on a link and states that the total number of vehicles on any link at the end of the current time interval is the net algebraic sum of vehicles on that link at the end of the previous time period, vehicles entering that link during the current period and vehicles exiting that link during the current period.

Constraints (4), (5) and (6) represent the time-dependent link-path incidence relationships which fundamentally characterize the dynamic assignment problem. Constraint (4) represents the dynamic relationship between the number of vehicles assigned to various paths and their aggregation on links. Constraint (5) illustrates the calculation of the path travel times using the dynamic link-path incidence variables. The number of time steps in which the dynamic incidence variable takes a value 1 implies the number of discrete time steps that a vehicle (or a group of vehicles) spent in the system, and multiplying with Δ gives the actual travel time in the system.

Constraint (6) states that the dynamic link-path incidence variables are a function of the assignment.

Constraints (7) and (8) are definitional constraints for the number of vehicles entering and exiting links in the various time intervals. Constraint (9) defines temporal correctness. Constraint (10) restricts the dynamic incidence variables to take values of 0 or 1. Constraint (11) represents the non-negativity requirement.

3.5 Rolling horizon formulation

This formulation represents a more realistic scenario of the information available to the controller. It assumes that information will be available for a "short" duration into the future. This provides the opportunity to use a rolling horizon approach with forecasted future O-D desires. The basic idea behind the rolling horizon approach is that current events will not be influenced by events "far" into the future, i.e. that vehicles currently assigned will not be influenced by vehicles assigned "far" into the future as the currently assigned vehicles will probably be out of the system by that time. The stage length h in the following diagram depicts that length of time (its value in actual problems is network specific). The roll period ℓ represents the short duration into the future for which O-D desires are available with reasonable certainty. To make an assignment of vehicles to various paths for the current period, the controller requires knowledge of O-D desires for the rest of the stage length as these are expected to influence current assignments. These O-D desires may be forecasted based on historical data and current information. O-D desires beyond the stage length h are assumed to be zero. The situation is now analogous to the complete information availability scenario, albeit, only for the duration covered by the stage length h. The system is solved for the duration of the stage length and O-D desires for the roll period (known with certainty) are assigned to the paths determined. The time frame is now "rolled" forward by the roll period and the above process is repeated till the end of the duration of interest. Hence, a series of optimizations are performed till the planning horizon is covered. The formulation is illustrated below:

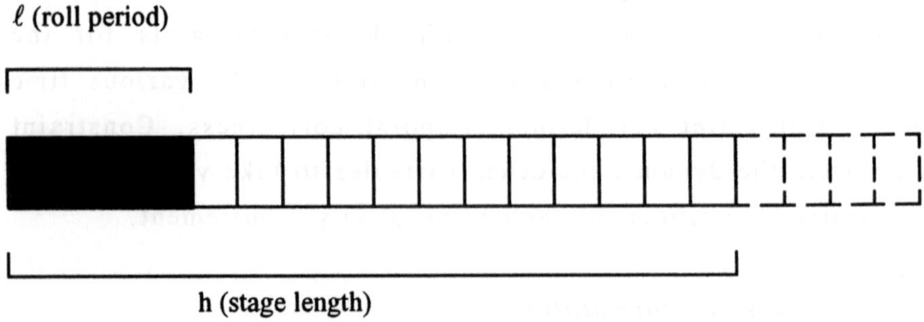

ℓ = The roll period(in number of time steps)

h = The stage length(in number of time steps)

η_ℓ = The current stage number

Given:

1. $r^{\tau}{}_{ij}$, $r^{\tau}{}_{ijk}$, $\forall$ i, j, k and $\tau = 1,2,\ldots\ldots\ldots\, \eta_\ell.\ell$, all O-D desires and assigned paths up to current period
2. $r^{\tau}{}_{ij}$, $\forall$ i, j and $\tau = \eta_\ell.\ell+1,\ldots\ldots\ldots\ldots,\ \eta_\ell.\ell+\ell$, O-D desires for a roll period of length ℓ for which assignment is to be made
3. $r^{\tau}{}_{ij}$, $\forall$ i, j and $\tau = \eta_\ell.\ell+\ell+1,\ldots\ldots\ldots\ldots,\eta_\ell.\ell+h$, O-D desires forecast for the rest of current stage based on historical and current information
4. $r^{\tau}{}_{ij} = 0$, $\forall$ i, j, $\tau > \eta_\ell.\ell+h$, current assignments are not affected by the O-D desires "far" into the future

Objective function:

$$\text{Min.} \sum_{\tau=\eta_\ell.\ell+1}^{\eta_\ell.\ell+h} \Sigma_i \Sigma_j \Sigma_k \left(r^{\tau}_{ijk} \cdot T^{\tau}_{ijk}\right)$$

Subject to:

constraints (1) - (11)

This formulation is identical to the deterministic scenario formulation except for assumptions on the amount of information available to controller and the time frame over which the objective function is evaluated. The formulation is shown for the current stage number η_ℓ .

The path assignments in each stage are determined for the entire stage, but implemented for only the roll period (as only the demand for this period is available with certainty). A number of pertinent questions arise at this point. How far is "far" ? What is the "optimal" stage length h? What is a good value for the roll period ℓ? How accurate are the forecasted values for future O-D desires? Is there a need for feedback to check if the assumptions made were realistic? How robust is the solution vis-a-vis the predicted O-D desires. These questions need to be addressed while implementing the solution methodology.

3.6 Stochastic formulation

Under this scenario, the controller possesses O-D trip desires for the current period only. O-D desires for the remaining assignment planning horizon are a vector of random variables with known distributions in each interval. A number of vectors are generated for each future interval and a Monte-Carlo simulation for various sequences of random variables is performed to obtain average values for the assignment of the O-D desires to various paths on the network for the current period. The additional variables used in this formulation are :

$\alpha(t)$ = optimal policy for the desired assignments in period t

x^t = vector containing the number of vehicles on each arc at the beginning of period t, based on decisions $\alpha(1)$, $\alpha(2)$,......, $\alpha(t-1)$ up to time t

$\psi(t \mid x^t)$ = The minimum total travel time from t to end of the planning horizon given x^t.

The formulation is as follows:

<u>Given:</u>

1. r^{τ}_{ij}, r^{τ}_{ijk}, $\forall$ i, j, k and τ = 1,2,........., t-1, O-D desires and assigned paths up to interval t-1
2. r^{t}_{ij} , $\forall$ i, j , current O-D desires
3. $x^t = \{ x^{ta}, \forall a\}$,current state of network
4. Known probability distribution for future O-D desires

<u>Objective function:</u>

$\psi(t \mid x^t) = \text{Min. } \{\Sigma_i \Sigma_j \Sigma_k (r^t{}_{ijk}(\alpha(t)) \;.\; E[T^t{}_{ijk}(\alpha(t))] + E(\psi(t+1|x^{t+1})) |x^t , \alpha(t))\}$

<u>Subject to:</u>

constraints (1) - (11)

The objective function $\psi(\cdot)$ consists of two terms. The first term is the product of the assignments in the current period based on the optimal policy $\alpha(t)$ obtained by minimizing the current objective function and the average travel time on the paths (obtained from the MC simulation). The second term is the expected value of the objective function for the next time period given the current O-D desires and the current optimal policy for assignment of O-D desires to paths. Hence, at each time step the objective function evaluates the best decision for the current period in such a way that the expected future objective function is also minimized. The constraints are identical to the constraints in the previous formulation. The objective function is a complicated nonlinear expression and the methodology for its evaluation needs further research.

In the remainder of the paper, we present a solution methodology for the first formulation (full information) which may also be applied for the second (rolling horizon).

4. Solution methodology

4.1 Simulation-assignment approach

We use a simulation based algorithm to solve the SO dynamic traffic assignment problem described in section 3.4. A traffic simulator is used to evaluate the objective function, ensuring consistency with realistic traffic behavior (FIFO, no holding back of traffic). The procedure assigns vehicles to various paths directly, obviating the need to infer a path assignment from the solution to a link-based formulation. The DYNASMART (DYnamic Network

Assignment-Simulation Model for Advanced Road Telematics) assignment-simulation model developed at The University of Texas at Austin is used to simulate traffic.

4.2 Description of the approach

The overall solution methodology is shown in Figure 1, for the formulation under which O-D desires are assumed known for the whole assignment duration. It can be suitably modified for the rolling horizon approach. The algorithm is a relatively straightforward extension of well-known solution methods for the static assignment problem, with key differences in each component of the algorithm and significant additional implementation challenges. A brief summary of the approach is as follows:

1. Set iteration counter I=0. Obtain historical paths for the entire duration (to obtain an initial solution, any other feasible approach could be used).
2. Assign O-D desires over the entire duration to the paths and simulate the resulting traffic patterns using DYNASMART.
3. Compute marginal link travel times using time-dependent experienced travel times and number of vehicles on links, as obtained at the end of the simulation.
4. Using time-dependent least cost path algorithm, compute the time-dependent shortest marginal paths (auxiliary paths) for each O-D pair, based on the marginal link travel times and the average link travel times.
5. Perform all-or-nothing assignment of O-D desires to the auxiliary least marginal time paths computed in step 4.
6. Update the paths and the number of vehicles assigned to the paths. Update of paths is done by checking if the auxiliary path is a previously existing path for that O-D pair and including it if it is not. The update of the number of vehicles (splitting of O-D desires among the various paths currently existing between the O-D after the path update) is done using the Method of Successive Averages.

7. Check for convergence using an Î-convergence criterion. Convergence can be achieved on the value of the objective function or on the number of vehicles assigned to various updated paths (the latter approach is preferred).

8. If convergence criterion is satisfied, stop. Otherwise, update the iteration counter I=I+1 and go to step 2 with the updated paths and number of vehicles.

4.3 Discussion of methodology

This section describes the various components of the SO solution methodology. After a brief introduction to DYNASMART, the approach to obtain the time-dependent marginal travel times. is illustrated. Next, the time-dependent least cost path algorithm is briefly addressed. Details of the updating mechanism are described, followed by the path assignment, procedure which interfaces the update mechanism with DYNASMART.

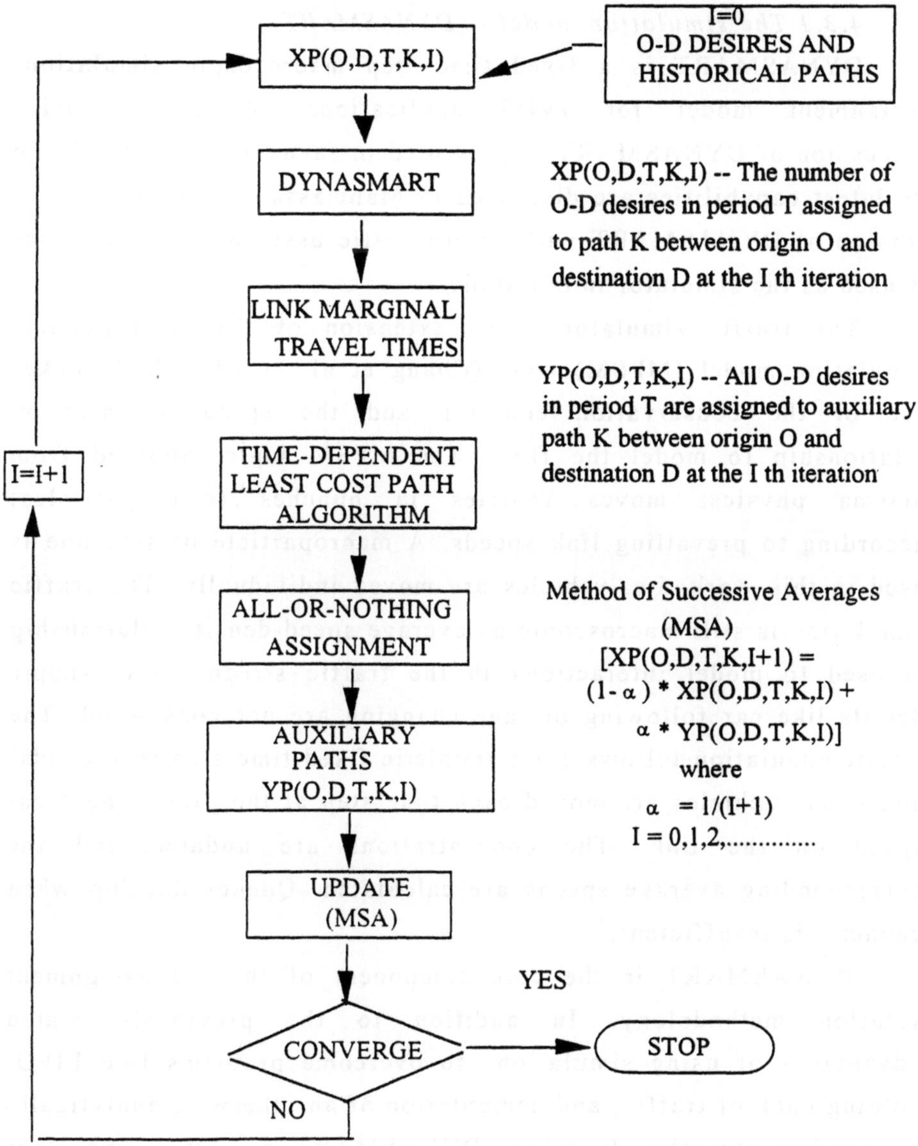

Figure 1. Solution Algorithm for the System Optimal Dynamic Traffic Assignment Problem

4.3.1 The simulation model -- DYNASMART

DYNASMART is a fixed time step macroscopic simulation-assignment model for IVHS applications. A comprehensive discussion of DYNASMART is presented in Jayakrishnan (1991) and its latest capabilities are discussed in Mahmassani et al. (1992). A version of DYNASMART with deterministic assignment capabilities is used as the simulator in our study.

The traffic simulator is an extension of the macroparticle simulation model (MPSM) code (Chang et al., 1985) which makes use of the conservation equation and the speed-concentration relationship to model the flow. The MPSM logic, adopted from plasma physics, moves vehicles in bunches (macroparticles) according to prevailing link speeds. A macroparticle of size one is used in this work, i.e. vehicles are moved individually. The traffic simulation is still macroscopic as average speed-density relationship is used to model interactions in the traffic stream. Microscopic details like car following or lane changing are not considered. The traffic simulation follows a deterministic fixed time step (6 seconds) approach. Vehicles are moved each time step at the prevailing local speed on the link. The concentrations are updated and the corresponding average speeds are calculated. Queues develop when capacity is insufficient.

DYNASMART is the core component of the SO assignment solution methodology. In addition to the previously stated advantages of using simulation to overcome problems like FIFO, holding back of traffic, and computation of an otherwise analytically intractable objective function, DYNASMART furnishes important post-simulation data, average travel times, predicted travel times, and number of vehicles on each link for each simulation interval for the remaining components of the algorithm. The above information is used in the marginal path, time-dependent shortest path, the update, and path assignment components.

4.3.2 Marginal travel times

This section discusses the computation of marginal travel times and their significance to the SO assignment problem.

4.3.2.1 Significance of marginals in SO assignment

The SO dynamic traffic assignment problem aims at minimizing the total system travel time. A global path marginal travel time denotes the travel time increment to the system by the addition of one vehicle to that path. Hence, the solution to the SO problem would entail assigning O-D desires to the time-dependent shortest global marginal paths. The computation of global marginals would entail computationally intensive brute force approaches to capture secondary effects that arise from network interactions over different time periods. The approach used here calculates the marginal costs in only an approximate manner that ignores some of the spatial and temporal interactions taking place in the network. In particular, the marginal cost imposed by an additional vehicle on a given path at a particular time is assumed here to be limited to impeding vehicles on the links constituting that particular path (still correctly recognizing the time-dependent incidence of that vehicle on the links). It may therefore be possible to improve on the solution by a more elaborate computation of the marginal costs.

The marginals are computed in our approach as follows. The marginal travel time on a link is the contribution of an additional traveler to the total travel time on that link. The path marginal total travel times are obtained by a summation of the time-dependent modified link travel times (which include the actual time-dependent link travel time and the time-dependent marginal contribution of an additional traveler) for all links on that path.

4.3.2.2 Methodology for obtaining marginals

The modified link travel times are obtained according to the following relationship:

For each O-D pair,

$$mltt(a,t) = tt(a,t) + itt(a,t) \cdot x(a,t)$$

where

mltt(a,t) = modified travel time in period t for link a

tt(a,t) = travel time (experienced/estimated) in period t for link a

itt(a,t) = increment in travel time in period t to traveler already on link a due to the additional traveler

x(a,t) = number of vehicles on link a at time t

The product of itt(a,t) and x(a,t) gives total link travel time increment due to an additional traveler on link a in period t. The difficulty lies in the evaluation of itt(a,t) which is the derivative of tt(a,t) with respect to x(a,t). The method to evaluate itt(a,t) is illustrated in Figure 2.

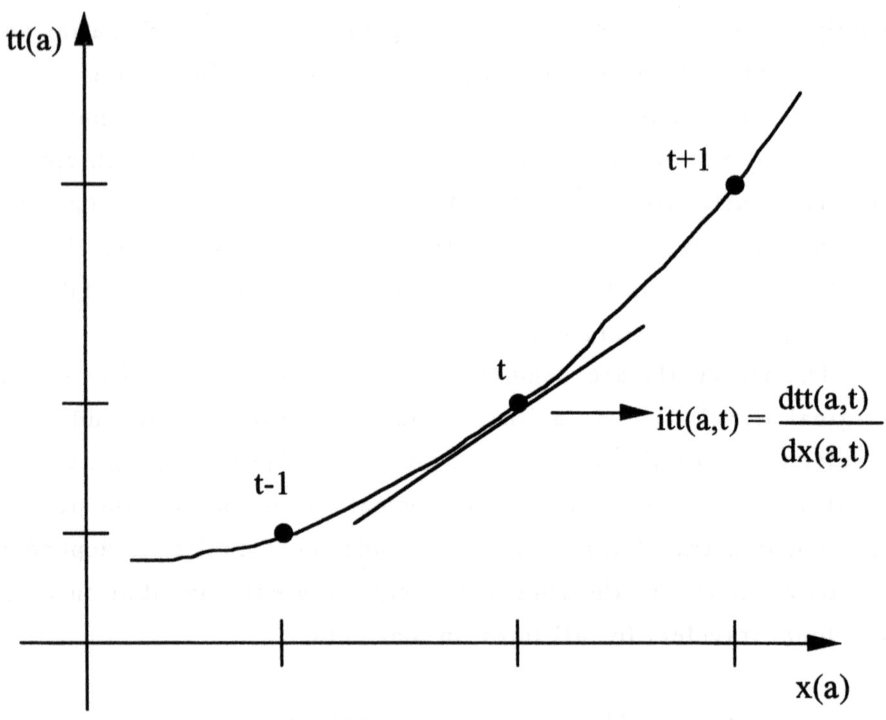

Figure 2. Computation of Marginals

Figure 2 shows a plot of travel time tt(a) versus the number of vehicles x(a) on link a. In the static case, the link travel times and flows are assumed constant. Hence, the calculation of the derivative using a static link-performance function is trivial. In the dynamic case, the travel times and the number of vehicles on a link are time-dependent. Consequently, the derivative is time-dependent and this makes its evaluation problematic.

The approach we use visualizes "time-dependent" link performance functions, which change gradually over time. If the time interval between successive evaluations of marginals is small, it appears reasonable to assume that three consecutive points in time are on the same curve. This assumption is made in the evaluation of itt(a,t), illustrated in the graph with three successive time points (t-1, t and t+1). A quadratic fit using the three points results in the time-dependent link performance curve at time t and the slope of this curve gives itt(a,t). The calculation of mltt(a,t) is then straightforward as the values of other variables are obtained from the simulation. The curves obtained with three successive time periods are relatively close to each other. However, small time intervals (in the order of a few seconds) may cause some instability in the curves because of sudden "jumps" in values of travel times and numbers of vehicles. There is a trade-off between their approximate correctness of the curves and the robustness.

The robustness of the marginal travel time values over time is important to the solution procedure, and is achieved through averaging techniques. Averaging can be done over time intervals, number of vehicles or both. The simulation time interval of 6 seconds in DYNASMART is too small for update of paths for a given O-D pair as no appreciable change takes place in the system in such a short time. An assignment interval of 3 to 5 minutes is used for updating paths. The marginal values are necessary for assignment intervals only and not for simulation intervals. One technique is to average the marginal time values over the simulation intervals in a given assignment interval and use this value for that assignment interval. Computing a moving average of marginals over the last n assignment time intervals is another averaging technique.

The marginal values can also be computed by taking an average of the marginal times for all vehicles present on a link during a given assignment interval.

Another issue is the value of travel time tt(a,t) that should be used to compute the marginals. Post-simulation data gives two types of travel times, average and experienced/estimated. The average travel time on a link (for a given interval) is based on an analytical model relating speed to the concentration on the link, but not including traffic phenomena like queue formation and discharge which can make the actual travel time experienced quite different from the average value. Hence, a more accurate representation of the marginal travel times is obtained by using the experienced/estimated travel times of vehicles as opposed to using analytically computed link travel times.

4.3.3 Time-dependent shortest marginal path computation

An essential problem consists of finding the shortest path from a node to all other nodes in a directed graph with time-dependent travel times. Dreyfus (1969) proposed to address the problems by using Dijkstra's algorithm in an expanded static representation of the time-varying problem for deterministic travel times. Kaufman and Smith (1990) made explicit an assumption that is sufficient for the validity of Dreyfus' approach and discussed this issue from an IVHS perspective.

The marginal travel times on paths are obtained by a summation of modified link times for all links on that path. Time dependency of these link travel times requires a time-dependent least cost path algorithm to calculate the paths. A new time-dependent least cost path algorithm developed at The University of Texas at Austin for use with DYNASMART is used to obtain the shortest marginal time-dependent (auxiliary) travel paths. The various capabilities of the algorithm including recognition of turn movements and very efficient data structures are discussed in Ziliaskopoulos and Mahmassani (1992a, 1992b).

Figure 3 details the procedure for the computation of the shortest marginal paths. The time-dependent least cost path algorithm requires average *and* marginal link travel times as inputs. A least marginal path calculation based solely on marginal link travel costs is incorrect because marginal link travel time does not have a physical interpretation, and vehicles move at the average (or experienced/estimated) travel times, as illustrated in the following example.

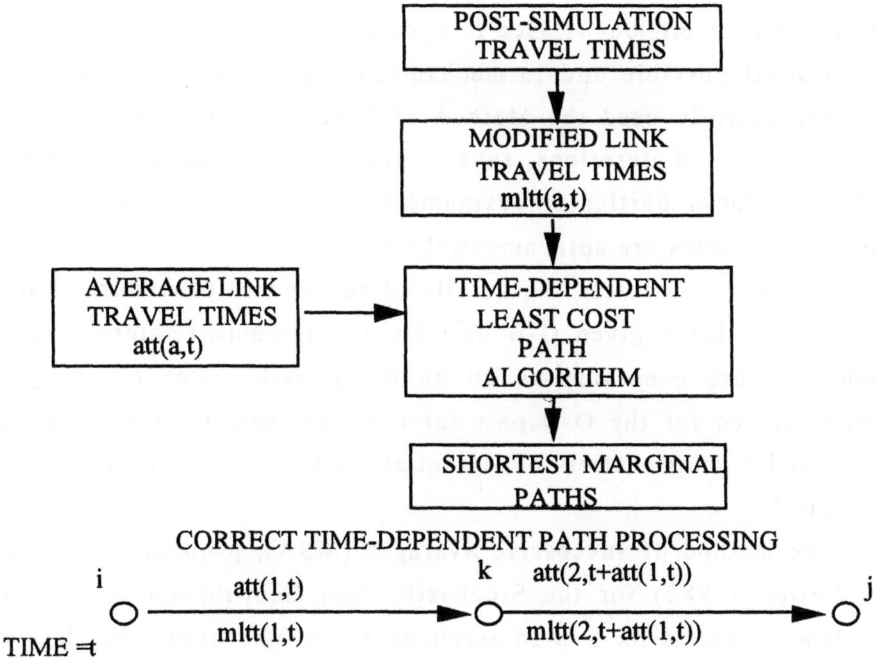

Figure 3. Computation of Shortest Marginal Paths

In Figure 3, consider the path from i to j. Starting at node i at time t, link 1 (i-k) is chosen as the next link on the path based on the link penalty mltt(1,t) the modified link travel time on link 1 at time t. However, node k is reached at time att(1,t) and *not* mltt(1,t) as att(1,t) is the actual time taken to move on link 1. Consequently, the modified link travel time considered on link 2 (k-j) is mltt(2,t+att(1,t)) and *not* mltt(2,t+mltt(1,t)).

Following all-or-nothing assignment is the update of paths described in the next section.

4.3.4 Update of paths and vehicle assignments

In the solution algorithm, auxiliary paths are generated and an all-or-nothing assignment of the O-D desires is made to these paths during each iteration as a means of obtaining an approximate descent direction (on average). However, the all-or-nothing assignment is not an end in itself, but is used to update the current solution.

Several possible update mechanisms may be used; to date, we have successfully used the Method of Successive Averages (MSA). Over a number of iterations, several paths are generated for a given O-D pair (for a particular assignment period), and the number of departing vehicles are split among these.

Figure 4 illustrates the details of the update procedure. Paths are included for a given O-D pair (in an assignment interval) only when they are generated as an auxiliary path, after checking if already stored for the O-D pair for that assignment interval. Thus, only partial enumeration of essential paths for the O-D pair is performed.

The method of successive averages (MSA), proposed by Sheffi and Powell (1982) for the Stochastic User Equilibrium assignment problem, is based on a predetermined move size along the descent direction. The move size α_n is determined a priori and not on the basis of the current solution. The move size α_n has to satisfy certain requirements for an algorithm to converge, and one of the simplest move-size sequences satisfying those requirements is used by MSA. At each iteration, MSA uses the inverse of the iteration number (n) as the move size. Hence, $\alpha_n = 1/n$.

The vehicle assignments for the next iteration are obtained by using this move size in the search procedure with the move direction

$d(n) = y_n - x_n$, where x_n and y_n represent the current solution values and auxiliary values respectively for iteration n. The solution for the next iteration x_{n+1} is obtained as $x_{n+1} = x_n + 1/n\ (y_n - x_n)$ which is equivalent to $x_{n+1} = (1 - 1/n)\ x_n + (1/n)\ y_n$. A comprehensive discussion on MSA is given in Sheffi (1985).

It is important to note that when the auxillary paths are obtained based on the appropriate local link marginal times, the resulting move direction is not guaranteed to be a descent direction.

A descent direction is generated only when least global marginal cost paths are obtained, as discussed earlier and highlighted by Ghali and Smith (1992)

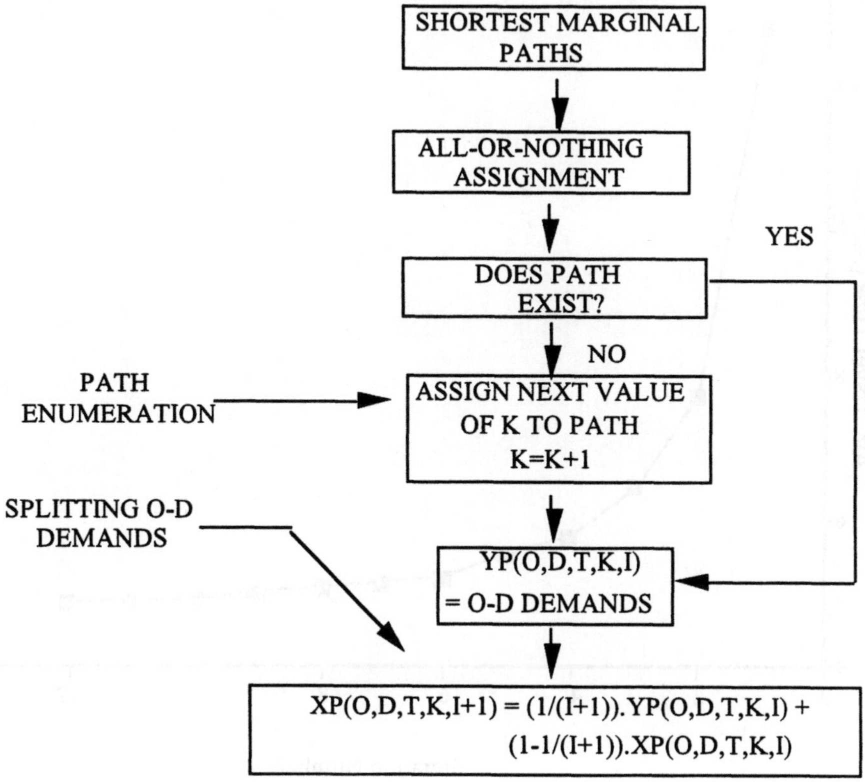

Figure 4. Update of Paths and Vehicle Assignments to Paths

5. Example application of the solution algorithm

The solution algorithm was applied to a typical test case consisting of a network with 50 nodes and 163 arcs, with around 38000 vehicles generated in a 35 minute peak period of traffic flow. Figure 5 shows the convergence pattern of the algorithm for the SO dynamic traffic assignment problem. The initial assignment results in an average trip time per vehicle of about 13 minutes. At convergence, the average trip time per vehicle is slightly less than 5 minutes.

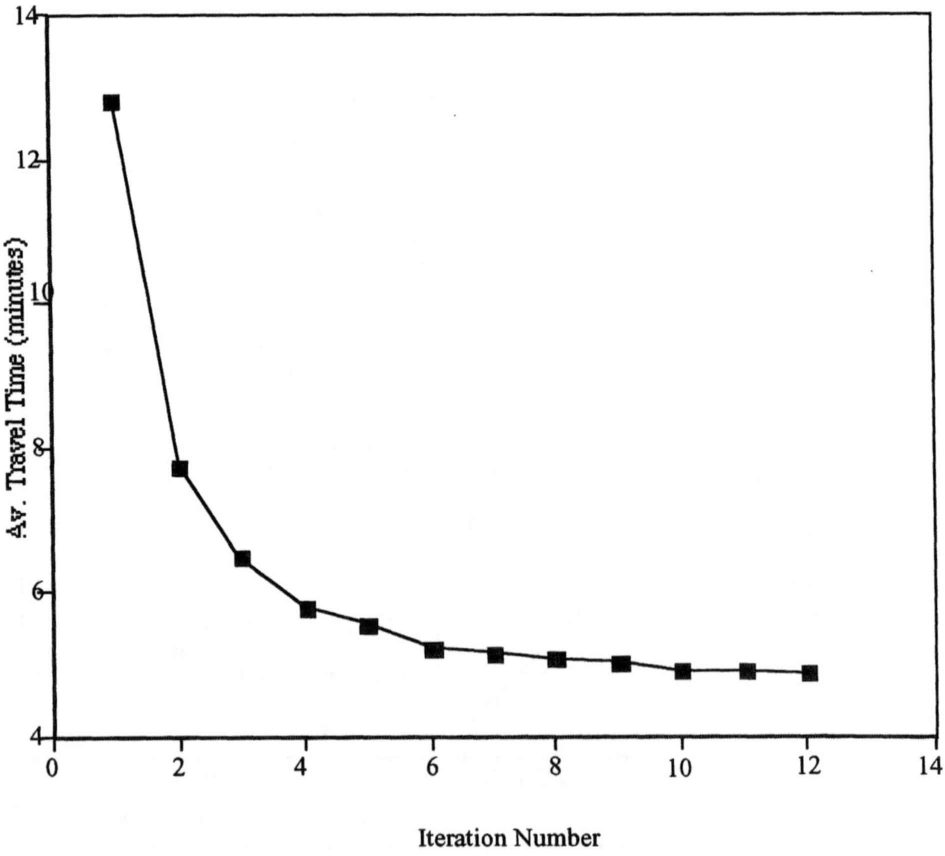

Figure 5. Typical Convergence Pattern for the Solution Algorithm

6. Concluding comments

The solution methodology has been implemented on a CRAY-YMP supercomputer. This elaborate implementation entails interfacing individual components through input-output data files as some of the individual components require considerable memory and computation time. This points to a need for efficient parallel and vector programming techniques for reducing computational time and efficient data structures for handling memory requirements.

A number of experiments have been conducted for various test networks to derive insights on the benefits obtained under alternative ATIS information supply strategies, and the results have been analyzed in Mahmassani and Peeta (1993). The complexity of traffic dynamic phenomena captured by the simulator when evaluating the objective function generally preclude the kind of well-behaved properties required to guarantee convergence of the solution algorithm in all cases. However, such convergence was achieved in all our experiments.

The algorithm in this paper addresses the problem where all vehicles are suitably equipped for one-way or two-way communication with a central controller and where users comply fully with the prescribed paths. The formulation and algorithms are being extended to represent more realistic scenarios where only a fraction of all vehicles are equipped, drivers may have access to different types of information, and may have different response behavior to the supplied information. Under such scenarios, some equipped drivers may comply with prescribed routes, others will make their own decisions based on current or predicted conditions, and yet others may act in a contrarian manner.

The solution algorithm in this paper also assumes that the controller has complete a priori information on O-D desires for the peak period. In reality, it is more likely that the controller will have information on O-D desires for only a short duration in the future. We address this problem using a rolling horizon approach which can also be used to implement the solution algorithm for the dynamic assignment problem in real-time.

Acknowledgments

This work was funded by the Federal Highway Administration under contract DTFH61-90-R-00074 titled "Traffic Modeling to Support Advanced Driver Information Systems (ADIS)". Additional support was provided by the U.S. Department of Transportation through the Southwest Region University Transportation Center. The authors are grateful to Ta-Yin Hu and Athanasios Ziliaskopoulos for collaboration on key aspects of the program, and to Drs. Richard Rothery, and Leon Lasdon for fruitful discussions. The work has benefited from the insightful comments of Al Santiago, Henry Lieu, and Ammar Kanaan at the Federal Highway Administration. The authors of course are solely responsible for the context of this paper.

References

[1] Boyce, D.E. (1991). *"Dynamic User-Optimal Traffic Assignment: A New Model and Solution Technique"*, submitted to Transportation Science.

[2] Carey, M. (1986). *"A Constraint Qualification for a Dynamic Traffic Assignment Model"*,Transportation Science, Vol. 20, pp 55-58.

[3] Carey, M. (1987). *"Optimal Time Varying Flows On Congested Networks"*, Operations Research, Vol.35, No.1, pp 58-69.

[4] Carey, M. (1992). *"Nonconvexity of the Dynamic Assignment Problem"*, Transportation Research, Vol. 26B, No. 2, pp 127-133.

[5] Dreyfus, S.E. (1969). *"An Appraisal of Some Shortest-Path Algorithms"*, Operations Research 17, pp 395-412.

[6] Friesz, T.L., Luque, J., Tobin, R.L. and Wie, B.W. (1989). *"Dynamic Network Traffic Assignment Considered as a Continuous*

Time Optimal Control Problem", Operations Research, Vol.37, No.6, pp 893-901.

[7] Ghali, M. O. and Smith, M.J. (1992). *"Optimal Dynamic Traffic-Assignment on a Congested City Network"*, presented at the 2nd International Capri Seminar on Urban Traffic Network, Capri, Italy.

[8] Ho, J.K. (1980). *"A Successive Linear Optimization Approach to the Dynamic Traffic Assignment Problem"*, Transportation Science 14, pp 295-305.

[9] Janson, B.N. (1991). *"Dynamic Traffic Assignment with Schedule Delay"*, Presented at the 71st TRB Annual Meeting, Washington D.C., 1992.

[10] Kaufman, D.E., Smith, R.L. (1990). *"Fastest Paths in Time-Dependent Networks for IVHS Application"*, IVHS Journal, Vol.1, No. 1, pp 1-11.

[11] Mahmassani, H.S., Peeta, S. (1993). *"Network Performance Under System Optimal and User Equilibrium Dynamic Assignments: Implications for ATIS"*, forthcoming in Transportation Research Record, TRB, 1993.

[12] Mahmassani, H.S., Peeta, S., Chang, G., Junchaya, T. (1992). *"A Review of Dynamic Assignment and Traffic Simulation Models for ADIS/ATMS Applications"*, Technical Report DTFH61-90-R-00074-1, Center for Transportation Research, The University of Texas at Austin.

[13] Mahmassani, H.S., Jayakrishnan, R. (1991). *"System Performance and User Response under Real-Time Information in a Congested Traffic Corridor"*, Transportation Research, Vol. 25A, No.5, pp 293-307 .

[14] Mahmassani, H.S., Hu, T, Jayakrishnan, R. (1992). *"Dynamic Traffic Assignment and Simulation for Advanced Network Informatics"*, The Proceedings of the Second International Capri Seminar on Urban Traffic Networks, Capri, Italy, July 5-8, 1992.

[15] Merchant, D.K., and Nemhauser, G.L. (1978). *"A Model and an Algorithm for the Dynamic Traffic Assignment Problems"*, Transportation Science, Vol. 12, No.3, August 1978, pp 183-199.

[16] Midler, J.L. (1969). *"A Stochastic Multiperiod Multimode Transportation Model"*, Transportation Science 3, pp 1-7.

[17] Papageorgiou, M., Messmer, A. and Senninger, H. (1990) *"Dynamic Network Traffic Assignment Via Feedback Regulation"*.

[18] Ran, B., and Shimazaki, T. (1989). *"A General Model and Algorithm for the Dynamic Traffic Assignment Problems"*, Proceedings of the Fifth World Conference on Transport Research, Yokohoma, Japan.

[19] Ran, B., Boyce, D.E. and LeBlanc, L.J. (1993). *"A New Class of Instantaneous Dynamic User-Optimal Traffic Assignment Models"*, Operations Research, Vol. 41, No. 1, 1993.

[20] Sheffi, Y. (1985). Urban Transportation Networks : *"Equilibrium Analysis with Mathematical Programming Methods"*, (Prentice-Hall, NJ).

[21] Sheffi, Y., Powell, W. (1982). *"An Algorithm for the Equilibrium Assignment Problem with Random Link Times"*, Networks 12, pp 191-207.

[22] Smith, M.J. (1991). *"A New Dynamic Traffic Model and the Existence and Calculation of Dynamic User Equilibria on Congested Capacity-Constrained Road Networks"*, Presented at the 71st Annual Meeting of TRB, Washington D.C., 1992.

[23] Wie, B.W. (1990). *"Dynamic Analysis of User Optimized Network Flows With Elastic Travel Demand"*, Presented at the 70th Annual Meeting of TRB, Washington, D.C., 1991.

[24] Ziliaskopoulos, A., Mahmassani, H.S. (1992). *"Design and Implementation of a Shortest Path Algorithm with Time-Dependent Arc Costs"*, Proceedings of the 5th Advanced Technology Conference, Washington D.C., 1992, pp 1072-1093.

[25] Ziliaskopoulos, A., Mahmassani, H.S. (1993). *"A Time-Dependent Shortest Path Algorithm For Real-Time Intelligent*

Vehicle/Highway Systems Applications", forthcoming in Transportation Research Record.

ASSIGNMENT IN THE INTEGRATION OF URBAN TRAFFIC CONTROL AND DYNAMIC ROUTE GUIDANCE

M.G.H. Bell, C.M. Shield, J.M. Anderson

TORG. Dept of Civil Engineering, University of Newcastle, UK

F. Büsch

Siemens AG , Traffic Control Systems, Munich D-81359, Germany

Higher levels of market penetration of dynamic route guidance systems, such as the EURO-SCOUT system from Siemens, require that some form of adjustment is made for the consequences of routeing recommendations. Ideally a procedure is required that generates route recommendations such that no guided driver could have taken a faster route. This is a form of Wardrop equilibrium, and necessitates a dynamic equilibrium assignment algorithm in order to obtain such a set of route recommendations. The determination of a dynamic equilibrium assignment requires the derivation of travel time functions. In this paper it is proposed that link travel time measurements from the dynamic route guidance system be combined with link flow measurements from a traffic signal control system, and that the coefficients of the link travel time functions be estimated recursively. A modified Frank-Wolfe algorithm for solving the dynamic equilibrium assignment problem is proposed. This requires the iterative use of a shortest path algorithm, such as that currently implemented in the EURO-SCOUT system. The properties of the algorithm are explored using a small numerical example.

1. Introduction

Higher levels of dynamic route guidance penetration require that some form of adjustment is made for the consequences of routeing recommendations. This implies some form of dynamic equilibrium assignment algorithm to find a set of route recommendations such that guided vehicles end up having taken best routes. This in turn requires the determination of travel time functions which allow for the impact on link delays, and therefore link travel times, of link traffic flows and, where appropriate, traffic responsive signal control. This paper concentrates on the EURO-SCOUT dynamic route guidance system from Siemens (see von Tomkevitsch, 1990, for a description of the system).

Substantial benefits are envisaged for the integration of traffic signal control and EURO-SCOUT (see CAR-GOES Deliverables 30 and 33). These arise from the complementarity of hardware (shared cables, posts, cabinets, etc.), from the complementarity of the measurements (the potential value of vehicle telegrams for traffic signal control and detector data for EURO-SCOUT), as well as from the improved performance of both systems through links at the control centre level. Dynamic equilibrium assignment constitutes an important aspect of the link between the control centres.

This paper considers dynamic equilibrium assignment from the point of view of dynamic route guidance. Interest therefore stems not so much from its realism as regards a description of unguided driver behaviour but from the point of view of user optimisation for guided drivers. In fact, as a description of unguided driver behaviour, dynamic equilibrium assignment is clearly unrealistic since it requires drivers to have sufficient information about future travel times or delays - the goal of dynamic route guidance.

The equilibrium referred to in this paper corresponds to the first principle set out in Wardrop (1952). It implies that *no guided driver could have taken a less costly alternative route*. It has been pointed out by Smith (1984) for the static case that, although a user optimum is a Wardrop equilibrium, the converse may not be true; the same presumably holds in the dynamic case (see Smith and Ghali, 1990). Situations can arise, and "Braess's paradox" is a good example of one, where actively rerouteing some drivers can reduced everyone's cost (see Murchland, 1970). Nonetheless, this paper focusses on Wardrop equilibrium for guided drivers in the belief that in practice a Wardrop equilibrium will lie considerably closer to the user optimum than to the unguided situation.

Wie et al (1990) have considered the situation where drivers are guided on the basis of current link costs such that at any moment an instantaneous Wardrop equilibrium obtains. This approach, however, leaves out of account the impact of current guidance on future traffic conditions. They identified three approaches to dynamic assignment:

- computer simulation, as in the case of CONTRAM (Leonard et al, 1978);
- the mathematical programming approach (the approach of this paper); and
- the optimal control theoretic approach (Papageorgiou, 1990, as well as Wie et al, 1990).

Consider the case where a dynamic route guidance system is coupled with an urban traffic control system in such a way that both measurements of link travel times and link flows are available (see Fig. 1). The problem is to obtain a set of route recommendations for guided vehicles such that guided vehicles, in so far as they comply with the recommendations, end up having taken best paths. The problem is that link costs may be influenced by the route recommendations, particularly in congested networks with higher levels of market penetration of the guidance system. A form of dynamic equilibrium assignment is therefore required, which in turn needs cost functions for the links.

It is assumed that link cost functions are available which capture sufficiently the behaviour of vehicle responsive traffic signal control. This implies non-separable cost functions, since for any intersection under vehicle responsive signal control, the green time allocated any stage will depend in part on the demand for competing stages. As an example, the SAEWA on-line strategy from Siemens seeks to modify the green time allocation to equalise the anticipated degree of saturation on all stages (see CAR-GOES Deliverable 1 for a description).

Equating cost with time, the parameters of these functions could in principal be estimated recursively from the incoming link travel time and flow measurements. Since time spent in queues constitutes a major component of link cost, and since queues evolve recursively, an autoregressive structure for the cost functions may well be most appropriate. Cybis (1992) notes that while flows in interval t can influence costs in interval t+1, the converse is not

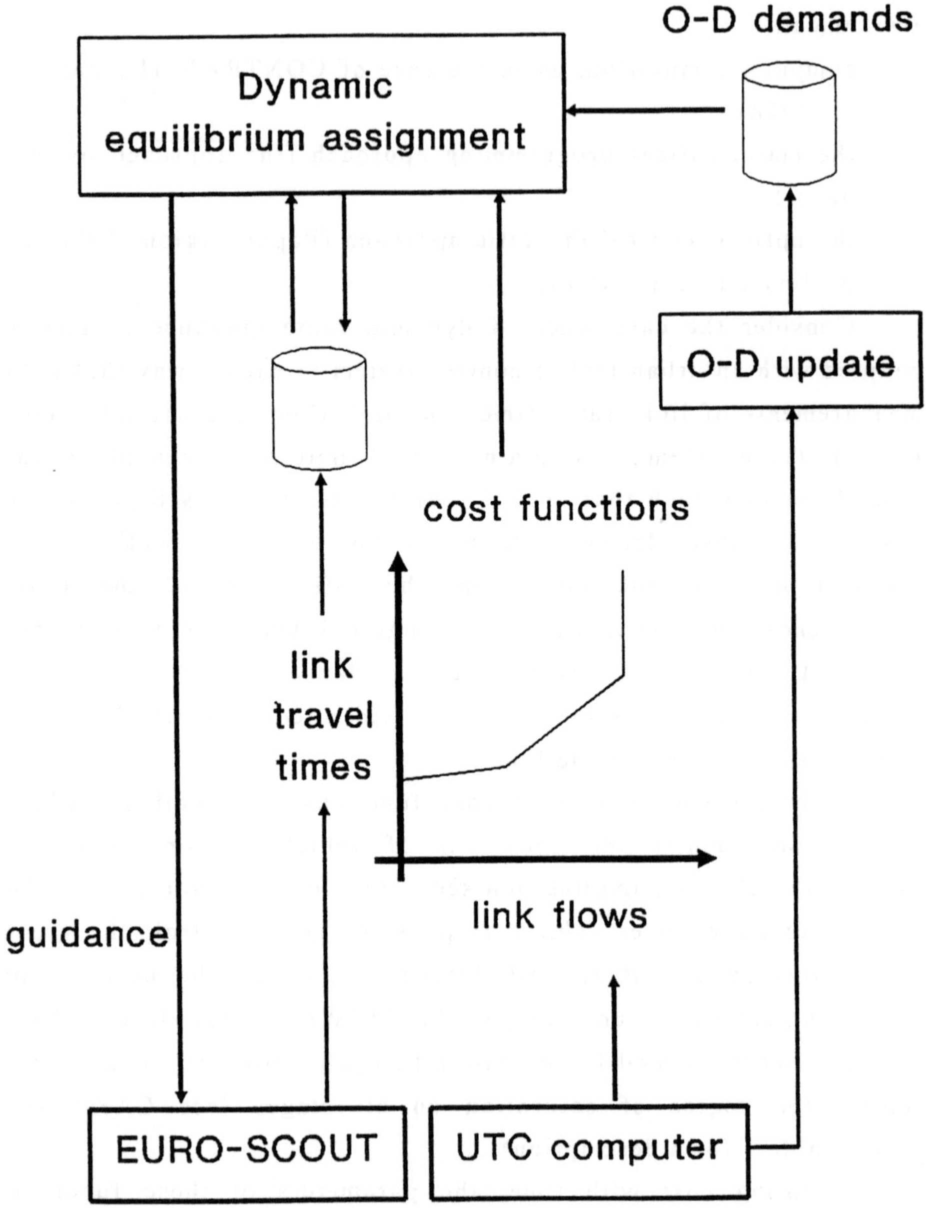

Fig. 1: The integration of DRG and UTC

true, again suggesting autoregressive cost functions.

The recursive estimation of the cost function parameters will be dealt with in another paper. In general, one would expect the relationship on any link between travel time and flow to be increasing with an asymptote at the maximum link flow (the link capacity). This non-linear relationship can be approximated by a piece-wise linear cost function. The cost functions thus estimated have applications other than dynamic route guidance, for example incident detection. When observations of flow and travel time arise which clearly do not conform to the relationship, an incident may have occurred. It would be possible to establish a library of cost functions dependent on the type of link, the form of traffic signal control and other factors; as the form of traffic signal control changes so would the cost function used in the dynamic equilibrium assignment.

2. Dynamic equilibrium assignment

In common with most modelling practice, a time slice approach is adopted here. Time is divided into a number of periods, each with its own set of O-D demands. A period of 5 minutes could be chosen to fit in with the operation of EURO-SCOUT.

Let

h_{ks} = flow on path k starting in period s;

c_{ks} = cost (travel time) on path k for a trip starting in period s;

$$\delta_{klst} = \begin{cases} 1 & \text{if a trip on path k starting in period s is to be found on link l in period t} \\ 0 & \text{otherwise;} \end{cases}$$

v_{lt} = flow on link l in period t;

g_{lt} = cost (travel time) on link l in period t; and

d_{js} = demand for O-D couple j in period s.

The flow conservation relationship states that

$$v_{lt} = \Sigma_k \Sigma_{s \leq t} \delta_{klst} h_{ks} \quad (1)$$

Path costs are given by

$$c_{ks} = \Sigma_l \Sigma_{t \geq s} \delta_{klst} g_{lt} \quad (2)$$

The difficulty of dynamic equilibrium assignment in comparison to steady state equilibrium assignment arises because δ_{klst} depends on link travel times which in turn depend on link flows.

In vector notation

$$\mathbf{v} = \mathbf{E}\ \mathbf{h} \quad (3)$$

and

$$\mathbf{c} = \mathbf{E}^T\ \mathbf{g} \quad (4)$$

The O-D demands impose constraints on the feasible set of path flows. In particular

$$d_{js} = \Sigma_{k \text{ connects } j} h_{ks} \quad (5)$$

Let D_s (referred to as the demand set) represent the set of feasible path flows for trips starting in period s. Subscript s will be dropped in the following.

If **h** is a Wardrop equilibrium then

$$\mathbf{c}^T(\mathbf{h}).(\mathbf{h} - \mathbf{u}) \leq 0 \qquad (6)$$

where **u** is any other set of feasible path flows. In other words, **c** is normal at **h** to D. This is the variational inequality formulation of Wardrop equilibrium originating from Smith (1979).

3. An equivalent optimisation problem

As noted by Cybis (1992), in the dynamic case we cannot as in the static case define an equivalent minimisation problem based on the standard objective function z (the sum of the integrals of the link cost functions). Consequently, an alternative objective function is required. This should be convex, or quasi-convex, and the constraints should also be convex.

Consider the gap function

$$F(\mathbf{h}) = \sup_{\mathbf{u} \text{ in } D} \mathbf{c}^T(\mathbf{h}).(\mathbf{h} - \mathbf{u}) = \mathbf{c}^T(\mathbf{h}).\mathbf{h} - \inf_{\mathbf{u} \text{ in } D} \mathbf{c}^T(\mathbf{h}).\mathbf{u} \qquad (7)$$

This is the maximum value of the expression on the left hand side of **(6)**. It has value 0 only when **h** is a Wardrop equilibrium, and is positive otherwise. Let **u** be the set of feasible path flows which minimise total system cost (in other words $\mathbf{u} = \arg\inf_{\mathbf{u} \text{ in } D} \mathbf{c}^T(\mathbf{h}).\mathbf{u}$). While $\mathbf{c}^T(\mathbf{h}).\mathbf{h}$ is convex if $\mathbf{c}(\mathbf{h})$ is monotone in **h**, $F(\mathbf{h})$ may unfortunately not be convex in **h**. Note that if shortest paths are unique, in the sense that there is only one for each O-D couple, then the gradient of F with respect to, and evaluated at, **h** is

$$\text{grad } F(\mathbf{h}) = \mathbf{c}^T(\mathbf{h}) + (\mathbf{h} - \mathbf{u})^T \mathbf{J}(\mathbf{h}) \qquad (8)$$

where $\mathbf{J}(\mathbf{h})$ is the Jacobian of **c** evaluated at **h**. If shortest paths are not unique, they could be combined (how this can be done in practice is not considered here). If $\mathbf{c}$ is monotone in $\mathbf{h}$ then $\mathbf{J}$ is positive definite, or semi-definite, and

$$\text{grad } F(\mathbf{h}).(\mathbf{u} - \mathbf{h}) \leq 0 \qquad (9)$$

Thus $\mathbf{u} - \mathbf{h}$ is a descent direction although $F(\mathbf{h})$ may not be convex. In this situation, F is quasi-convex in the sense that, when $\mathbf{h}$ is not a Wardrop equilibrium, there is always a descent direction, namely $\mathbf{u} - \mathbf{h}$.

Smith and Ghali (1990a) have demonstrated the monotonicity of delay at the link level, however this does not imply monotonicity of the route cost function $\mathbf{c}(\mathbf{h})$. They did not succeed in proving the existence of a dynamic Wardrop equilibrium. However, if the current set of fastest paths is unique (one path per O-D couple) and if we assume that infinitessimal changes in link travel times do not change $\mathbf{E}$, then

$$\mathbf{J}(\mathbf{h}) = \mathbf{E}^T\mathbf{G}(\mathbf{v})\mathbf{E} \qquad (10)$$

where $\mathbf{v} = \mathbf{E}\ \mathbf{h}$. If $\mathbf{G}$ is positive definite then so is $\mathbf{J}$ (provided there are as many linearly independent path flows as there are link flows), ensuring that the path cost functions are monotonic, at least locally.

4. An iterative algorithm

Consider the case where time is divided into intervals of say 5 minutes. From now into the future, a travel time is associated with each link for each 5-minute interval. There is a database of past link travel times, as in the case of the EURO-SCOUT dynamic route guidance system. This is operated on in an iterative way by the dynamic equilibrium assignment algorithm to produce predictions of future link travel times (see Fig. 2).

Smith and Ghali (1990a) and Cybis (1992) have suggested the

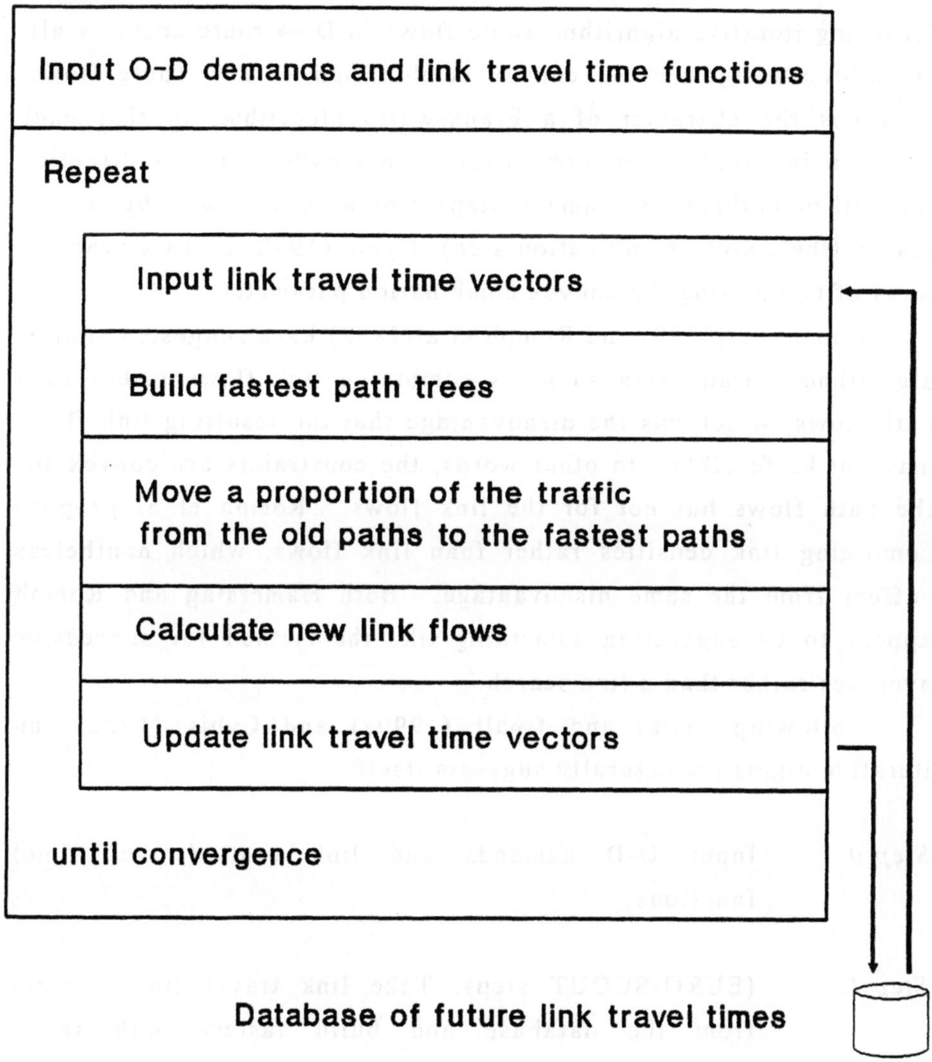

Fig. 2: Dynamic equilibrium assignment algorithm

following iterative algorithm: route flows in D $\rightarrow$ route costs $\rightarrow$ all-or-nothing assignment $\rightarrow$ convex combination $\rightarrow$ new route flows. This has the character of a Frank-wolfe algorithm, in that each iteration involves the determination of a feasible descent direction (the all-or-nothing assignment step) which is followed by a line search (the convex combination step). Cybis (1992) looks at various ways of calculating the convex combination parameter.

Hamerslag(1989) and Romph et al(1992) have suggested similar algorithms. Hamerslag suggests combining link flows rather than path flows, which has the disadvantage that the resulting link flows may not be feasible. In other words, the constraints are convex for the path flows but not for the link flows. Romph et al propose combining link densities rather than link flows, which nontheless suffers from the same disadvantage. Both Hamerslag and Romph appear to be suggesting something like the method for successive averages rather than a line search.

Following Smith and Ghali (1990a) and Cybis (1992), an iterative algorithm naturally suggests itself:

Step 0: Input O-D demands and link cost (travel time) functions.

Step 1: (EURO-SCOUT step). Take link travel time vectors from the database and build fastest path trees. Dijkstra's algorithm may be used. Label new paths.

Step 2: (line search). Move traffic from the current paths to the fastest paths identified in Step 1 so as to minimise F (the gap function). Golden Section search may be used.

Step 3: (load path flows). Convert path flows into link flows using the current link travel time vectors from the database.

Step 4: (update database). Calculate new link travel times on the basis of link flows derived in Step 3. Update the database.

Step 5: (stopping criterion). Test for convergence. If not yet sufficiently converged, return to Step 1; otherwise produce link choice proportions p_{ja} (proportion of traffic between O-D couple j to use link a). The gap function F may be used to assess convergence.

The principal difficulty regarding convergence of this algorithm lies in Steps 3 and 4, because the allocation of path flows to particular links in particular time intervals depends on the link travel times and therefore on the link flows themselves. This means that the link travel times used in Step 3 may not be consistent with the link travel times calculated in Step 4. The algorithm is currently under investigation. It is to be hoped that any inconsistency diminishes with the iterations.

5. Numerical example

To explore the behaviour of the algorithm described in the preceding section and to compare its performance to that of the steady state Frank-Wolfe algorithm, the network shown in Fig. 3 has been adopted. There are 12 links, 1 origin, 1 destination, 7 nodes and 6 paths. Linear travel time functions of the following linear form have been assumed:

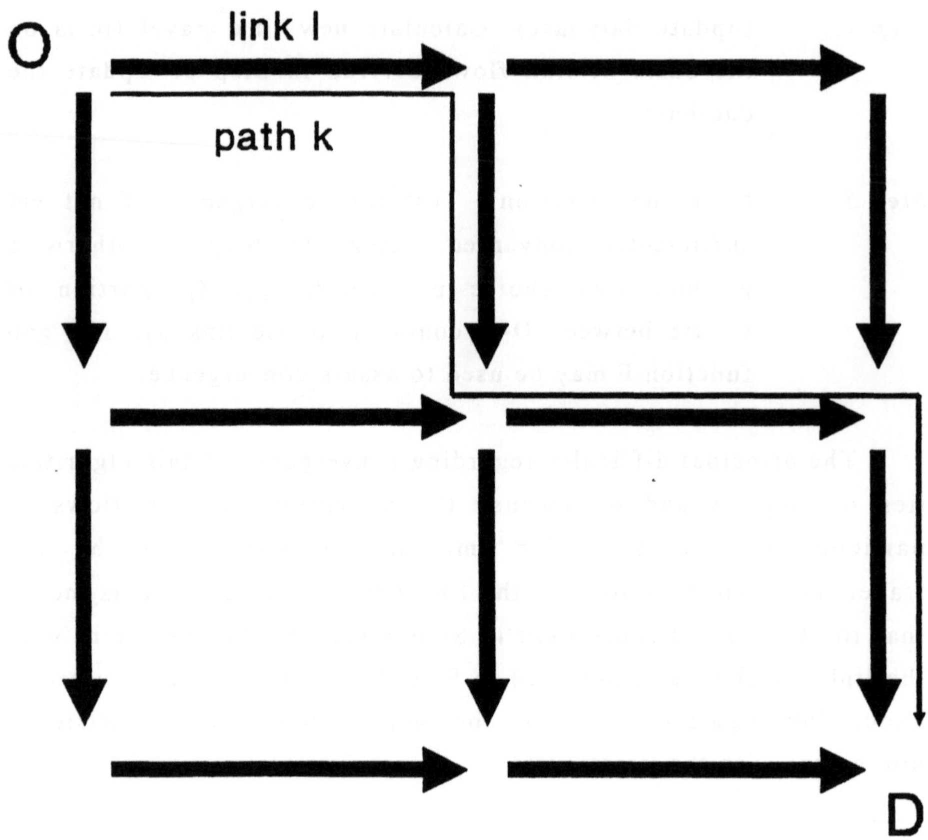

Fig. 3: The example network

$$
\begin{aligned}
|y_1| &= |10| + |0.3\ 0.0\ 0.0\ 0.0\ 0.0\ 0.0\ 0.0\ 0.0\ 0.0\ 0.0\ 0.0\ 0.0|\ |x_1| \\
|y_2| &= |10| + |0.0\ 0.3\ 0.0\ 0.0\ 0.0\ 0.0\ 0.0\ 0.0\ 0.0\ 0.0\ 0.0\ 0.0|\ |x_2| \\
|y_3| &= |10| + |0.0\ 0.0\ 0.3\ 0.0\ 0.0\ 0.0\ 0.0\ 0.0\ 0.0\ 0.0\ 0.0\ 0.0|\ |x_3| \\
|y_4| &= |10| + |0.0\ 0.0\ 0.0\ 0.3\ 0.0\ 0.9\ 0.0\ 0.0\ 0.0\ 0.0\ 0.0\ 0.0|\ |x_4| \\
|y_5| &= |10| + |0.0\ 0.0\ 0.0\ 0.0\ 0.3\ 0.0\ 0.3\ 0.0\ 0.0\ 0.0\ 0.0\ 0.0|\ |x_5| \\
|y_6| &= |10| + |0.0\ 0.0\ 0.0\ 0.3\ 0.0\ 0.3\ 0.0\ 0.0\ 0.0\ 0.0\ 0.0\ 0.0|\ |x_6| \\
|y_7| &= |10| + |0.0\ 0.0\ 0.0\ 0.0\ 0.3\ 0.0\ 0.3\ 0.0\ 0.0\ 0.0\ 0.0\ 0.0|\ |x_7| \\
|y_8| &= |10| + |0.0\ 0.0\ 0.0\ 0.0\ 0.0\ 0.0\ 0.0\ 0.3\ 0.0\ 0.0\ 0.0\ 0.0|\ |x_8| \\
|y_9| &= |10| + |0.0\ 0.0\ 0.0\ 0.0\ 0.0\ 0.0\ 0.0\ 0.0\ 0.3\ 0.0\ 0.3\ 0.0|\ |x_9| \\
|y_{10}| &= |10| + |0.0\ 0.0\ 0.0\ 0.0\ 0.0\ 0.0\ 0.0\ 0.0\ 0.0\ 0.3\ 0.0\ 0.3|\ |x_{10}| \\
|y_{11}| &= |10| + |0.0\ 0.0\ 0.0\ 0.0\ 0.0\ 0.0\ 0.0\ 0.0\ 0.3\ 0.0\ 0.3\ 0.0|\ |x_{11}| \\
|y_{12}| &= |10| + |0.0\ 0.0\ 0.0\ 0.0\ 0.0\ 0.0\ 0.0\ 0.0\ 0.0\ 0.3\ 0.0\ 0.3|\ |x_{12}|
\end{aligned}
$$

The y-values are the fitted travel times, the intercept terms correspond to the empty link travel times, the x-values are the link flows and the matrix of coefficients indicate the incremental effect of link flow on link travel time. Note that the functions are not separable. In practice, given link travel time observations from a dynamic route guidance system such as EURO-SCOUT and traffic flow measurements from a traffic signal control system, these coefficients can be estimated recursively. As indicated in Section 1, linear functions are unlikely to be sufficiently accurate in practice.

A pulse of 100 trips are assigned to the network. In the first iteration a path is selected through the network using Dijkstra's algorithm. The network demand is loaded onto the path, the flow on the path's links are calculated and the link travel times are adjusted to match with the new link flows. Whilst the travel times are adjusted, the program ensures that each node is reached in the correct time slice. The details of the initial path through nodes 1,2,4,7 and 3 are as shown in Table 1.

Link	Flow	Cost	Start Time Slice
1 - 2	100	40	0
2 - 4	100	40	8
4 - 7	100	40	16
7 - 3	100	40	24

Table 1. The initial path after one iteration

In the second iteration, a new fastest path is determined. In this iteration, all the flow is not placed on the new path but spread across the two existing paths in a ratio descibed by w in the equation:

$$f(w) = (\mathbf{h} - \mathbf{u})^T \mathbf{c}((1-w)\mathbf{h} + w\mathbf{u}) \qquad \mathbf{(11)}$$

Parameter w is determined by solving f(w) = 0 by, for example, the bisection method.

After the second iteration, there are two paths; one through 1, 2, 4, 7 and 3 and the other through 1, 5, 6, 7, and 3. These are shown in Table 2.

Link	Flow	Cost	Start Time Slice
1 - 2	50	25	0
1 - 5	50	25	0
2 - 4	50	25	5
4 - 7	50	40	10
5 - 6	50	25	5
6 - 7	50	40	10
7 - 3	100	40	18

Table 2. Link flows and their costs after two iterations

Subsequent iterations proceed in a similar fashion. Solving f(w) = 0 becomes more complex. After the twentieth iteration the latest fastest path is through 1,5,8,9 and 3. Link costs and flows are shown in Table 3.

Link	Flow	Cost	Start Time Slice
1 - 2	39.5	21.8	0
1 - 5	60.5	28.1	0
2 - 4	39.5	21.8	4
4 - 7	39.7	26.4	8
5 - 6	29.7	18.9	5
5 - 8	30.8	19.2	5
6 - 7	15.2	26.4	8
6 - 9	14.5	23.6	8
7 - 3	54.7	40.0	13
8 - 9	30.7	23.6	8
9 - 3	45.3	40.0	12

Table 3. The links and their costs after twenty iterations

It is noticeable that after several iterations the flows and corresponding travel times become balanced across the links. A high flow on a particular link will reflect its inclusion in several paths. The algorithm appears to converge satisfactorally.

6. Conclusions

Higher levels of market penetration of dynamic route guidance systems, such as the EURO-SCOUT system from Siemens, will require that some form of adjustment is made for the consequences of routeing recommendations. Ideally a procedure is required that generates route recommendations such that *no guided driver could*

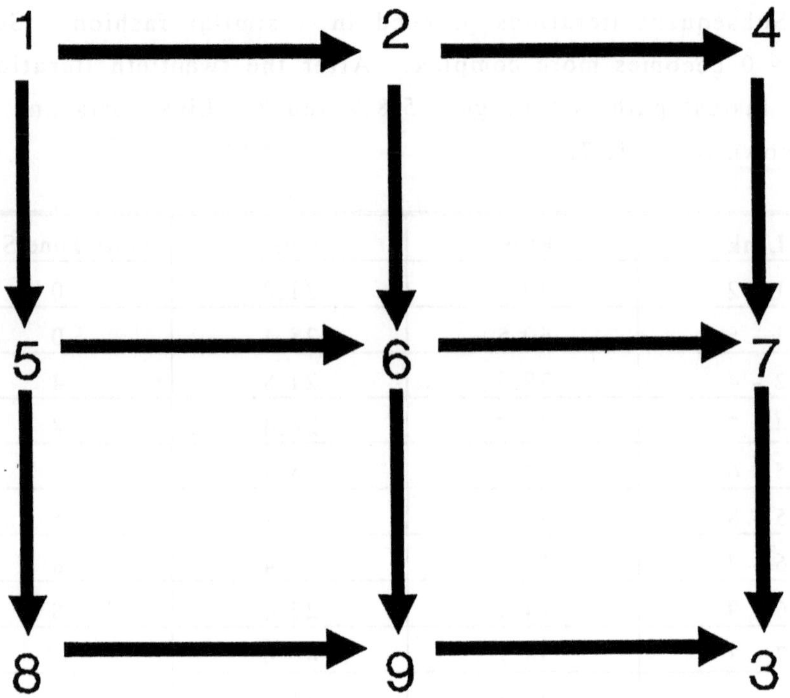

Fig. 4: Network used in numerical example

have taken a faster route. This is a form of user equilibrium, and necessitates a *dynamic equilibrium assignment algorithm* in order to obtain such a set of route recommendations.

The determination of a dynamic equilibrium assignment requires the derivation of travel time functions. In this paper it is proposed that link travel time measurements from the dynamic route guidance system be combined with link flow measurements from a traffic signal control system, and that the coefficients of the link travel time functions be estimated recursively. The effects of traffic responsive signal control may be allowed for by including in the functions the demands for competing stages. For any link, an autoregressive term can be included to allow for temporary over-saturation.

The Frank-Wolfe algorithm is generally used to find a steady state equilibrium assignment. In this paper, a modified Frank-Wolfe algorithm for solving the dynamic equilibrium assignment problem is proposed. This requires the iterative use of a shortest path algorithm, such as that currently implemented in the EURO-SCOUT system. The properties of the algorithm are explored using a small numerical example. The convergence requirements of the algorithm need further investigation.

The solution of the dynamic equilibrium assignment problem is seen by industry as being of great importance for the next generation of dynamic route guidance system, which will be integrated with urban traffic control. In subsequent work, it is intended to test procedures on a small network under urban traffic control in Berlin or Köln.

References

[1] Bell, M.G.H. & Busch, F. (1992) *"The integration of urban traffic control and dynamic route guidance: The role of dynamic assignment"*. Vehicle Navigation and Information Systems, Oslo,

September.

[2] CAR-GOES Deliverable 1: *"Report on existing traffic control systems"*. May 1989.

[3] CAR-GOES Deliverable 30: *"Formulation of integrated strategies"*. July 1991.

[4] CAR-GOES Deliverable 33: *"Final report"*. September 1991.

[5] Cybis, H.B.B. (1992) *"A dynamic user equilibrium assignment model"*, Proceedings of the UTSG Conference, Newcastle, January.

[6] Hamerslag, R. (1989) *"Dynamic assignment in three-dimensional timespace"*. Transportation Research Record 1220.

[7] Ho, J.K. (1990) *"Solving the dynamic traffic assignment problem on a hypercube multicomputer"*. Transportation Research, Vol. 24B, 443-451.

[8] Leonard, D.R., Tough, J.B. and Bagueley, P.J. (1978) *"CONTRAM, a traffic assignment program for predicting flows and queues in peak periods"*. TRRL Report 841.

[9] Merchant, D. & Nemhauser, G. (1978) *"A model and algorithm for the dynamic traffic assignment problem"*. Transportation Science, Vol. 12(3), 183-199.

[10 Murchland, J.D. (1970) *"Braess paradox of traffic flow"*. Transportation Research, Vol. 4, 391-394.

[11] Papageorgiou, M. (1990) *"Dynamic modelling, assignment and route guidance in traffic networks"*. Transportation Research, Vol 24B, 471-495.

[12] Romph, E. de, Grol, H.J.M van & Hamerslag, R. (1992) *"A dynamic assignment model"*. Paper presented to the Second International Seminar on Urban Traffic Networks, Capri, July.

[13] Smith, M.J. (1979) *"The existence, uniqueness and stability of traffic equilibrium"*. Transportation Research, Vol. 13B, 295-304.

[14 Smith, M.J. (1984) *"Two alternative definitions of traffic equilibrium"* Transportation Research, Vol. 18B, 63-65.

[15] Smith, M.J. & Ghali, M. (1990a) *"Dynamic traffic assignment and dynamic traffic control"*. Proceedings of the 11th International Symposium on Transportation and Traffic Theory, Yokohama, July.

[16] Smith, M.J. & Ghali, M. (1990b) *"The dynamics of traffic assignment and traffic control: A theoretical study"* Transportation Research, Vol. 24B, 409-422.

[17] Wie, B. Friesz, T. and Tobin, R. (1990) *"Dynamic user optimal traffic assignment on congested multidestination networks"* Transportation Research, Vol. 24B, 431-442.

[18] von Tomkevitsch, R. (1990) *"Evolution of road/vehicle communications for road transport informatics"*. Proceedings of the International Conference on the Highway/Vehicle Link for Road Users, Paris.

[19] Wardrop, J.G. (1952) *"Some theoretical aspects of road traffic research"*. Proceedings of the Institution of Civil Engineers, Part II, Vol. 1(36), 325-362.

DESIGN ASPECTS OF ADVANCED TRAVELER INFORMATION SYSTEMS

Isam Kaysi

Faculty of Engineering and Architecture
American University of Beirut
P.O. Box 11-0236 Beirut Lebanon

Moshe Ben-Akiva

Department of Civil and Environmental Engineering
Massachusetts Institute of Technology
Room 1-181
Cambridge MA 02139

Andre de Palma

University of Geneva, Dept. COMIN
102 Boulevard Carl-Vogt, CH-1211
Geneve 4 Suisse

The acceptance of route guidance advice by motorists is expected to occur only whenever such advice is experienced to be valid and reliable. Two factors may cause route guidance systems to provide motorists with unreliable advice: (1) the traffic information constituting the basis for guidance advice is inaccurate; and, (2) the impact of a large fraction of motorists responding to the guidance, and the subsequent overreaction that occurs, is ignored whenever the guidance advice is being set.

In this paper, the impact of the above factors is investigated for a small prototypical network using simulation. Factors in the design of traveler information systems that are likely to provide guidance with higher validity and reduce (or eliminate) the occurrence of overreaction are identified. It is concluded that need exists for predicting traffic conditions as the basis for setting guidance advice, and that distributing traffic over a number of alternate routes or a high temporal update frequency can be used to reduce or eliminate overreaction. Finally, the manner in which the quality of predictions affects the benefits to be gained from guidance is analyzed.

1. Introduction

The increasing level of urban traffic congestion and the detrimental effect it has on mobility, accessibility, and safety in urban areas has become a commonly observed phenomenon. To alleviate this situation, transportation planners and traffic managers are looking for ways to improve the utilization of existing facilities with enhanced traffic management and operations schemes. It has been suggested that during the coming decades, efficient operation of existing road networks will be achieved through dynamic traffic management schemes which make use of the available and anticipated advanced technologies [13]. Within this context, Intelligent Vehicle-Highway Systems (IVHS) are currently being developed. These IVHS systems envision the linking of road infrastructure, vehicles, and drivers using advanced communication technology, computers, information display equipment, and traffic control systems.

Advanced Traveler Information Systems

Two IVHS schemes which are expected to provide such improvements are Advanced Traveler Information Systems and Advanced Traffic Management Systems. Other developments within the IVHS scope include Automated Vehicle Control Systems which are expected to help drivers perform certain vehicle control functions. The implementation of these schemes using advanced technologies is envisioned to produce an integrated vehicle/roadway transportation system.

In this context, Advanced Traveler Information Systems (ATIS) based on modern information technology is expected to play an important role in reducing traffic congestion and improving traffic flows and safety. ATIS have the potential to reduce delays due to both incident and recurrent congestion by providing information to motorists with respect to alternative paths to their destinations and to actual traffic conditions on links of interest using a combination of roadside signals and onboard systems. However, the level of benefits to be derived from such advanced traveler information systems depends to a large degree on the quality of the guidance or information being provided.

Guidance provided to drivers may conceptually be based on one of the following three categories of traffic information [3]:

- **Historical Information** - information which describes the state of the transportation system during previous time periods.
- **Current Information** - the most up-to-date information about current traffic conditions.
- **Predictive Information** - information concerning expected traffic conditions during subsequent time periods when travel can occur.

As will become clear later, these different categories have a significant impact on the quality of guidance that is provided to drivers. Most existing systems rely on historical or current information while primitive predictions of traffic conditions are attempted in a few others.

Potential Adverse Impacts of Guidance

In considering the potential benefits of alternative traveler information systems, it is also essential to evaluate the potential adverse effects that more accurate predictive information may induce. Several behavioral phenomena may negate some of the beneficial effects of improved information, prime among them being overreaction [3].

Consider a situation in which a substantial fraction of drivers receive *information on traffic conditions*. *Overreaction* occurs when drivers' reactions to traffic information cause congestion to transfer from one road to another. It may also generate oscillations in road usage. Overreaction may happen if too many drivers respond to current information on traffic conditions.

Several instances of this type of overreaction have been reported. For example, a French radio service "Bison Fute" which advises drivers going on holiday how to plan their itinerary and optimal day for departure has experienced overreaction in past years. Another instance is provided by Fujii [7], discussed in Boyce [6]: "If we use a simple scheme such as shortest-time path guidance (when most of the vehicles are under control of the system), it might cause oscillation of traffic flow among alternative routes. In order to avoid this situation, an advanced scheme such as route distributing guidance should be prepared." Using a simulation model, Kobayashi [11] found that oscillatory behavior may occur when road guidance by shortest path route is extended to a large proportion of drivers (90%).

Therefore, it becomes clear that overreaction is mainly due to a large fraction of motorists responding to guidance advice. This results in oscillations in road usage and shifting of congestion from one road to another. These oscillations, in turn, are likely to result in invalid guidance being provided to motorists

The evidence of overreaction in prior research indicates that the two main features of this phenomenon, as depicted in Figure 1, are:

- the existence of an "optimal fraction" of guided drivers that provides highest total benefits
- the occurrence of disbenefits (or reduced benefits) if the guided fraction exceeds a certain threshold

Figure 1. Overreaction Features

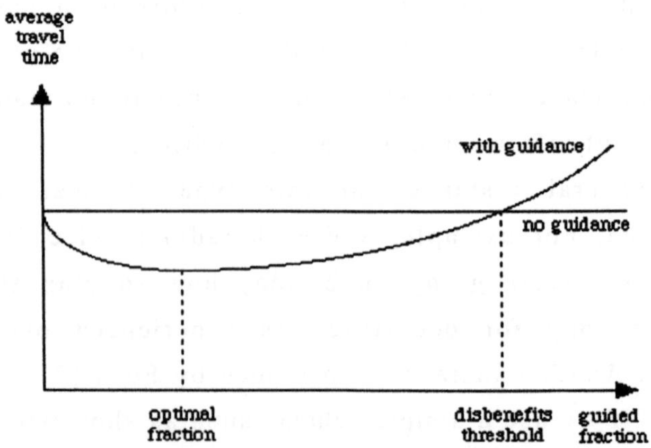

Desirable Properties of Route Guidance

ATIS are expected to generate and disseminate traveler information that can be used for real-time diversion of traffic. Guidance provided to drivers has to possess a number of *desirable properties* if it is to be useful for dynamic traffic management and if it is to be adopted by drivers.

To be useful for dynamic traffic management, guidance has to be able to *respond in real-time* to emerging traffic conditions and *anticipated* demand levels in such a way that it *reduces the overall extent of congestion*. In addition, since traveler information may prove to be counterproductive in some cases, guidance should be provided in such a way that it *reduces the potential adverse impacts of more accurate information*.

From the drivers' point of view, guidance should be *valid* in the sense that, for a given departure time, the travel time a driver experiences on the route he was guided to should be shorter than that on alternate routes. Moreover, guidance has to improve the *reliability* of travel times experienced by drivers by reducing variability and eliminating unusually long travel times.

If drivers receive guidance which is not characterized by these desirable properties, they will not have confidence in the information system. As a result, drivers may not comply with the guidance being provided, in which case the dynamic traffic management potential of the information system would not materialize.

The objectives of this paper are: (1) to analyze ATIS design features that will be capable of achieving the above-mentioned properties; and (2) to conduct prototypical evaluation analyses of the value of some of the design elements. This second objective will be accomplished using simulation tools.

2. ATIS Design Issues

The actual benefits realized from traveler information systems will depend heavily on the quality of the traffic information provided to drivers [5]. For instance, the occurrence of adverse impacts due to improved information may not be avoided if the provision of driver information is not based on a well-designed implementation framework (see [10]). Moreover, the unavailability of accurate traffic information may result in poor guidance directives being provided to drivers. The likely impact would be a gradual loss of confidence in the system and its eventual abandonment. For example, a recent survey of users of the ALI-SCOUT demonstration project (Janko, [8]) found a very low compliance rate among commuters receiving guidance who, after some experimentation, came to believe that they could do better than the guidance being provided by the system. A possible reason for this poor system performance is the fact that guidance is sometimes being based on overly inaccurate predictions of traffic conditions.

Such observations support the need for an ATIS implementation scheme that is designed to provide guidance which is characterized by the desirable properties listed above.

The achievement of all the desirable properties associated with guidance being provided to drivers represents a significant challenge to developers of ATIS for several reasons. First, the implementation of route guidance necessitates a systems-based treatment of the guidance problem including the identification of modular functions that have to be developed and the options available for each, as well as the isolation of the many parameters that enter into the problem. Moreover, modeling tools required to implement the modular functions referred to above have to be specified and developed.

ATIS Design Features

The design features that should be embodied in any proper ATIS implementation are:

Feature 1: Projection of Traffic Conditions

A dependable look-ahead capability is central to the success of guidance systems since it permits the advice to be based on traffic conditions that are most likely to materialize in the future. This advice is expected to be more responsive to such conditions and providing more opportunities for reducing overall congestion levels. Guidance with such a capability has the best potential for improving the travel time reliability of drivers. In contrast, guidance based solely on the use of historical data or available measurements from surveillance equipment is not able to anticipate future variations in demand levels or the buildup of congestion at specific locations in the network.

Feature 2: Nature of Projection Methodology

There exists a need for a congestion prediction capability that provides accurate projections of future traffic conditions. Such a module would make use of current traffic conditions, predicted demand levels, guidance being provided, and anticipated driver response to guidance in making its predictions of future congestion in the network. A congestion prediction capability results in guidance which is more responsive to (a) future shifts in demand patterns, (b) upcoming variations in network capacity due to traffic control actions, and (c) the potential adjustments in driver travel decisions. Therefore, the system would be in a better position to guide drivers to routes that will be less heavily congested when they use them.

Feature 3: Consistency Requirements

Consistency has to be maintained between the guidance being provided to drivers and the predicted traffic conditions. That is, the guidance being provided is *based on a prediction of the future traffic conditions* which in turn will depend on the guidance being provided. Achieving consistency should prove to be quite helpful in maintaining high levels of guidance validity. The consistency issue may be depicted as shown in Figure 2.

Guidance Strategies

When specifying the guidance strategy to be adopted in the ATIS, three major issues have to be addressed, namely:

(1) What *logic* should the guidance strategy adopt in order to provide drivers with guidance advice? Should *information* or *route directives* be provided to drivers? Should route directives be based on *shortest path* guidance or on *route distributive* guidance? How should route distributive guidance be implemented? How can the guidance logic ensure guidance/projection consistency?

(2) Given that guidance provided at specific locations of the network can be updated periodically, what is the impact of *temporal update frequency* on the effectiveness of routing strategies?

(3) Given that drivers may receive guidance updates at various locations as the move through the network towards their destinations, what is the impact of *spatial update frequency* on the effectiveness of routing strategies?

Figure 2. Consistency Requirements

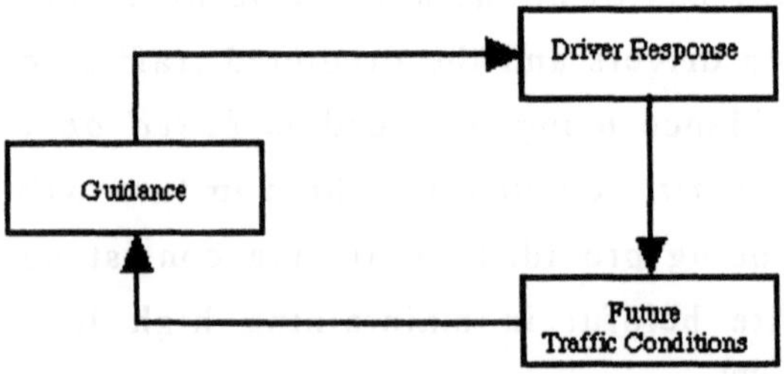

Dealing with Overreaction

Next we discuss possible schemes within the overall ATIS design to reduce the potential adverse impact of overreaction.

Effect of Type of Information: When *current* traffic information is being provided, overreaction is likely to take place if drivers fail to consider or underestimate the potential responses of other drivers. However, in such a case overreaction may be avoided if drivers' decisions are based on correct expectations of other drivers' reactions. For example, some drivers may not shift to a reportedly faster route because they anticipate a rush of the other drivers to that route.

Predictive traffic information that is based on a simple statistical projection is not able to foresee the occurrence of overreaction since driver response to guidance does not enter into the predictions. In a proper ATIS design, guidance should be based on *predictive* traffic information which utilizes explicit models of *dynamic driver travel decision making* in response to guidance. In such a case, overreaction may be avoided if the congestion prediction capability accurately predicts driver behavior and reaction to information, including the fraction of drivers who will follow the guidance advice.

Effect of Guidance Logic: Using shortest path guidance is likely to fail if the guidance strategy ignores the impact of drivers' reactions to information. This scheme does not perform well in congested networks since the single route to which drivers are guided will become congested quickly as drivers switch to it and it would no longer represent the best route. This results in an inconsistency between predicted and realized traffic conditions and leads the guidance system to suggest (or drivers to switch to) other routes thus resulting in an oscillatory behavior which is characteristic of overreaction cases.

A "consistency" feature in the guidance logic should ensure that an alternate guidance mechanism (i.e. distributive route guidance) is resorted to whenever the occurrence of overreaction cannot be avoided using single route guidance.

Effect of Temporal Update Frequency:

More frequent guidance updates are also likely to reduce the potential occurrence of overreaction since that would limit the number of vehicles that are exposed (and react) to any specific guidance message.

In summary, we expect that the occurrence and severity of overreaction will be reduced either by (a) basing the guidance on predictive traffic information that accounts for driver reaction, by (b) providing guidance in such a way that traffic is distributed over several reasonable paths if single route guidance is likely to cause overreaction, or by (c) performing more frequent guidance updates. In what follows we investigate these ideas using a small prototypical network.

3. The Sample Network

Network Structure

We now introduce the network that was used in the evaluation study. The network was chosen to be simple but at the same time to include features that will enable an analysis of the factors involved in the design of traveler information systems. The network is presented in Figure 3 and consists of 4 links (labeled 1 to 4). Each of these four links has a potential bottleneck at its end.

Figure 3. Prototypical Network

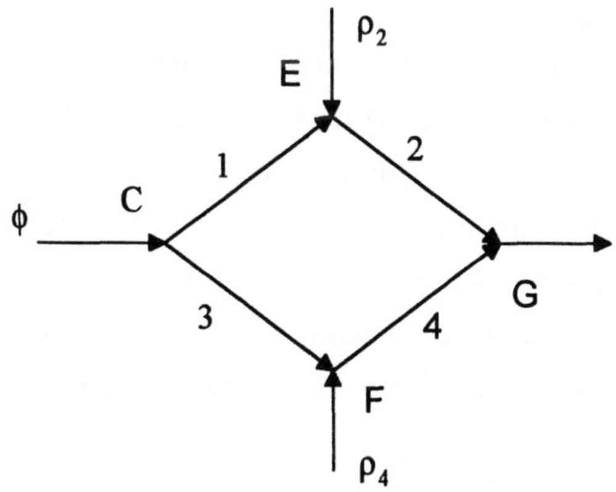

O-D Flows

Associated with the network are 3 O-D flows all of which are destined to node G, as follows:

- $\Phi(t)$ entering at node C
- $\rho_2(t)$ entering link 2 at node E
- $\rho_4(t)$ entering link 4 at node F

Flow $\Phi(t)$ has only two possible routes to the destination. Moreover, flows $\rho_2(t)$ and $\rho_4(t)$ have no route choice as they have to follow links 2 and 4, respectively. These latter flows represent "external demands" that are loaded onto the network and affect its traffic conditions but are outside the scope of the guidance system in the sense that they can not be controlled or diverted.

Guidance Provision

The network being used consists mainly of two routes in parallel. The guidance provision setup that will be used in the evaluation study consists of drivers receiving route guidance at point C concerning which of the two routes they should follow to their destination, point G.

In the network being used, $\Phi(t)$ represents the incoming flow which is to be guided and which arrives at time t at point C. $\Phi_1(t)$ and $\Phi_3(t)$ are the flows departing from C at time t and following routes 1 and 3, respectively ($\Phi(t) = \Phi_1(t) + \Phi_3(t)$).

If traffic to be guided constituted the only flow on the two routes (no side flows) then information on the current traffic conditions would be sufficient to compute future traffic conditions and to guide traffic properly, in the absence of accidents or any other factors that could vary route capacity. However, for the network adopted here the side flows need to be taken into consideration when providing guidance at C since future traffic conditions will depend on the future arrival of side flows in the network.

Network Performance

The link performance functions used to determine the travel times on links, or the time a vehicle spends on a specific link, assume that each link consists of a running section that has a constant free flow travel time and a bottleneck at the end. The travel time on the running sections of the four links is set to zero.

As such, we will consider delays at bottlenecks as the only measure of travel disutility faced by traffic. Moreover, each bottleneck is modeled as a deterministic queue which possesses a specific discharge capacity. The network has four bottlenecks in total (of capacities s_1, s_2, s_3, and s_4), two on each of the parallel routes. Queues in the bottlenecks build up or dissipate based on the level of the incoming flows relative to the bottleneck discharge capacity.

Topology

The network being used in the case study represents an extreme case of network topology as it relates to spatial update frequency. This network represents the case in which traffic that is guided at point C cannot switch from the route selected at C. As such, *no possibilities* exist for adapting to evolving traffic conditions or for recovering from mistakes that were made when guiding vehicles at C.

4. Simulation Study

Network Performance

The flows on the network are subdivided into packets with the size of each packet determined by the incoming flows during each interval of time. The movement of the first and last vehicles in each packet is simulated. Variations in link capacities are assumed to occur as a result of simulated accidents. Accidents are described by their frequency of occurrence, the affected links, and their duration (Kaysi, [9]).

Driver Response to Guidance

The behavior of both guided and unguided drivers was accommodated within the case study since both categories of drivers affect the overall traffic conditions on the network. The behavior of each of these categories is described below.

Guided Vehicles: It is assumed that drivers of guided vehicles will have a specific probability of complying with the guidance. This probability of compliance, together with the percent of equipped vehicles, will determine the probability that any vehicle will be considered to be "guided". This "guided probability" constitutes an important parameter of the case study and can be interpreted as follows:

guided probability = probability vehicle is equipped * probability of compliance given that the vehicle is equipped;

Unguided Vehicles: In order to model the behavior of unguided drivers, a stochastic user equilibrium solution was obtained for the network being used. Such a solution assumes that the drivers' perceptions of travel times on the different links in the network are random (see Sheffi, [12]) and assigns the O-D demand among alternative routes in such a way that the *perceived* travel time on the various routes between the O-D pair is equal.

The stochastic equilibrium conditions were obtained using *average* O-D demand patterns. Route choices made by drivers are described by a binary choice logit model where the path travel time represents the only factor in the driver's disutility function. The time-of-day-dependent patterns of path flows that correspond to stochastic user equilibrium were obtained for each network (see [2]).

Based on these patterns, the fractions of vehicles that follow each of the two routes leaving node C for different times of the day were obtained. These "stochastic equilibrium fractions" simulate the average behavior of drivers for a no-information, average-conditions case.

In the case study, drivers who are unguided at point C are assumed to follow each of the two routes with a probability equal to the stochastic equilibrium fractions. In other words, the unguided vehicles are assumed to distribute themselves, on average, according to the same fractions as the ones that occur in the stochastic equilibrium case. The rationale behind this assumption is that unguided drivers are expected to follow their "habitual behavior" which responds to average traffic conditions and, therefore, cannot adjust to day-specific random effects such as accidents or demand variations. Moreover, such drivers are not expected to be able to discriminate fully between traffic conditions that vary by time of day. In the case study it is assumed that unguided drivers are only able to discriminate amongst three periods of the day which experience significantly different traffic volumes (uncongested, congested, and moderately congested). Therefore, the stochastic equilibrium probabilities are averaged over these three periods of the day. These averages will be referred to as "average" probabilities.

Aggregate Driver Behavior: For traffic leaving C during a specific time interval t, and assuming the guidance message indicates that drivers should follow route 1 (i.e. link 1), the probability that any individual vehicle will follow route 1 is:

$P_1(t)$ = Prob. [vehicle is guided] * Prob. [compliance | guided] +

Prob. [vehicle is unguided] * Prob. [follows route 1 | unguided]

$P_1(t)$ = Guided Probability + (1 - Guided Probability) * $AP_1(t)$

where $AP_1(t)$ is the "average" probability that a driver will follow route 1 during time interval t. The two terms in the equation refer to guided and unguided vehicles, respectively. A "guided" vehicle will follow the guidance since the compliance rate is already factored into the "guided probability". For an unguided vehicle, the probability that it follows route 1 is equal to the "average" probability of following route 1 during that time interval.

Given that each vehicle departing C during time interval t has a probability $P_1(t)$ of following route 1, the total number of vehicles following route 1 during that interval of time may be represented by a binomial distribution. The reasonably large number of cars leaving during each interval permitted the use of the normal approximation to the binomial in the case study to determine the number of vehicles following each route during that interval.

Experimental Factors

The case study involves an analysis of the impact of the following experimental factors:

Guided Probability: The relation between guidance effectiveness and the probability of vehicles being guided is investigated in the case study to determine whether adverse impacts of information materialize as the guided probability increases.

Information Basis for Guidance: The impact of two major issues related to the basis for guidance provision is evaluated in the case study (see [9] for a more detailed discussion).

- Prediction of Traffic Conditions: The use of "predicted" traffic conditions as opposed to other types of traffic information (such as "instantaneous" measurements of queue lengths) for setting driver guidance is evaluated.
- Prediction Quality: In order to provide projections of future traffic conditions, an O-D prediction capability may be

required. The simulation study evaluates the impact of the quality of such predictions on the guidance effectiveness based on a comparison of two cases: (1) perfect O-D predictions are available ("perfect"), and, (2) O-D predictors are capable of predicting the overall day to day variation, but not the specific within day period-by-period variations ("day").

Measures of Effectiveness

In order to evaluate the effectiveness of guidance scenarios under various operating conditions and guidance parameters, measures of effectiveness are required. The following measures were selected:

Average Delay: The average delay of all drivers (both guided and unguided) who travel between C and G during the analysis period is used as one measure of effectiveness of the guidance scenarios compared to each other and to the no guidance case.

Guidance Validity: The percent of drivers who are "correctly guided" conveys a measure of how valid the guidance was. Specifically, this measure indicates the percent of guided drivers who experienced travel times that were shorter than the ones experienced by drivers who left during the same time interval but followed the other route to the destination. The importance of this measure lies in the fact that driver compliance with guidance is expected to depend to a large degree on whether the guidance they receive is valid or not, in the sense indicated above.

Travel Time Reliability: The ability of ATIS to provide guidance that will allow guided vehicles to avoid long delays is expected to be one of the major advantages of such systems. The case study evaluates this aspect by assessing the maximum travel time experienced by guided and unguided vehicles.

Simulation Results

Simulation results are presented in Figures 4 to 8. The first 3 figures compare the performance of the no guidance, instantaneous, and predictive guidance cases based on average delay, maximum delay of guided vehicles, and guidance validity for a guidance update frequency of 2.5 minutes. The figures indicate that the predictive (day) case performs consistently better than the instantaneous case based on all measures. More specifically, considering Figure 4, it can be observed that all guided probability levels in the predictive (day) case provide reductions in average delay relative to the instantaneous case. In addition, no disbenefits occur with predictive guidance even at high levels of guided probability. As far as maximum delay is concerned (Figure 5), predictive guidance offers reductions of close to 10% relative to guidance based on instantaneous information. Moving to Figure 6, it can be observed that while guidance validity improves when guidance is based on predictive information, the validity levels sustained at higher levels of guided probability (close to 0.6) are not likely to be sufficient for travelers to develop confidence in the guidance system.

Figure 4. Average Delay
2.5 minute updates

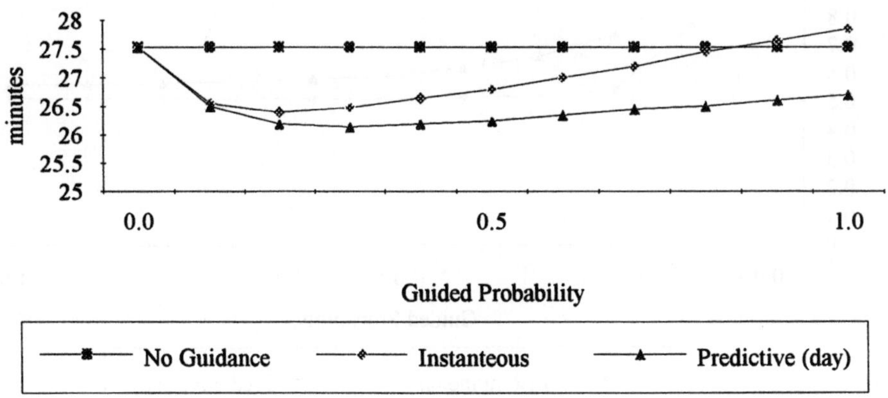

Figure 5. Maximum Delay / Guided
2.5 minute updates

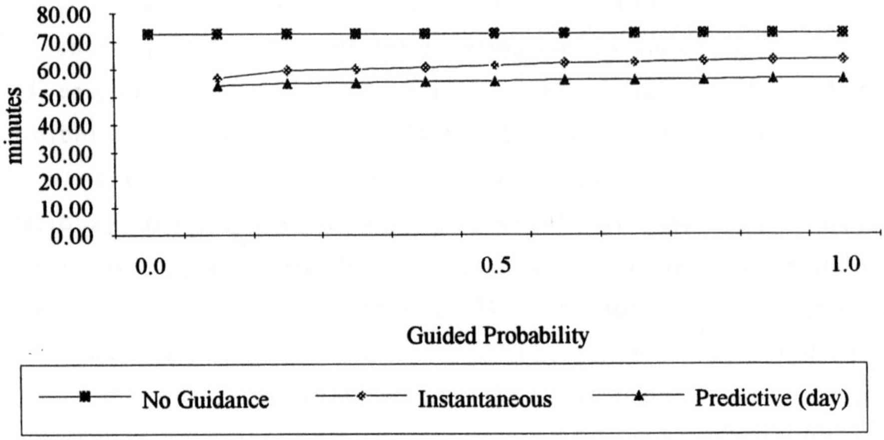

Figure 6. Validity
2.5 minute updates

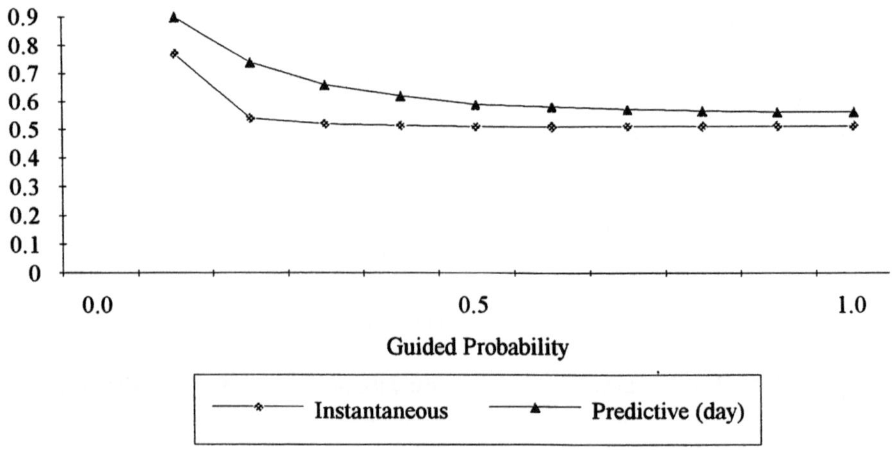

Figures 7 and 8 indicate that when guidance based on predictive traffic information was implemented at 30 second guidance update intervals, *overreaction was totally eliminated* and *high validity* was achieved even for higher guided probabilities. Moreover, Figures 7 and 8 indicate that the quality of O-D projections significantly affects guidance validity, while average delay is affected to a much lesser degree. This demonstrates that very good O-D predictors are needed to achieve validity levels that are necessary for sustained operation of the guidance system.

In general, simulation results indicate that only small reductions in average delays were achieved through guidance. On the other hand, *more significant reductions in maximum delays* were achieved. This emphasizes the potential of guidance systems in improving travel time reliability.

Figure 7. Average Delay
predictive / 30 second updates

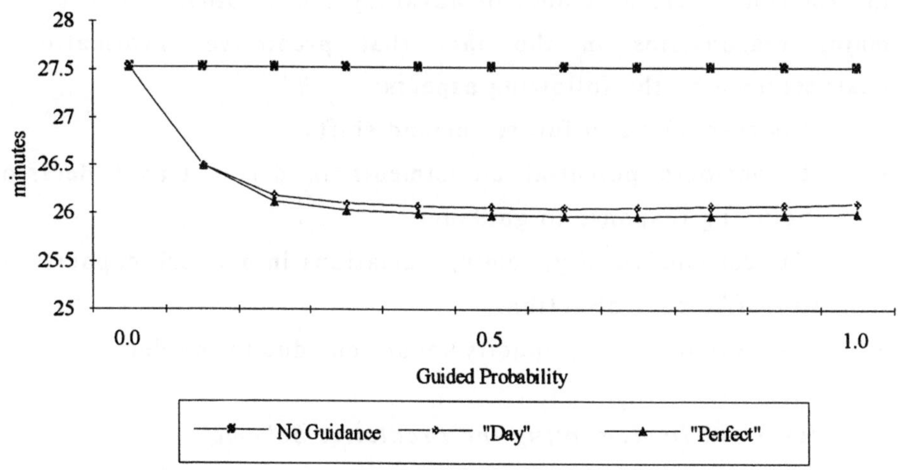

Figure 8. Validity
predictive / 30 second updates

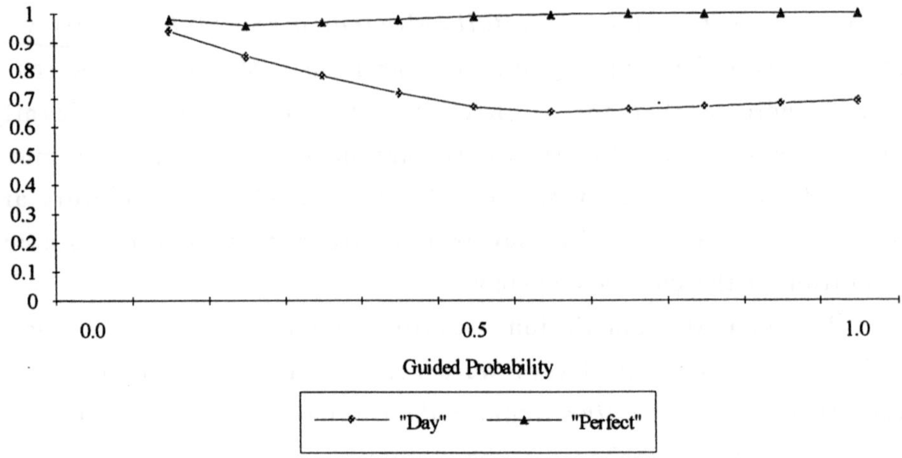

Potential Advantages of Guidance Based on Predicted Traffic Conditions

As demonstrated above, basing route guidance on predictive information offers a number of advantages over other schemes. The major reason lies in the fact that predictive information is characterized by the following aspects:

- it is responsive to future demand shifts
- it considers potential adjustments in driver travel decisions including response to guidance
- it is responsive to upcoming variations in network capacity due to traffic control actions
- it is responsive to capacity reductions due to incidents

Issues in Implementing the Predictive Scheme

Given that predictive information generally offers improved guidance, there are several issues that have to be kept in mind as far as the implementation of such a scheme is concerned. First, good predictors of O-D levels are required. Moreover, the performance of the predictive scheme improves if incident detection capabilities are available. Finally, its superiority is most prevalent in networks which do not offer "cheap" opportunities for re-routing. Therefore, the implementation of such a scheme requires an analysis of the nature of the network where the installation of a guidance system is being considered as well as an assessment of the cost/benefit relations inherent in the utilization of more advanced O-D predictors and/or incident detection capabilities.

5. Conclusions

The success of any ATIS implementation will clearly depend on the quality of the guidance it provides to travelers. One of the major issues that has to be considered in the design stage of ATIS systems is the utmost importance of guidance validity. In fact, a guidance system is not likely to sustain operation with low validity levels as travelers are likely to lose confidence in the system and abandon it quickly. The case study presented above indicated that basic ATIS designs are unlikely to provide adequate validity levels and that a more advanced system design is a necessity. The major characteristics of such advanced system designs include: (a) basing guidance on predictive traffic information that accounts for driver reaction, (b) providing guidance in such a way that traffic is distributed over several reasonable paths if single route guidance is likely to cause overreaction, and (c) performing more frequent guidance updates.

For a system design which encompasses the above elements and which is most likely to provide guidance characterized with the desirable properties outlined at the outset of this paper, we propose an ATIS implementation framework that consists of: (a) a surveillance system, (b) a congestion prediction module, and (c) advanced control and routing capabilities. The elements of such a framework are fully illustrated in [9] and [10]. Other research work into developing components of the framework is reported in [1] and [4]. This ATIS implementation framework is based on a dynamic network modeling approach which is needed to assess accurately network performance as well as to forecast traffic conditions that may exist in the near future in order to develop real-time diversion strategies to alleviate both recurring and non-recurring congestion conditions.

References

[1] Ashok, K. and M. Ben-Akiva (1993). *"Dynamic Origin Destination Matrix Estimation for Real-Time Traffic Management Systems"*, forthcoming in ISTT Proceedings, Berkeley, California, July 1993.

[2] Ben-Akiva, M. and A. de Palma (1987). *"Dynamic Models of Transportation Networks"*, Proceedings of PTRC Transport and Planning Summer Annual Meeting, University of Bath, England.

[3] Ben-Akiva, M., A. de Palma, and I. Kaysi (1991). *"Dynamic Network Models and Driver Information Systems"*, Transportation Research, Vol. 25A, No. 5.

[4] Ben-Akiva, M., H. Koutsopoulos, and A. Mukundan (1992). *"A Dynamic Traffic Model System"*, paper presented at the Second International Capri Seminar on Urban Traffic Networks, Capri, 1992.

[5] Bonsall, P.W. and A.D. May (1989). *"Evaluation of In-Vehicle Route Guidance Systems"*, paper presented at the 5th WCTR, Yokohama, Japan.

[6] Boyce, D. (1988). *"Contributions of Transportation Network Modeling to the Development of a Real-Time Route Guidance System"*, paper presented at the first International Conference on Transportation for the Future, Sweden, May 1989.

[7] Fujii, M. (1986). *"The CACS Project"*, paper presented at the 65th Annual Meeting of the Transportation Research Board, Washington, DC.

[8] Janko, J. (1991). *"LISB - The Berlin Route Guidance System"*, paper presented at the 70th Annual Meeting of the Transportation Research Board, Washington, D.C., January 1991.

[9] Kaysi, I. (1991). *"Framework and Models for the Provision of Real-Time Driver Information"*, Unpublished Ph.D. Dissertation, Department of Civil Engineering, MIT, November 1991.

[10] Kaysi, I., M. Ben-Akiva, and H. Koutsopoulos (1993). *"An Integrated Approach to Vehicle Routing and Congestion Prediction for Real-Time Driver Guidance"*, forthcoming, Transportation Research Record.

[11] Kobayashi, F. (1979). *"Feasibility Study of Route Guidance System"*, Transportation Research Record, 737, pp. 107-112.

[12] Sheffi, Y. (1985). *"Urban Transportation Networks"*, Prentice-Hall, Englewood Cliffs, N.J.

[13] U.S. Department of Transportation (1989). *"Discussion Paper on Intelligent Vehicle-Highway Systems"*.

AN ON-LINE DYNAMIC TRAFFIC PREDICTION MODEL FOR AN INTER-URBAN MOTORWAY NETWORK

Ben-Akiva, M.,

Massachussets Institute of Technology, USA
Hague Consulting Group, Netherlands

Cascetta, E.,

University of Naples, Italy

Gunn, H.,

Hague Consulting Group, Netherlands

This paper specifies the overall structure of an on-line Dynamic Traffic Model System for an inter-urban motorway network. The model predicts traffic conditions on a motorway network in real time, and can be used to provide information on (expected) congestion to a motorway traffic control centre.
The model receives traffic data such as traffic flow, average speed and average detector occupancy in real-time from a motorway monitoring system. Additionally, a database of historic traffic information will be used.
The model system contains several submodels : a *statistical* model for filtering noisy traffic data, and for very short term predictions (prediction horizon : 5-15 min.); a *dynamic traffic assignment* model for short term predictions (prediction horizon : 15-60 min.); a real-time *O/D matrix prediction* model to provide the input flows to the statistical and assignment models. Predictions from the two models are merged where their respective time horizons overlap. The present and predicted traffic conditions will be displayed for evaluation by the traffic operator. The real-time aspect of the system is reflected in the need for timing and reliable synchronisation of the various model components.

1. Introduction

Steadily increasing traffic loads on Western European roads, especially on motorways, suggest the possibility that the performance of road systems might be seriously compromised in the next 15 years or so. The expected growth in demand (even after stringent pricing policies) may cause drastic increases in traffic delays due to congestion, may seriously reduce traffic safety and may lead to increased energy consumption and emissions. In order to reduce these undesirable effects traffic management measures are required.

Recent research initiatives in the USA, Japan, and the EC (the DRIVE I and II programmes) are testing the use of Advanced Transport Telematics (ATT) to improve road safety, to increase general transport efficiency, and in general to use the existing road infrastructure more efficiently. In view of the massive investments necessary to expand the infrastructure, and the equally massive costs caused by traffic delays, use of ATT technologies to improve the efficiency of existing road infrastructure must be explored.

The use of ATT for dynamic traffic management and control can take many different forms, for which a variety of instruments is available. A recent publication on dynamic traffic management in the Netherlands by the Ministry of Transport ([13]), for instance, mentions:

- road-side based systems: signalling, ramp metering, variable message signs, fog detection and warning systems
- incident management: rapid detection, warning, and road clearance
- priority measures: High Occupancy Vehicle lanes, Heavy Goods Vehicles lanes, bus lanes
- in-vehicle systems: general broadcast services, RDS-TMC, other static and dynamic route guidance and travel information systems for drivers
- travel information: multi modal or public transport information
- vehicle control systems.

Most of these instruments can be used to apply two different types of control strategies: "myopic" ones which base their actions exclusively on the current state of the system and "predictive" or anticipatory strategies which take forecasts of the future states of the system into consideration. While the latter approach seems more appealing from a theoretical point of view, the state of practice generally follows the former.

A key requirement of dynamic traffic management is accurate real-time data on the actual as well as the expected short term future traffic state. Dynamic monitoring systems, currently based on inductive loops embedded in the pavement, are capable of providing the required information about actual traffic conditions, with only minor delay (typically 1 or 2 minutes). Information on expected future traffic conditions 30 or 60 minutes ahead can only be provided by means of a prediction tool, capable of predicting traffic conditions in real-time. Such a tool is currently not available for inter-urban motorways.

The DYNA project, which is part of the DRIVE II programme of the EC, deals with the development, implementation and testing of a real-time traffic prediction model system for inter urban motorway networks including an initial application in a pilot area around Rotterdam. The foundations for the design of the DYNA system are laid out in earlier research (see [4], [5], [6], and [12]).

The main input for the system will consist of real-time measurements of traffic flows, speeds and occupancies to be provided by loop detectors. Additional information will be obtained from historic traffic flow databases and descriptions of the road network infrastructure. Also the immediate user of the system in the traffic control centre, will provide updates of relevant information, such as temporary capacity reductions due to accidents or roadworks, extreme weather conditions, special events and so forth.

To achieve the goal of determining the current traffic state and predicting future traffic states between 1 and 30 or 60 minutes ahead the system requires practical and accurate prediction methodologies which are capable of working in real-time. Instead of relying on one single methodology, it was decided to apply two different methods in parallel, whose results will be merged:

- a statistical model, using a state-space/Kalman filter approach with explicit feedback of prediction errors; it is expected that this model will be most accurate for very short time periods, eg. 1 to 10 or 15 minutes ahead
- a dynamic traffic assignment model, working in connection with a dynamic O/D prediction method; it is expected that this

method will be more accurate for prediction horizons in the range of 15 to 30 or 60 minutes ahead.

Both models will incorporate the possibility of simulating the effects of dynamic traffic management measures, including the provision of information on delays and travel times, route advice, variable message signs, ramp metering, and other measures. In the statistical model this will be done by extending the number of states of the system, in the dynamic assignment approach this will be done by explicitly simulating drivers' responses in the route choice procedure and/or link performance functions.

An important feature is that the system will analyze a network rather than a series of individual and independent links. This means that the question of route-choice enters the modelling process. This is particularly relevant for assessing the impact of possible traffic control measures on traffic flows and speeds, and ultimately for allowing the traffic operator to evaluate the predicted impacts of alternative control strategies. A network-wide analysis enables the traffic operator to optimise global network performance by evaluating network-wide implications. Without due regard to such implications there is no guarantee that local measures taken to optimise traffic flow on one part of the network will not disproportionally penalise performance elsewhere (see e.g. [1]).

The purpose of this paper is to present the overall design of the DYNA project and the architecture of the DYNA traffic prediction model. In section 2 of this paper the objectives of the DYNA project are described. In section 3 the functionality of the system is presented, together with the available inputs and the required outputs and operational constraints. In section 4 the system architecture is described: briefly addressing the theoretical and modelling approaches which will be used and how they are related to the state of the art.

2. Objectives and requirements of the traffic prediction system

The overall objective of the DYNA project is to develop prediction capability of expected future traffic states for an interurban motorway network. The DYNA system would be employed by a traffic control centre to generate optimal control strategies. It should provide, with minimal delay, a clear picture of what would happen in terms of traffic performance under alternative scenarios of traffic control measures. As such a real-time traffic prediction system such as the one outlined in this paper is an essential part of a more general dynamic traffic management system.

The goal of the DYNA project is to develop, implement and apply a fully operational dynamic traffic model system for real-time applications. The system will provide short-term forecasts (1 to 30 or 60 minutes ahead) of traffic flows and travel times.

More specifically, the system will need to provide for each link for each prediction period, forecasts of:

- traffic flow
- (average) speed
- (average) travel time
- whether or not there is a queue expected
- if there is a queue: number of vehicles in queue
- if there is a queue: (average) waiting time
- if there is a queue: expected duration.

The forecasts need to be suitable for use in designing and evaluating alternative control strategies. Although the provision of information to the drivers or potential drivers is not part of the current Pilot project, the predictions need to be sufficiently accurate to be able to do so in future extensions.

The system should not only be able to function well and provide sufficiently accurate forecasts under "ordinary" traffic conditions (both congestion free and congested), but also under more unusual traffic conditions (accidents, bad weather, special events).

The model system is to be applied at real scale and evaluated in a Pilot Project. This will be an application to an inter-urban motorway network in The Netherlands, in the Rotterdam area.

3. System environment and interfaces

This section describes the environment in which the DYNA system will operate. The system environment diagram is shown in Figure 6.

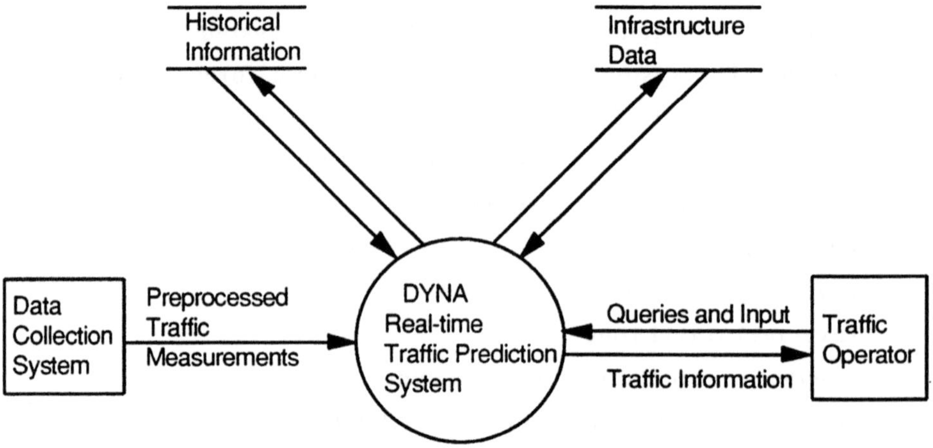

Figure 1. Environment of the traffic prediction system

The diagram illustrates how the prediction system interacts with the supporting data bases and with the Traffic Operator using the system. As can be seen, there are four separate entities in the environment of the system:

- the Data Collection System
- the Historical Information
- the Infrastructure Data
- the Traffic Operator.

The Data Collection System feeds the prediction system directly with the necessary real-time data, and through that system it also updates and extends the Historical Information. It may also supply data to update the Infrastructure Data, through a dynamic capacity estimator (see for example [18]).

The Historical Information simply feeds into the prediction system, and is kept up-to-date using data from the Data Collection System and revised estimates from the traffic prediction system.

The Infrastructure Data is mainly time invariant, but some of its information is dynamically generated, such as the dynamic capacity estimates, which are used to automatically update the Infrastructure Data. It feeds the prediction system with network information, and it receives information updates from the Traffic Operator, for instance when temporary or permanent changes occur in the network.

The Traffic Operator also interacts directly with the prediction system: in the simplest form of use the Operator reviews the forecasts and acts when unusual situations are anticipated. He may also have straightforward queries for the system (eg: zoom in on a bottleneck), and in the envisaged future extensions of DYNA the Operator will be able to specify packages of control measures which will be simulated and evaluated by the system.

3.1 Inputs to the system

3.1.1 Real-time traffic data

The real-time traffic data will be provided by a motorway monitoring system. In the Netherlands, where the Pilot Project will be implemented, two systems will be operational:

- the Motorway Control and Signalling System (MCSS). This system, in existence on certain sections of the network since the eighties, consists of buried loop detectors, connected via a communication network to a central processor. The detectors are placed at intervals of about 500 metres along the motorway.

Local processing units are used to pre-process and transmit the traffic data.

- the Monitoring System (MS) will be implemented on those parts of the motorway network which are not covered by the MCSS. Buried loop detectors will be placed at intervals of between 4 and 8 kilometres on motorway sections, and at all on-ramps and off-ramps. All loop information is pre-processed and aggregated into 1-minute averages by roadside processors.

The data from the two systems are being integrated in a unified computer system. All roadside processors of both the MCSS and the MS systems are being connected to a central minicomputer located in a traffic control centre, which will collect and store all the 1-minute aggregate data.

The Pilot Study area, where the initial application of the DYNA project will take place, is situated in the area surrounding the city of Rotterdam (see Figures 2 and 3 below). The Pilot network comprises about 200 Kilometres of motorway, some 12 motorway intersections, and 50 on-off ramps. Installation of the MS is expected to be complete by early 1994.

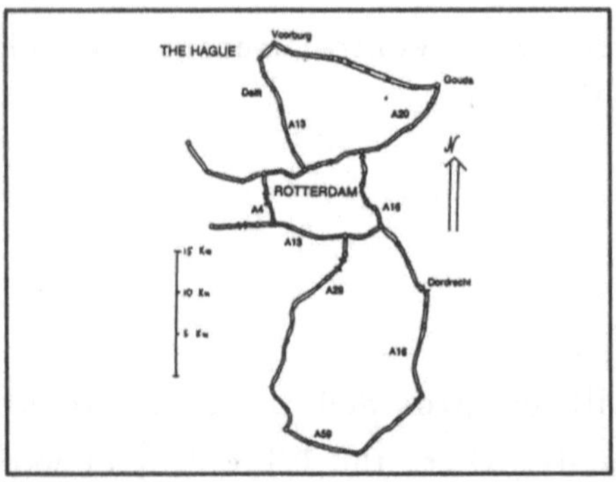

Figure 2. Pilot Area

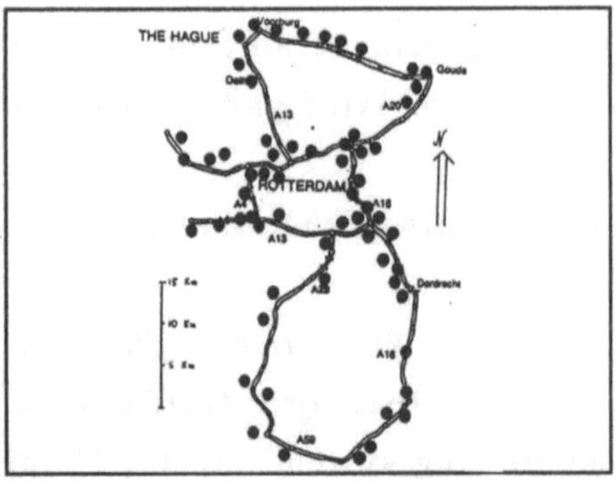

Figure 3. Approximate location of the road-side processors

On some sections of the Pilot network, notably the A13 just north of Rotterdam, the existing MCSS system is already operational. On those sections the MCSS data will be used, on the other sections the data to be provided by the MS system will be used. The traffic measurements supplied by the MCSS are somewhat less detailed then those from the planned MS, but the spatial density is somewhat higher.

The MCSS system provides on a 1-minute basis :

- traffic flow intensity (veh/hr)
- mean traffic speed (km/hr).

Additionally, the MCSS system provides a simple congestion indicator (congestion yes/no), derived from observed sudden drops in speed or observed speeds falling below a preset threshold. Also, the system transmits some additional data on the reliability of the measurements (incl. detector breakdown).

Note that MCSS does not provide information on:

- vehicle category (only totals across all vehicle types are produced)
- separate lanes (only total traffic flow summed across all lanes per direction are produced)
- standard deviations of speeds.

The MS system will provide on a 1-minute basis :

- traffic flow intensity (veh/hr) per lane per vehicle category (5 categories are distinguished)
- mean traffic speed (km/hr) per lane per vehicle category
- standard deviation of traffic speed (km/hr) per lane per vehicle category
- mean detector occupancy (%) per lane.

The MS system provides the same simple congestion indicator as the MCSS, as well as the same additional data on reliability.

Like all measuring systems, the detectors are prone to measurement error. We distinguish two kinds of measurement error through detector malfunction:

- complete detector breakdown, and
- operational detectors returning incorrect measurements.

To safeguard against detector malfunction, diagnostic

information is provided through the monitoring systems along with the measurements. This information is produced by the roadside processors, which detect and signal complete detector breakdown, and which also provide the percentage of measurements that is identified as possibly erroneous.

3.1.2 Historical traffic data

Two broad categories of historical information are distinguished:

- historic traffic measurements, derived from the monitoring systems
- historic O/D information, relevant for the real-time estimation and prediction of the origin-destination matrices.

The historic traffic measurements are generated as part of the continuous flow of incoming data which are temporarily stored by the computer of the traffic surveillance system. In addition DYNA will generate its own historical database consisting primarily of O/D matrices based on historical data. The historic O/D information is required for the estimation and prediction of the O/D matrices required by the Dynamic Traffic Assignment model. O/D estimation relies on a mix of a-priori information and information contained in traffic flow data. As the information content of the traffic flow data is typically fairly limited, the availability of good quality a-priori information is an important condition to obtain accurate estimates of the O/D pattern. The a-priori information is to be provided by the historic O/D patterns.

Historic O/D patterns could be derived where possible from observation (one or more dedicated license plate surveys to establish actual O/D patterns during certain periods of days). Alternatively, synthetic estimates of time varying O/D matrices may be generated by using the large scale transport planning model; off-line simulations can be used to generate a large number of a-priori matrices for different times of the day, days of the week, congestion levels, etc.

These initial matrices will be continuously updated using the O/D patterns that were previously estimated by the on-line prediction system, either running in off-line or in on-line mode; this process will generate updated a-priori matrices for different times of the day and days of the week, but also for extreme weather conditions and special events such as sporting and cultural events, etc.

3.1.3 Infrastructure data

The infrastructure data provides the prediction system with the necessary description of the network. This includes the following elements:

- network topology consisting of links and entry and exit ramps. Links can have a length of between 100 and 2000 M.; arcs between two intersections may contain more then one link.
- any circulation restrictions that might apply (e.g. prohibited turns)
- the geometric specifications of the network elements (such as link length, width, curvature)
- outflow capacity of each link
- storage capacity of each link
- speed limit that applies to a link
- any type of lane control that might be applied
- state of any control devices that might be installed alongside the network, e.g. variable message signs at diversion points inside the network.

It also comprises continuously updated information on link closures (due to accidents or road works).

It is known from previous research that outflow capacity varies with time (see [16]), and recently a dynamic capacity predictor has been developed (see [18]). It is based upon the real-time data supplied by the Monitoring System, and will be available for use in the DYNA project.

3.1.4 Input from the traffic operator

The traffic operator, should be able to provide certain inputs to the prediction system, such as:

- Accident reports: following the rapid penetration of car-phones, accidents are nowadays reported much quicker by drivers with car-phones than through any other source. The traffic operator, once warned about an accident, can provide the relevant information to the prediction system, which in turn can simulate or predict the traffic problems that it may cause.
- Temporary capacity reductions may be caused by incidents or by temporary changes to the network due to roadworks. For the last category the exact starting and ending times and the magnitude of the capacity reduction can be foreseen by the operator, and should be pre-entered into the system database. For the first category, caused by incidents, the occurrence, the duration and the magnitude cannot be foreseen, but it is important to inform the prediction system as quickly as possible about such event and when possible to also provide an estimate of expected duration and capacity reduction.
- Special events that are likely to affect traffic demand, when known in advance, should be entered into the system by the traffic operator. Such events might include special days (e.g. holidays), school vacations, national events, special exhibitions and so on. Ideally a library of prototype historical O/D matrices should be available for such occasions, from which the traffic operator would be able to select one according to the information available to him.
- Weather conditions are known to have a significant influence on motorway operations. Thus some indication of the type of weather is required. At a minimum this could be a dry/rain indicator, which could be generated automatically, but indications of expected heavy rain, snow or fog might be usefully supplied by the traffic operator.

3.2 Outputs from the system

The real-time prediction system will have one direct user: the traffic operator responsible for the inter-urban motorway network for which the forecasts are provided. Under ordinary smooth flow conditions, the traffic operator will probably want to monitor the network's performance passively. As bottlenecks develop, a more active role can be expected from the operator: the police may be warned, and drivers in the network and those drivers planning to enter the network may also be informed about the (expected) congestion. Furthermore the operator may consider dynamic traffic management measures aiming to improve the network flows: for instance variable message signs may be switched on, informing drivers approaching possible diversion points about the (expected) delays, and possibly suggesting the use of an alternative route.

The system outputs will include the following variables for each link in the network :

- traffic intensity [Veh/Hr]
- (average) speed [Km/Hr]
- (average) traffic density [Veh/Km]
- congestion alert (for detected congestion)
  (location, queue length, expected delay, expected duration)
- warning of expected congestion
  (location, queue length, expected delay, expected duration)
- (average) travel time
- an indication of the reliability of the prediction

3.3 Operational requirements

There are a number of requirements which are of particular importance when the design of required prediction system is considered.

The first, and obviously single most important requirement is real-time operation. The fact that forecasts are to be produced in real-time implies that all necessary data processing and computing needs to be completed within a time-scale which is (considerably) shorter than the prediction horizon. Another relevant consideration here is the update frequency of the forecasts: ideally the system should provide a new, updated forecast each time new information has become available. In practise that means every minute, given the one-minute intervals at which the real-time traffic flow data are produced. The required speed of operation of the prediction system puts a number of serious limitations on:

- the complexity and speed of the algorithms to be used: clearly a compromise between accuracy and simplicity needs to be sought
- computer speed: a very fast computer is required, within the available resources; theoretically a super computer might be attractive, in practice alternatives exist that have a much higher performance/cost ratio
- data access: a very fast access medium with sufficient storage capacity is required for all the input data required by the prediction system; also the amount of input data needs careful consideration: data reduction techniques should be used where possible to avoid excessive amounts of stored data.

A second, also very important requirement, is that the prediction system should be able to simulate the effects of a wide range of possible traffic management and control measures.

4. System architecture

Within the system environment shown in the previous section, the system itself must accommodate the input requirements and the output potential of the two basic prediction approaches that are used inside the system: the Statistical Model (STM) and the Dynamic Traffic Assignment (DTA). The overall system structure is depicted in Figure 4 below:

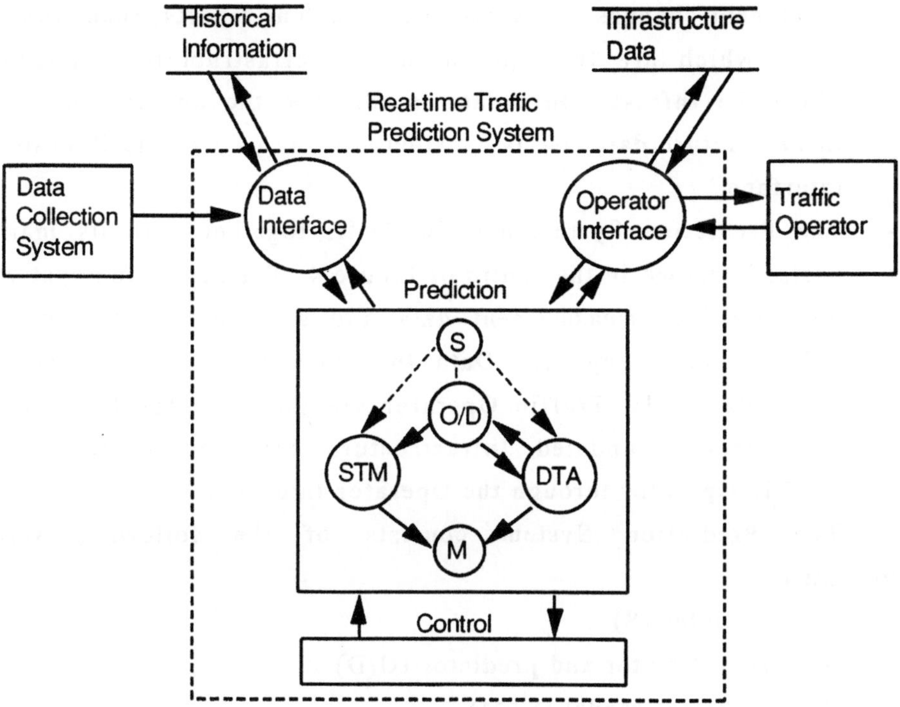

Figure 4. Overall System Architecture

This figure is an enlargement of Figure 1, now showing the major elements of the prediction system itself (inside the dotted box). The Real- time Traffic Prediction System is enlarged to show the following key components:

- The Data Interface in the top left corner is a separate module that maintains contact with and accepts measurement data from the Data Collection System, and makes these available to the other modules of the prediction system. Access to the Historical Information is likewise channelled through the Data Interface.
- In the top-right corner is the Operator Interface. The Traffic Operator communicates with the system only through this interface. It provides the Traffic Operator with a running display of the actual and predicted traffic situations. Furthermore, the Operator Interface specifically enables the Traffic Operator to inform the system about temporary or more

permanent changes in the network (accident reports, roadworks, etc.), which are then stored in the infrastructure database. Since the Infrastructure data is editable, the updating of the Infrastructure database is considered to be part of the Operator interface.

- The Prediction System module itself, together with its main parts, is shown in the centre of Figure 4. The prediction system both receives measurement data and accesses the historical information through the Data Interface. It makes its results available to the Traffic Operator via the Operator Interface, and receives updated Infrastructure Information from the Traffic Operator through the Operator Interface.

The Prediction System consists of the following five subsystems:

* the Scheduler (S)
* the O/D estimator and predictor (O/D)
* the Statistical Model (STM)
* the Dynamic Traffic Assignment (DTA)
* the Merger (M).

The main modelling elements are: O/D estimation, very-short-term traffic flow prediction through the statistical model, and short-term traffic flow prediction through Dynamic Traffic Assignment.

The actions of these models are synchronised by the Scheduler, and the predictions of both models are combined by the Merger (where their prediction horizons overlap) into a single prediction. It is this combined prediction that is sent to the Operator Interface.

The three main models are discussed in some more detail below.

- At the bottom is the Control System : a separate module (to be implemented later) to evaluate control policies and recommend control measures. This represents the "ultimate" system, as it is currently envisaged.

4.1 Monitored versus entire network

The DYNA system is driven by the real-time traffic measurements. These data are available for the motorway network that is equipped with the monitoring system, and are not available for the more extended road network that may be used by the drivers. As a consequence the real-time prediction system can only cover the (equipped) inter-urban motorways. We call this the monitored network.

The real-time input data covers only cars that are on the monitored network. The parts of the trips between the real origins and the entry-points and the parts between exit points and the final destinations are not observed (see figure 5). Thus for the monitored network, the "origins" of the cars on the network are the entry points, and the "destinations" are the exit points.

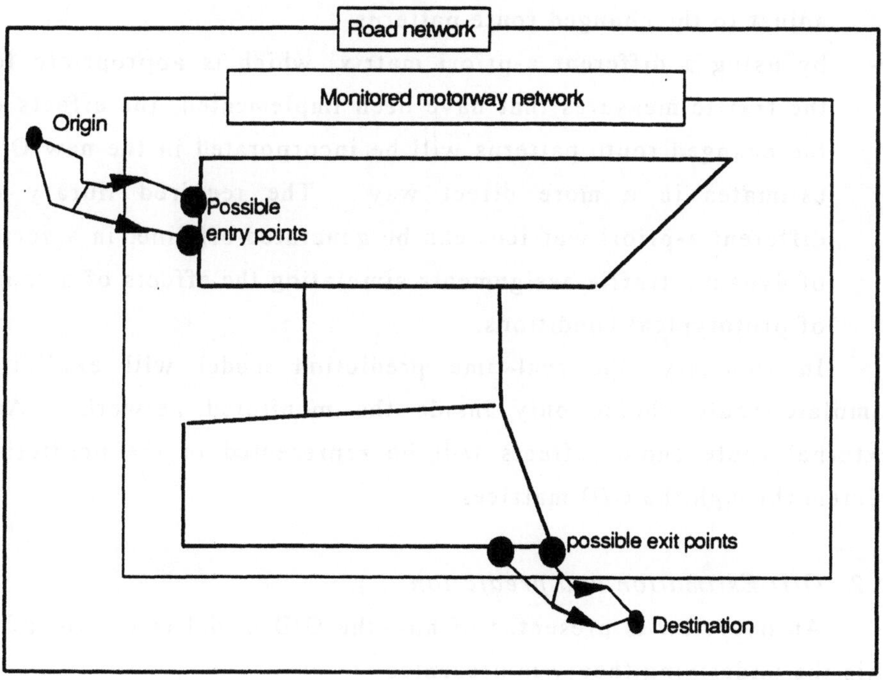

Figure 5. Road network versus Monitored network

The dynamic, real-time O/D estimation and traffic assignment are therefore limited to estimating Entry/Exit matrices and to simulating route choice within the monitored network. As a consequence the real-time dynamic traffic assignment will only simulate route choice decisions by drivers on the monitored network, given their entry points and desired exit points. In practice drivers might, in response to specific traffic management measures (e.g., on board route guidance systems) change their routes outside the monitored network, which might bring them to different entry and exit points. In the prediction system this problem is addressed by the O/D estimation model: any observed or expected changes in routes outside the monitored network will be reflected in the predicted O/D matrices:

- through the observed changes in the traffic flows at the entry and exit links the predicted O/D matrices will (with some lag) adjust to the changed route patterns
- by using a different a-priori matrix, which is appropriate for the traffic measures that have been implemented, the effects of the changed route patterns will be incorporated in the new O/D estimates in a more direct way. The required library of different a-priori matrices can be generated off-line, in a series of dynamic traffic assignments simulating the effects of a range of prototypical conditions.

In summary, the real-time prediction model will explicitly simulate route choice only inside the monitored network. Any external route choice effects will be represented in the prediction system through the O/D matrices.

4.2 O/D Estimation and prediction

An overview is presented of how the O/D model is set up; after this the underlying theory is presented.

4.2.1 Schematic view

Figure 6 shows the O/D subsystem in greater detail. Traffic

counts and Historical O/D matrices are made available through the Data Interface. The most recent Assignment Matrix is made available by the DTA module. This Assignment matrix maps the (unobservable) O/D flows into the observable link flows, and constitutes a key input to the subsystem. The O/D estimation module takes the counts, the historic O/D matrices, and the most recent assignment matrix to produce an estimate of O/D flows which are on the network. This estimated O/D matrix is used to obtain a predicted O/D matrix for the subsequent time-intervals.

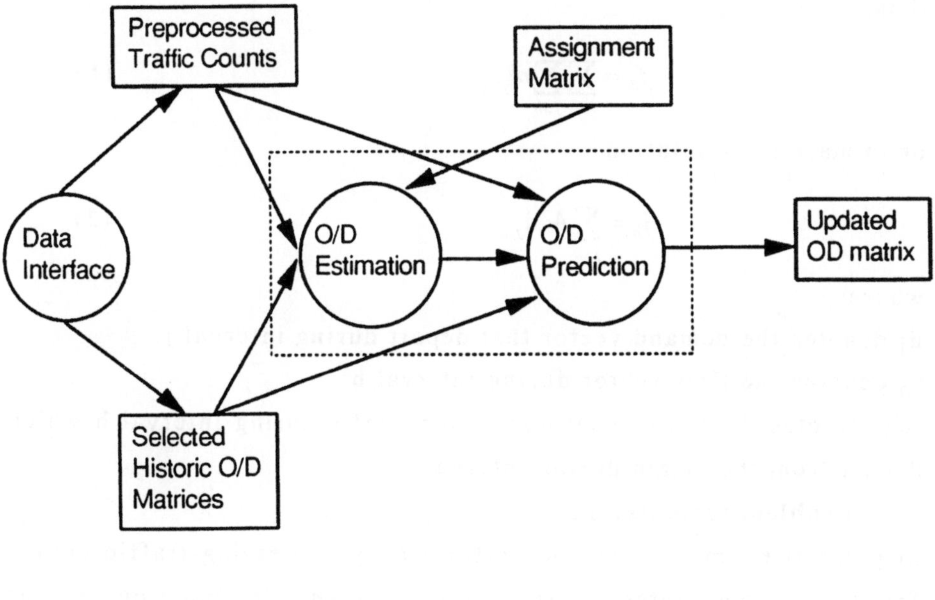

Figure 6. Expanded view of O/D estimation and prediction

4.2.2 Theoretical framework

The estimation of dynamic O/D matrices from traffic counts can be approached in two different ways. *Simultaneous estimation* can be used to obtain the O/D matrix for multiple time intervals. *Sequential estimators* iteratively produce the O/D matrix per time interval using as input the traffic counts for that time interval and the estimated O/D matrices of previous time intervals. The latter approach, which is computationally less demanding is the only possible option for real-time applications. Discrete time intervals are used throughout.

Definitions and notation

Let:

d_{ij} be the demand for O/D pair i which depart from the origin during time interval j

f_{lh} be the flow at measuring point l during interval h

a^{ij}_{lh} be the fraction of d_{ij} crossing measuring point l during interval h

then:

$$f_{lh} = \sum_{i} \sum_{j \le h} a^{ij}_{lh} d_{ij} \quad (1)$$

or in matrix formulation :

$$\mathbf{f}_h = \sum_{j \le h} \mathbf{A}^j_h \mathbf{d}_j \quad (2)$$

where:

$\mathbf{d}_j$ denotes the demand vector that depart during interval j

$\mathbf{f}_h$ denotes the flow vector during interval h

$\mathbf{A}^j_h$ denotes the assignment matrix for traffic during interval h which depart from the origin during interval j

Problem formulation

Sequential estimation can be performed by expressing traffic counts for the current interval as a linear stochastic function of the unknown demand for the same interval, starting from relationship (2) between demand and traffic flows:

$$\mathbf{v}_h = \sum_{j \le h-1} \mathbf{A}^j_h \mathbf{d}^*_j + \mathbf{A}^h_h \mathbf{d}_h + \vartheta_h \quad (3)$$

where:

$\mathbf{v}_h$ is the traffic count vector for the current interval h

$\mathbf{d}_j^*$ is the estimated demand vector for interval $j<h$

d_h is the unknown demand vector for the current interval h
ϑ_h is a random error

The crossing fractions contained in the assignment matrices should be computed through a dynamic network loading procedure that is dependent on the demand vector, leading to a fixed-point problem. For real-time applications an O/D estimation for interval h is performed at the end of interval h, based on the estimate for previous time intervals up to h-1 and the prediction for time interval h, which can be used to estimate crossing fraction matrices for interval h.

Through relationship (3) the dynamic estimation problem can be formulated as a minimisation problem:
Let d_h^0 be the a-priori estimate of the demand vector for interval h (d_{h-1}^* can be used if nothing else is available). The estimator of the demand vector for interval h is:

$$\mathbf{d}_h^* = \arg\min[\phi_1(d_h, d_h^0) + \phi_2(d_h, v_h, d_1^* \cdots d_{h-1}^*)], \qquad d_h > 0 \qquad \textbf{(4)}$$

where $\phi_1()$ is the error function for the a-priori demand vector and $\phi_2()$ is the error function for the available traffic counts during interval h. Several formulations can be adopted for $\phi_1()$ and $\phi_2()$ based on different statistical assumptions about the random errors. For more detail on the estimation of dynamic O/D matrices see [9].

O/D prediction will be based on a simplification of the approach developed in [2]. It is based on an autoregressive process describing the pattern of deviations between the a-priori matrix d_h^0 and the estimated matrix d_h^*. Given the calculated deviations for intervals ....,h-3, h-2, h-1. h the a-priori matrices for intervals h+1, h+2, ... are adjusted using the coefficients of a the previously estimated autoregressive process.

4.3 *Dynamic traffic assignment*

For a given O/D matrix, a Dynamic Traffic Assignment (DTA) model predicts the flows and travel times on the links of the network. A schematic overview of the structure of DYNA's DTA is given below, followed by a brief outline of the theoretical framework underlying the dynamic assignment approach.

4.3.1 Schematic view

Figure 7 shows a detailed view of the DTA subsystem. Central to this subsystem is a module that keeps track of the past, current, and latest predicted 'State of the Network'. The DTA predictions feed into the Merger subsystem. The Data Interface supplies traffic measurements to this process, which enables it to update current travel times. Up-to date Infrastructure Data is used, reflecting any changes made by the Traffic Operator through the Operator Interface. Directly from the Operator Interface is an arrow, indicating that the Traffic Operator may access this 'state of the network' directly.

The previously described O/D prediction subsystem produces an updated dynamic O/D matrix. This O/D matrix enters the Path Choice module, where given the observed (past and current) and projected (future) States of the Network a set of Path Flows is produced. The Path Flows and the State of the Network are used by the Dynamic Network Loading (DNL) module to derive the Link Flows. The Link Performance module uses the Link Flows calculated in this way by the DNL module, and the current link attributes, to calculate the travel times of all links. These, in turn, are used to update the State of the Network.

The DNL module is based on a "continuous packet" approach, where the dynamic O/D flows are represented during the simulation by a number of "packets" of cars. Separate packets are generated for each combination of: O/D pair, user-type (e.g. whether or not equipped with on-board information systems), selected path through the network, and departure time. Continuous packets have variable length with explicit location of the front and back ends of each packet on the network. For more detail see [6] and [10]).

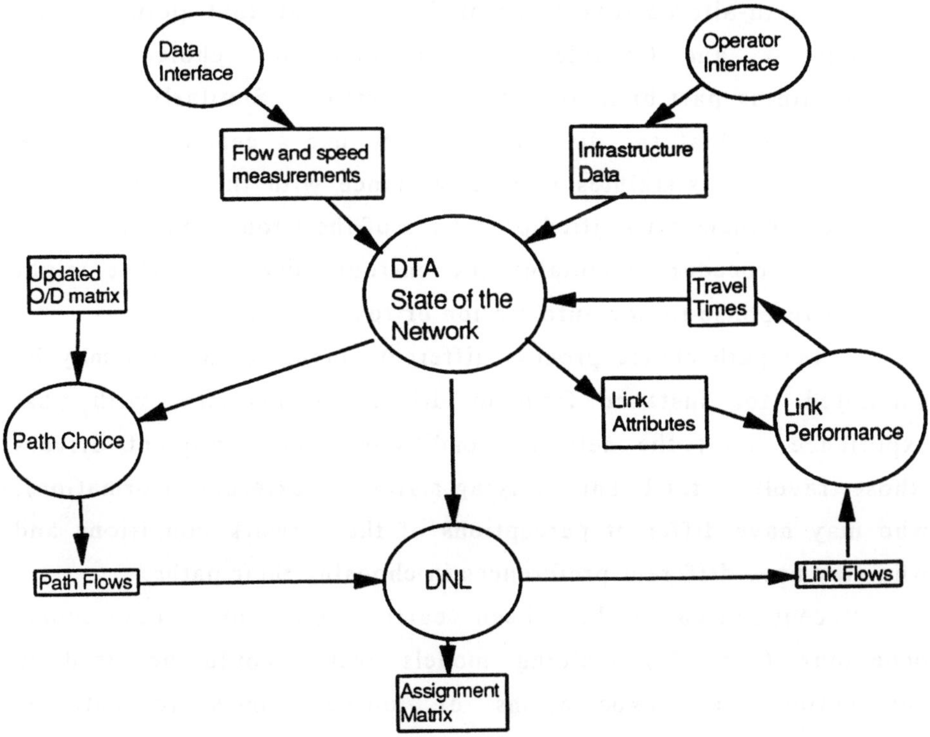

Figure 7. Expanded view of DTA

Link Performance will be simulated using link-specific speed-flow curves calibrated on observed flows. A two-regime model will be used, with speed-flow curves fitted in off-line mode being used during uncongested operation, and additional waiting times being modelled using a D/D/1 queuing model once congested operation has been reached (see also [17] and [15]).

The Path Choice module simulates path choice behaviour of the drivers using a procedure which can be summarised as follows:

- Drivers entering the network have an initial choice set, containing their feasible paths given their entry and exit points.
- On the basis of the information available to them at that moment, their path choice at network entry is modelled using a probabilistic discrete choice model.
- At each diversion point in the network, the drivers (may) given up-to-date information about the network conditions, consider

the path-alternatives relevant for them at that point in the network, and (if relevant) reconsider their choice for the remaining part of their trip on the network. Similarly they may be advised at the diversion node to use a certain route; the model then simulates their compliance with the given advice, and produces a modified prediction of their route choice.

This procedure simulates the en-route diversion that might occur in response to new information or route-advice.

In the path choice process different classes of drivers may be simulated, for instance frequent drivers (commuters, with past experience about the network conditions) and infrequent drivers (those travelling for leisure, relying mainly on external information), who may have different perceptions of the network conditions and who may have different preferences in choosing their paths.

Recent research has been carried out into route-choice behaviour (see [3]) yielding models which could be used in conjunction with expectations of journey times to estimate proportions of drivers using the different paths. In addition, a large-scale market-research exercise has been undertaken to investigate 'values-of-travel-time-savings', differentiating between user groups and contexts (see [11]).

For the prototype DYNA system, simple models will be assembled from these sources. When the control systems are actually operational, it will be feasible to sample travellers and obtain data on the basis of actual decisions (compliance rates with advice, evaluation and use of information). This will be the most useful basis for operational models, but will not be available for some time. It is to be expected that such a sampling process will be needed on a regular basis, at least in the early years of a new information system, as familiarity with the potential of the system grows.

4.3.2 Theoretical framework

Traffic assignment models are used to simulate link flows on transportation networks and the resulting link performances, such as travel times, congestion, pollution and energy consumption. Traffic assignment models are based on a *demand model* predicting user behaviour and a *supply model* representing network performance.

If travel demand and supply are assumed to be uniform over the simulated period (e.g. the morning peak), the assignment model is called *within-day or intra-periodic static.* Time dependent demand (for instance during the build-up of the rush hour) and/or changes in supply (due for instance to weather conditions or accidents) generally lead to time-dependent flows and queues which can only be simulated through *within-day or intra-periodic dynamic* models. Such models can also take account of the effects of real-time control measures (such as variable message signs or traffic information broadcasts).

Demand model

Users are grouped into n_i classes that are homogeneous in all behavioural characteristics. In particular they travel between the same O/D pair (possibly for the same purpose), and they have a common set K_i of available paths. It is assumed that on each day, d_i users of class i travel between the O/D pair. Elasticity of demand can be dealt with through the introduction of a fictitious alternative (e.g. a dummy path for within-day static models, and a dummy departure time for within-day dynamic models) corresponding to the choice of not travelling at all. In this case d_i is the number of potential travellers in user class i, and **d** is the corresponding O/D demand vector.

The reference period is divided into n_j intervals, which can be assumed, without loss of generality, to have equal length T; F_{jk} denotes the path flow, i.e. the number of users that depart during interval j and embark on path k (where k denotes the actual path followed to the final destination when all en-route diversions have been taken into account). Denote the corresponding path cost by C_{jk}, the path flow vector by **F** and the path cost vector by **C**.

The *demand model* describes user behaviour by expressing path flows as a function of path cost. Let $p_{i,jk}$ denote the probability that a user of class i selects departure interval j and path k, and denote the corresponding n_i x (n_j x n_k) choice matrix (usually modelled through random-utility discrete-choice. models) by **P**. Assuming that **P** directly depends on the path cost vector **C** (see [7], [8] for more general models) leads to:

$$\mathbf{F} = P(\mathbf{C}).d \qquad (5)$$

Supply model

The supply model describes the relationship between path costs and path flows. In the within-day static case the supply model can be easily formulated using the link-path incidence matrix. In the within-day dynamic case a more complex approach is needed, which is described below.

In traffic assignment models the performance of a link in the network is related to a variable associated with each link, usually called "link flow". Unlike the static case, in a within-day dynamic network the definition of link-flow is not unique since time and space averages need not coincide. Different definitions can be adopted corresponding to different modelling approaches. Denoting the flow (however defined) on link l during time interval h by f_{lh}, and the corresponding link flow vector by **f**, the relationship between link and path flows can be formally expressed through the Dynamic Network Loading (DNL) map as follows:

$$\mathbf{f} = \Phi[\mathbf{F}] \qquad (6)$$

In general this map is non-linear, and cannot be expressed analytically as in the within-day static case.

Let **r** denote the link performance (such as running speed or queuing delay) vector, which in congested networks generally is a function of the link flow vector:

$$\mathbf{r} = r(\mathbf{f}) \qquad (7)$$

In the within-day dynamic case both the time needed to traverse a link and the generalised link cost (if different from link travel time) depend on link-performance as defined by the link performance function and the flow. Since this flow can be time-varying, the cost of traversing a link is determined by arrival time at the link. Once the link flows are known through the DNL map, the link performance vector can be determined and the arrival times at each of the links of the path can be computed for any departure time. Then, the corresponding link costs can be defined as functions of both link performances and arrival times.

As above, the generalised transportation cost on path k, averaged across all users leaving during interval j and following path k is denoted by C_{jk}, and the corresponding path cost vector by **C**. Path costs for a given departure interval j can be computed from the corresponding link costs, so that path costs depend on the link performance vector:

$$\mathbf{C} = \Gamma(\mathbf{r}) \tag{8}$$

A within-day dynamic supply model is obtained once both the DNL maps and the link-performance functions are specified:

$$\mathbf{C} = \Gamma[r(\Phi[\mathbf{F}])] \tag{9}$$

Dynamic Network Loading

In the following the DNL map is examined in more detail. A straightforward generalisation of the within-day static formulations is not possible since, as mentioned before, the definition of link flow is not unique. Therefore link and path flows must be redefined in a consistent way. Additionally a fixed point problem arises in the definition of the DNL map since the time-dependent link flows depend on the time-dependent link performance, which in turn depend on the links flows.

Denoting the fraction of the path flow F_{jk} contributing to link flow f_{lh} by *crossing fraction* b^{jk}_{lh}, the flow on link l can be expressed as:

$$\mathbf{f}_{lh} = \sum_{j \le h} \sum_{k} b^{jk}_{lh} F_{jk} \tag{10}$$

Note that the crossing fraction is equal to zero when link l does not belong to path k, the departure interval j is later than h (j>h), or when path flow F_{jk} does not cross link l during interval h. The crossing matrix, denoted by **B**, is a lower triangular (n_l x n_j) x (n_k x n_j) block matrix, and the following holds:

$$\mathbf{f} = \mathbf{BF} \tag{11}$$

Crossing fractions $b^{jk}{}_{lh}$ consistent with the definition adopted for link flows depend on the network topology and the time needed to reach link l, travelling on path k and departing during interval j.

Therefore they depend on the travel times on all links preceding link l on path k, which are functions of the link performance vector r, so that:

$$\mathbf{B} = B[\mathbf{r}] \tag{12}$$

Therefore the relationship between link flows and path flows can be expressed in a general way as:

$$\mathbf{f} = B[\mathbf{r}].\mathbf{F} \tag{13}$$

On the other hand, as stated above, link performance in congested networks depends on the link flows so that by combining (7) and (13) a fixed point problem arises in the definition of the DNL map (6):

$$\mathbf{f}^* = B[r(\mathbf{f}^*)].\mathbf{F} \tag{14}$$

Once link-performance functions (7) have been specified, a dynamic network loading method essentially reduces to the computation of the crossing fractions, i.e. the $b^{jk}{}_{lh} = b^{jk}{}_{lh}[\mathbf{f}]$ that satisfy (14). It is worth noting that in non-congested networks where travel times are constant and independent of link flows, the link-path crossing matrix does not depend on link flows and the fixed point problem reduces to $\mathbf{f}^* = \mathbf{BF}$, formally analogous to the within-day static case.

For on-line application with relatively long time intervals the most effective approach towards the solution of the DNL problem seems to be the continuous packet method. Several approaches can be adopted for defining link flows f_{lh} and for solving the fixed-point problem (14). Real-time applications may require the use of a simple link-flow definition (e.g. the number of vehicles crossing a given point of the link per time unit) and of a heuristic algorithm for solving (14).

4.4 *Statistical model*

The statistical model is a loose abbreviation to denote a mechanism for modelling and predicting traffic patterns on a network of motorways that is based on statistically manifest relationships between traffic variables.

4.4.1 Schematic view

Figure 8 presents more detail of how the statistical model (STM) will operate inside the prediction system. The statistical model will be formulated in terms of a state-space model (see [19]) that describes the traffic situation on the network (depicted as 'STM State of Network').

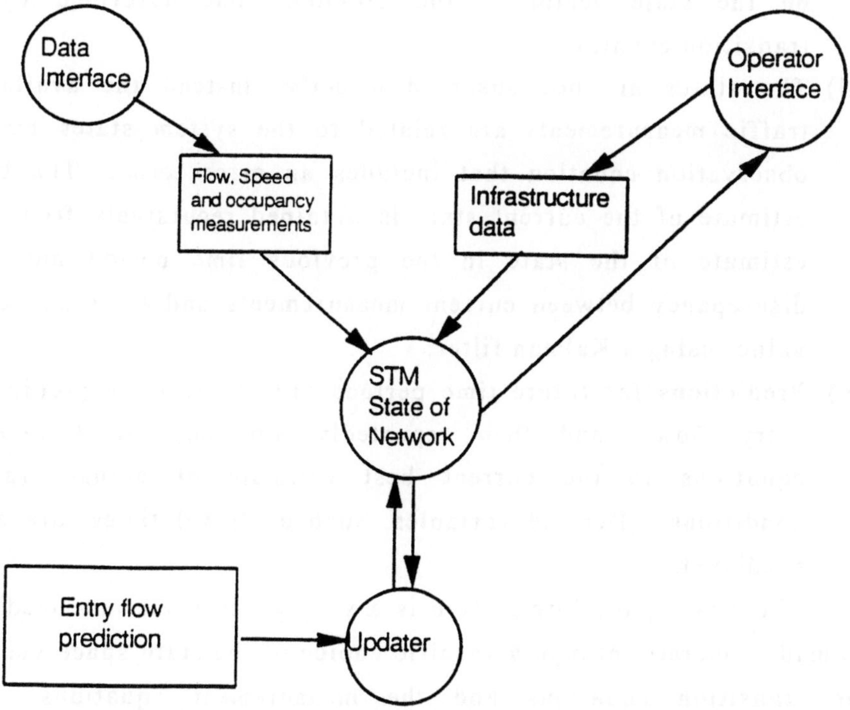

Figure 8. Expanded view of the Statistical Model

The component parts of the STM are:

(i) A network representation model that uses the infrastructure data to incorporate the topology of the network (its links, on- and off-ramps), and their interactions. These dependencies are formulated as a version of well known traffic flow models based on a continuous fluid approximation. Parameters in the model are updated by comparing predicted with actual flows. The entry flows predicted by the O/D model are input to this model.

(ii) A methodology for extracting the best state space approximation to this representation at a variety of network operating conditions. The state space model is built on equations described in [19]. The true state of the system, that is the state of traffic on the motorway network is described by a finite dimensional state vector for t=1,2,.... The system moves stochastically, and the state $\mathbf{x}_{t+1}$ at time t+1 depends only on the state vector at the previous time described by a transition equation.

(iii) The states are not observed directly; instead the available traffic measurements are related to the system states by an observation equation that includes an error term. The best estimate of the current state is obtained recursively from the estimate of the state in the previous time period and the discrepancy between current measurements and their expected value, using a Kalman filter.

(iv) Predictions for future time periods are obtained by prediction entry flows, and then repeatedly applying the transition equations to the current best estimate of actual traffic conditions. Derived variables, such as travel times, are also predicted.

The state-space formulation is a very general one and needs to be made concrete through a specific choice of the state space vector, the transition equations and the measurement equations. The observable quantities are traffic counts, traffic speeds, detector occupancy. Other system variables such as number of vehicles on a link, flow during a period, traffic density, whether or not there is congestion are not available as measurements, and must be estimated.

4.4.2 Theoretical framework

State space formulation

The true state of the system, i.e. the state of traffic on a motorway network is described by a p-dimensional vector $\mathbf{x}_t$ for t=1,2,.... The states $\mathbf{x}_t$ are not observed directly; instead quantities $\mathbf{y}_t$ are observed.

The observable quantities $\mathbf{y}_t$ are related to the system states $\mathbf{x}_t$ by the *observation equation*:

$$\mathbf{y}_t = \mathbf{H}_t \mathbf{x}_t + \mathbf{v}_t \quad (15)$$

where $\mathbf{v}_t$ is a error term, and $\mathbf{H}_t$ is a matrix mapping the state vector onto the vector of observable quantities.

The evolution of the system over time is modelled by the *transition equation*:

$$\mathbf{x}_{t+1} = \mathbf{F}_t . \mathbf{x}_t + \mathbf{w}_t \quad (16)$$

where $\mathbf{w}_t$ an error term and $\mathbf{F}_t$ is a matrix indicating the evolution of the state vector over time.

It is assumed that $\mathbf{v}_t$ and $\mathbf{w}_t$ are independent random variables.

Since the true state of the system $\mathbf{x}_t$ is not observed, it has to be estimated from the available observations. This estimate will have to be revised each time that a new set of measurements $\mathbf{y}_t$ becomes available.

Kalman filter

A Kalman filter is used to reursively update the best estimate $\hat{\mathbf{x}}_t$ of $\mathbf{x}_t$, given the estimate for the previous state $\hat{\mathbf{x}}_{t-1}$, and the current measurement $\mathbf{y}_t$.

A special feature of the Kalman filter is that it takes account of the difference between the most recent measurements $\mathbf{y}_t$ and the *expected* observations $\mathbf{y}_{t|t-1}$ to adjust its next estimate of $\mathbf{x}_t$ as follows:

- the estimate x_{t-1} is substituted in the prediction equation, giving an estimate $x_{t|t-1}$ of the current state of x
- next this result is substituted in the observation equation, giving an estimate of the expected measurement $y_{t|t-1}$ based on our best knowledge of both x_{t-1} and the transition mechanism
- the difference between the expected observation $y_{t|t-1}$ and the real observation y_t is then taken into account in the calculation of an updated estimate $\hat{x}_t$ of the real state x_t.

Predictions for future states can in principle be obtained by repeatedly applying the transition equations to the current best estimate of actual traffic conditions x_t.

Choice of variables

The DYNA Statistical Model will use the simplest state space model that will give acceptable results. For this reason initial tests will be conducted with a simple choice of state spaces, for instance:

$$\mathbf{x}_{s,t} = [r_{s,t}, q_{s,t}^{out}]^T \tag{17}$$

where:

$r_{s,t}$ = the density of vehicles on link s at time t

$q^{out}_{s,t}$ = the outflow from link s at time t

The speed $v^{out}_{s,t}$ at which traffic leaves link s is modelled as $v_{s,t} = v(r_{s,t})$ where V() is a speed/density relationship. An initial choice of the transition equations for a simple link will be:

$$r_{s,t+1} = r_{s,t} - (q_{s,t}^{out} - q_{s-1,t}^{out}) / l_s \tag{18}$$

where l_s is the length of the link

4.5 Temporal aspects

In [5] an outline is given of the architecture of an Advanced Driver Information System (ADIS), both regarding the system structure and the temporal aspects of the real-time congestion prediction system. The temporal aspects of any real-time prediction system are depicted in figure 9 below:

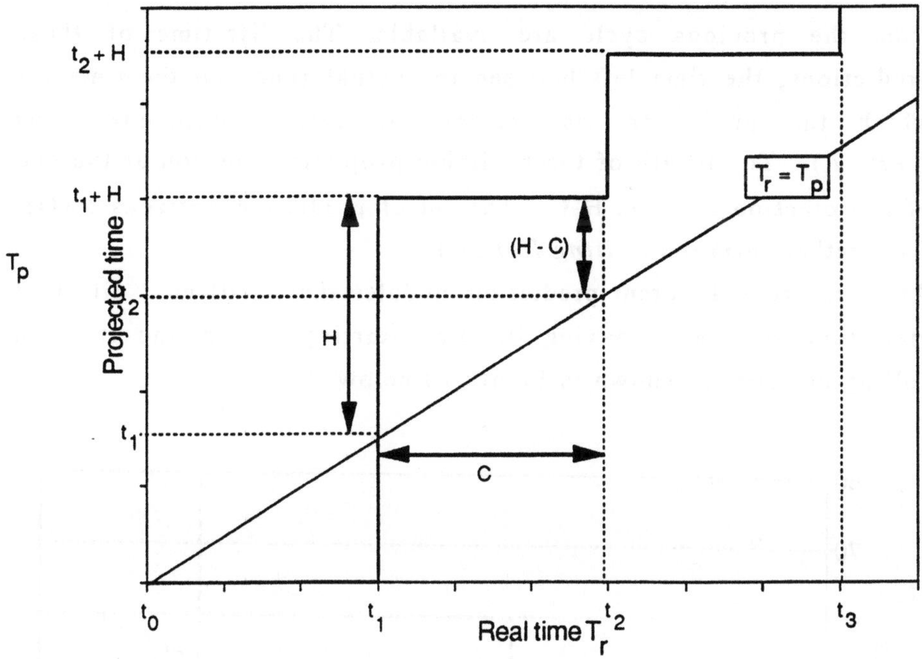

Figure 9. The Rolling Horizon prediction

On the horizontal axis the real-time T_r is set out; on the vertical axis is the projected time T_p for which predictions are made.

Predictions are updated at times t_0, t_1, t_2, etc., i.e. when the previous computational cycle ends and a new one begins. Predictions started at t_i become available at time t_{i+1}.

The step function denotes the computing activity, the prediction horizon, and the availability of new predictions. The T_p value of the horizontal lines denote the prediction horizon for the prediction currently being computed. Between times t_i and t_{i+1} (where the step function is constant) the system is busy computing the next prediction. At the step discontinuities t_i, the current prediction becomes available, and a new prediction cycle is started.

The prediction is made for some time in the future, the *Prediction Horizon* H. The duration of a computational cycle is denoted as C. During a computational cycle only the predictions from the previous cycle are available. The life-time of these predictions, the time left between the actual time and the time for which the prediction was meant, decreases continuously (as $t_i + H - T_r$). The length of the available projection horizon at the end of a projection cycle is H-C. The set of prediction horizons t_i+(H-C) together form the *Rolling Horizon*.

As there are 2 different prediction modules there will be 2 different step functions, each having its own starting points and its own rolling horizon, as shown in Figure 10 below:

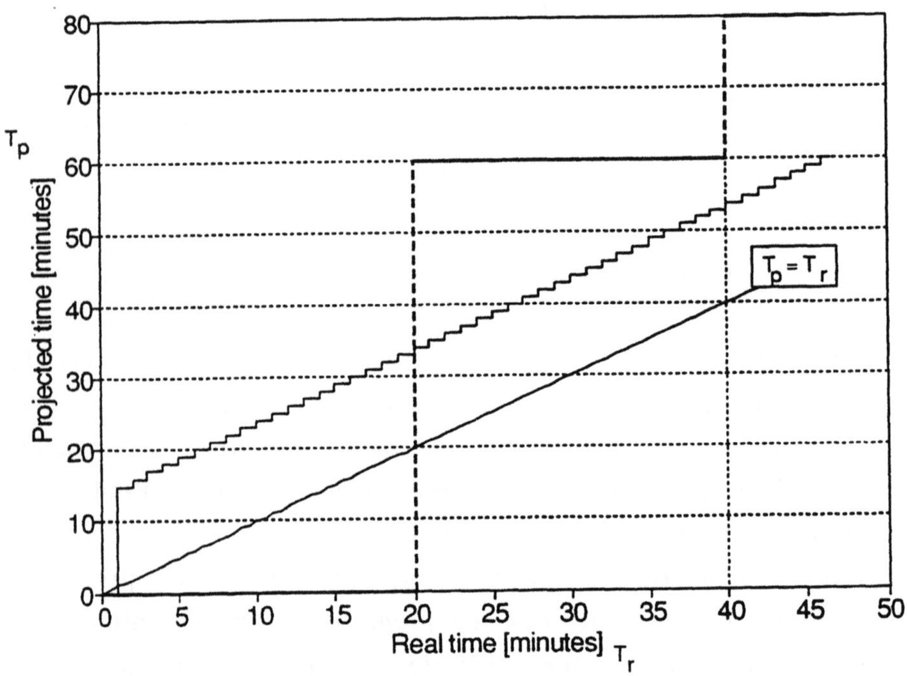

Figure 10. Two Rolling Horizons

The scale is in minutes, the very-short-term prediction is assumed to have a run-time of 1 minute, and a maximum prediction horizon of 15 minutes (drawn as solid step function), the short-term prediction is assumed to have a run-time of 20 minutes, a maximum prediction horizon of 60 minutes (drawn as dotted step function), and a minimum prediction horizon of 20 minutes.

A consequence of the fact that predictions start at fixed times, is that the start of an incident at any time t_x between the start of a prediction module at t_i, and the time t_{i+1} when the prediction started at t_i will be available can only be incorporated in the predictions after t_{i+1}. When a severe incident is detected by the STM, or when the operator enters an accident alert, the DTA will be notified. It will then redo the calculations using observed rather then estimated link-performance.

Various modules need information produced by other modules, so there must be some kind of scheduling of and synchronisation between modules.

The following are examples of the sudden availability of relevant information that must be taken into account quickly, and therefore pose the need for communication and synchronisation of modules :

- the Operator may inform the system that one or more lanes, or a whole road will be shut down (e.g. to reflect accidents reported by the police or road works). This information will have to be taken into account by all prediction modules.
- the Monitoring System has a new set of measurements ready for the Statistical module and the O/D estimation module to process. These data should be received without delay.
- an incident has been detected. Both the Statistical module and the DTA module should take this into account as soon as possible. It might even be worthwhile to interrupt the DTA procedure and take the incident into consideration, although this has the disadvantage that there is no longer any guarantee that the DTA predictions will be available in a fixed time interval.

In general the DTA module continuously uses the latest updated O/D matrix produced by the O/D module, and perhaps incident warnings. This implies a constant need for communication between the Incident Detection Module, the Statistical model, the O/D model and the Dynamic Traffic Assignment model.

4.6 Merging of very short and short term traffic predictions

There are several reasons to combine the predictions produced by the statistical model and the dynamic traffic assignment:

- it is to be expected that the statistical models can be evaluated faster than the dynamic traffic assignment.
- over very short time periods (5-15 min) driver's route choice can be expected to play only a modest role since there are very few alternatives to choose from.
- the Kalman Filter provides a natural way to include both the prediction, and the feedback of prediction error to improve the subsequent prediction.
- over periods of 30 min. or longer, choice behaviour of drivers can be expected to play an important role, which is difficult to capture in a statistical model.
- especially when the effect of control measures is to be evaluated, behavioural aspects can be expected to be of central importance; the dynamic traffic assignment approach provides a good framework into which other behavioural models can be integrated.

To obtain traffic flow forecasts for 5, 15, 30, and 60 minutes ahead, the predictions from both the Statistical and the Dynamic Traffic Assignment models may be merged in some way. When the predictions overlap, more weight will be given to the statistical model for shorter prediction horizons, and more to the Dynamic Traffic Assignment model for longer prediction horizons.

The merged predictions could be updated every minute to take the latest predictions from the statistical module into account. As shown in the (hypothetical) example, the prediction horizon of the slower prediction module may be approached (and possible overtaken) by the rolling horizon of the faster prediction module. It is evident that for each combination of predictions a different set of weights is needed.

One of the questions to be solved is how to integrate these predictions to minimise the system prediction error. As this depends on actual performance of the various models, including their actual runtime, it is proposed to weight the models predictions using a variance minimisation criterion that can be estimated as the model is running.

4.7 Self calibration through feedback of prediction errors

Another important aspect is the requirement that the models should be self-calibrating, so that a prediction error at any point in space or time will be damped out instead of amplified by subsequent applications of the model. This requirement can be met for the various models as follows :

1) O/D estimation model: At each cycle of the O/D estimation the model uses the most recent set of observed flows

2) Statistical Model: Feedback of prediction error is included in a natural way by adopting an adaptive prediction method that explicitly adjusts new forecasts on the basis of observed errors in previous forecasts

3) Dynamic Traffic Assignment: The DTA will, wherever possible, use the most up-to-date observed information to ensure that each new prediction is always firmly rooted in observed conditions; this will ultimately prevent the system from drifting away from actual conditions.

5. Conclusions

In this paper an outline has been presented of the structure of a dynamic traffic model for real-time applications as conceived in the DRIVE II project "DYNA", and some of the theory on which the models are founded.

Work is in progress on detailed specification of the algorithms, the detailed design of the computer software, and the testing of the models with actual measurements. The model structure as described in this review is a synopsis of the system to be built, and serves as a guideline for model development. It does not contain 'final' model specifications but instead indicates the framework within which model development must take place. Additionally it provides a sketch of the functional specification, and indeed specifies to some extent the basic design of the system.

Acknowledgements

The work reported in this paper was undertaken as part of the DYNA project supported by the European Commission within the DRIVE II programme. The partners in the project are Hague Consulting Group, Centro Studi sui Sistemi di Transporto, Elasis, the University of Lancaster, the University of Napoli, the Université Libre de Bruxelles and the Netherlands Rijkswaterstaat. In preparing this paper we have benefited from the contribution of: Giulio Cantarella, Massimo Di Gangi, Domenico Inaudi, Eric Kroes, Karel Lindveld, Andre de Palma, Stef Smulders, and Joe Whittaker.

References

[1] Arnot, R., de Palma, A., Lindsey, R. (1993) *"Information and Usage of Free-Access congestible facilities"*, Technical Reports, COMIN, University of Geneva, 1993.14

[2] Ashok, K., Ben-Akiva, M. (1993) *"Dynamic origin-destination matrix estimation for real-time traffic management systems"*. Proceedings of the ISTTT, Univ. of California, Berkeley.

[3] Ben-Akiva, M., Bergman, M., Daly, A., Ramaswamy, R. (1984) *"Modelling inter-urban route choice behaviour. Proceedings Ninth International Symposium on Transportation and Traffic Theory"*, VNU Science Press.

[4] Ben-Akiva, M., de Palma, A., Kaysi, I. (1991) *"Dynamic Network models and Driver Information Systems"*. Transpn. Res. 25**A**,5,251-266.

[5] Ben-Akiva, M. and Kaysi, I. (1991) *"Congestion Prediction using Dynamic Traffic Assignment"*. Working paper August 1991; dept. of Civil Engineering, Massachussets Institute of Technology, Cambridge Massachussets

[6] Cascetta, E., Cantarella, G.E. (1991) *"A day-to-day and within-day dynamic stochastic assignment model"*. Transpn. Res. 25**A**,5,277-291

[7] Cascetta, E., Cantarella, G.E. (1992a) *"Non linear dynamics and equilibrium in transportation networks. Submitted to Advances in Equilibrium modelling, analysis and computation"* - Annals of operation research, A. Nagurney Ed.

[8] Cascetta, E., Cantarella, G.E. (1992b) *"Modelling dynamics in transportation networks: part I the within-day static case"*. Presented at Triennial Symposium on Transportation Engineering, Univ. Naples, Italy.

[9] Cascetta, E., Inaudi, D., Marquis, G.(1993) *"Dynamic O/D estimators using traffic counts"*. Transportation Science, forthcoming.

[10] Di Gangi, M. (1992) *"An algorithm for continuous-flow dynamic*

network loading". Presented at the 6th. World Conference on Transportation Research, Lyon, France.

[11] HCG (1990) *"The Netherlands Value of Time Study: Final Report"*. Study commissioned by the Dienst Verkeerskunde, Rijkswaterstaat to Hague Consulting Group, Final report February 1990.

[12] Kaysi, I., Ben-Akiva, M. (1992) *"An integrated approach to vehicle routing and congestion prediction for real-time driver guidance"*. Paper prepared for presentation at the TRB annual meeting.

[13] Ministry of Transport (1992) *"Dynamic Traffic Management in the Netherlands Transportation and Traffic Research Division"*, Directorate-General for Public Works and Water Management, Ministry of Transport, Public Works and Water Management.

[14] ODIN (1991) *"Drive I project V1047 'ODIN' : O/D estimation in Urban and Extra Urban networks"*. Commission of the European Communities : Directorate General XIII Telecommunications, Information Industries and Innovation Directorate F Race Programme and Development of Advanced Telematics Services

[15] Transpute (1992) *"Vervolgonderzoek algorithmen voor schatten van vertragingstijden en file-omvang bij verkeersmonitoring"*. Transpute report to Dienst Verkeerskunde, Rijkswaterstaat CXT91068.RAP

[16] Toorenburg, J.v. (1986) *"Praktijkwaarden voor de capaciteit"*. Rijkswaterstaat, Dienst Verkeerskunde - March 1986

[17] Toorenburg, J.v. (1989) *"Dynamisch schatten van vertragingstijden en file-omvang"*. Transpute report to Dienst Verkeerskunde, Rijkswaterstaat

[18] v.d. Vlist, M., Arem, B. v., Smulders,S. (1992) *"An on-line procedure for estimating capacity under prevailing roadway and traffic conditions"*.

[19] Whittaker, J. (1991) *"A Kalman filter for network travel time prediction"*. Technical report: Mathematics Department, Lancaster University.

NETWORK DESIGN MODELS AND METHODS FOR URBAN TRAFFIC MANAGEMENT

Giulio E. Cantarella

Dip. di Ingegneria dei Sistemi e Matematica
Università degli Studi di Reggio Calabria
Via E. Cuzzocrea 69, 89100 Reggio Calabria

Antonio Sforza

Dipartimento di Informatica e Sistemistica
Università degli Studi di Napoli "Federico II"
Via Claudio 21, 80125 Napoli

In urban traffic planning and management the determination of network configuration (topology and link capacities), signal setting parameters (cycle time, green times and offsets) and link flow pattern is generally required. To this aim a set of models, belonging to the family of network design models, has to be defined. In these models the path choice user behaviour has to be taken into account, in order to define the link flow pattern. The signal setting contributes to the definition of the user costs (travel and delay costs) and assumes a relevant role in the path choice. For this reason the signal setting parameters (data and variables) should be explicitly included in the model formulation.
This paper is divided in two parts, the first one related to the network design models which can be formulated for urban traffic problems where the traffic signal setting assumes a relevant role, the second one related to the experimentation of an iterative procedure suitable for the computation of the signal setting parameters of a network with an equilibrium flow pattern.

1. Introduction

In urban traffic management problems the determination of the following elements is generally required:

-network configuration (topology and link capacity)

-signal setting parameters (cycle, green times and offsets)

-link flow pattern.

To this aim a set of models, belonging to the family of network design models, has to be defined. In these models the path choice user behaviour has to be taken into account, in order to define the link flow pattern. The signal setting contributes to the definition of the user costs (travel and delay costs) and assumes a relevant role in the path choice. For this reason the signal setting parameters (data and variables) should be explicitly included in the model formulation ([2], [3], [16], [17], [28], [29], [30], [31]).

In the following the network design models which can be formulated for urban traffic management are described. Some of them were already presented in literature in the last fifteen years. Exact and approximate algorithms were proposed for their solution, but a wide experimentation on real problems does not exist. Other models are quite innovative and still require solution algorithms and a wide experimentation on test and real network. Starting from the classical Equilibrium Network Design model (END), the Equilibrium Network Signal Setting (ENSS) and the Equilibrium Network Design and Signal Setting (ENDSS) will be presented.

Then the experimentation of the iterative procedure ENETS (Equilibrium NEtwork Traffic Signal setting) for the ENSS problem will be described. This procedure shows some features different from other procedures proposed in literature. In particular, the traffic signal setting is performed in two successive steps: single junction signal setting and network coordination. This way a more efficient method for network traffic signal setting is obtained, and the green time scheduling, that is stage composition and sequence, can be included in the decision process. The results of the application of this procedure to a test network are presented. They show good performances in term of convergence to a good feasible solution with an acceptable computation time, depending on the required precision.

2. Network Design Models

In the the field of traffic and transportation management and planning, Network Design models are devoted to the computation of the network topology (starting from a reference configuration) and the link capacities (that is the maximum number of vehicles which can cross a section in a time unit), with the objective to optimize a system performance function expressing, for example, the user total transportation cost or the total emission of pollutants. Network design models assumes traditionally the following set of input parameters:

i) a reference configuration of the network topology,
ii) travel demand,
iii) maximum capacity and investment cost for each link,
iv) free-flow travel cost and cost-flow function (depending on link capacity) for each link.

The variables of the models can be decisional or descriptive and can be divided between continuous and discrete variables [4]. The decisional variables are associated to the link capacities, depending on network topology and geometric characteristics of the road (vector **y**). They can be the width of the road section (continuous variables) or the lane number (discrete variables, 0/1 variables if there is only one possibility for the road section lay-out). The link flows (vector **q**) are descriptive continuous variables.

In urban traffic problems the link capacity is also influenced by the adopted signal setting. Moreover the cost-flow functions are greatly affected by the network coordination. For this reason also the following input parameters are needed for signal setting:

v) crossing incompatibilities for each junction,
vi) saturation flow (depending on geometrical characteristics) and lost time for each approach,
vii) distance and free flow speed between each pair of adjacent junction, consistent with (iv).

For the same reason the set of decisional variables has to be enlarged, to include the signal setting parameters (vector **p**):

- cycle time, green timing and scheduling for each junction,
- offset for each pair of adjacent junctions.

With this set of input parameters and variables the objective function related to the minimization of the travel cost can be expressed as:

$$\mathrm{Min}_{\mathbf{y},\mathbf{p}}\ Z\ (\mathbf{y},\mathbf{p},\mathbf{q}), \text{ where } \mathbf{q} = \phi(\mathbf{y},\mathbf{p})$$

In these models the user path choice behaviour, expressed by the function $\mathbf{q}=\phi(\mathbf{y},\mathbf{p})$, can be taken into account in several ways, depending on the representation adopted for the link costs ([9], [11], [27]). If these costs are constant, that is the network is assumed not congested, each o/d travel demand can be loaded on the network to the shortest path (all or nothing assignment) or to a set of efficient paths in a proportional way (probabilistic-proportional assignment). If the link costs vary with flow, that is the network is assumed congested, each o/d travel demand is distributed in equilibrium among the available paths (equilibrium assignment) in a deterministic or stochastic way. In particular, if a deterministic equilibrium is used each o/d demand will be distributed on the available paths in such a way that each user will experience the same cost on each one of the used paths and cannot experience a lesser cost on a unused path (I Wardrop principle, [35]).For separable cost flow functions this equilibrium flow pattern can be determined through the minimization of the Integral Cost Function:

$$\mathrm{Min}_{\mathbf{q}}\ I(\mathbf{y},\mathbf{p},\mathbf{q}) = \Sigma_{ij} \int_0^{q_{ij}} c_{ij}(\mathbf{y},\mathbf{p},t)dt$$

where

$c_{ij}(\mathbf{y},\mathbf{p},t)$ is the cost flow function for link ij, depending on network topology **y** and signal setting **p**

q_{ij} is the flow on link ij.

For other approaches including user behaviour in network design models see [8].

2.1. Equilibrium Network Design (END)

The most known network design model used in transportation problem is the so-called Equilibrium Network Design model (END) [25]. The decisional variables are the link capacities, only depending on the geometric characteristics of the link. These variables can be continuous or discrete, so generating the continuous END and the discrete END. The link flows are descriptive variables. The model has the following formulation:

$$\begin{aligned}
& \mathrm{Min}_y\ Z(y, q^*) \\
\text{s.t.}\quad & y \in Y \\
& q^* \in Q, \quad q^* = \phi(y) \\
& h^T y \leq B
\end{aligned}$$

where:

Y is the set of feasible networks in terms of topology and link capacities,

Q is the set of feasible link flow patterns,

$q^* = \phi(y)$ is the flow vector, result of a traffic assignment model, giving the equilibrium flow pattern for a network y,

h is the vector of link investment costs

$h^T y \leq B$ is a budget constraint.

This formulation may lead to a bilevel optimization model with decisional variables y at the outer level and the decisional variables q at the inner level [24]. Any traffic assignment model for determining link flow pattern can be used. As said before, a symmetric deterministic equilibrium model is used, $q^*=\phi(y)$ is such that $I(y,q^*) = \mathrm{Min}_q\ I(y,q)$, with $q \in Q$.

A particular case of the discrete END arises for binary decisional variables. In this case, assuming the link costs as a constant, each o/d demand, according to a deterministic user behaviour, will use the shortest path. Thus the problem becames the well known combinatorial discrete 0/1 network design problem with a budget constraint.

The equilibrium network design models are used in extra-urban transportation problems where the signal setting can be assumed not relevant and the network topology can be modified adding new links. This is not the case of urban management problems where the signal setting has to be analysed to reach a correct simulation of the system and moreover most of the network topology changes refers share of the total road width between the two link ways.

2.2. Equilibrium Network Signal Setting (ENSS)

As above mentioned Signal Setting assumes a relevant role in urban traffic problems and contributes to the definition of the link capacities together with the geometric characteristics of the link. It is possible to formulate a model where the decisional variables are not explicitly related to the link capacities, but rather to the signal setting parameters (green times and offsets), still taking into account the user path choice behaviour. The model, called Equilibrium Network Signal Setting, assumes fixed the network topology and the link geometric characteristics:

$$\mathrm{Min}_p\ Z(p,q^*)$$

s.t.

$$p \in P$$

$$q^* \in Q, \quad q^* = \phi(p)$$

where:

P is the set of feasible values for signal setting parameters,

$q^* = \phi(p)$ is the equilibrium flow pattern for the given network with a signal setting p, result of a traffic assignment model.

A graphical representation of the objective functions of the outer and inner optimization problems is described in [6], together with the analysis of lower bound solutions [19]. This graphical representation is briefly summarized in the following for the simple network sketched in fig.1, with 5 nodes and 5 links, proposed in [14]. Two origin-destination pairs (A-B, which can use two alternative paths, and C-D, on a single path) are considered. Junction E is the only signalized junction with a fixed cycle time.

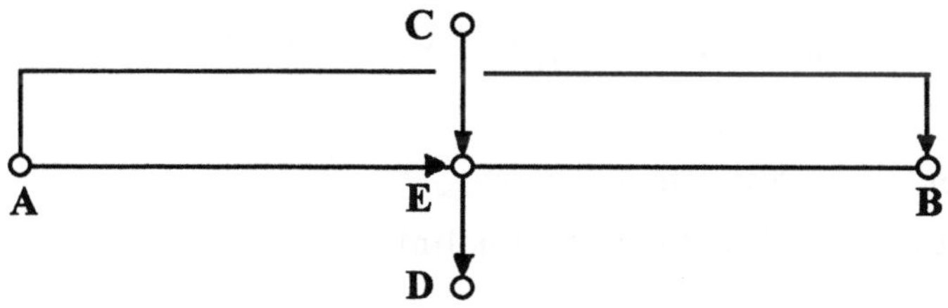

Fig.1. Example network

The parameters of such a simple problem, that is green times at the junction E and link flows, can be graphically represented in a two-dimensional plane (p,q). Indeed, flows on paths A-B and A-E-B, that is q_{AB} and $q_{AE}=q_{EB}$, are complementary with respect to the demand value between A and B. Green times at approaches of junction E (g_{1E} on link A-E and g_{2E} on link C-E) are complementary with respect to the cycle time. Moreover flow on path C-E-D, that is $q_{CE}=q_{ED}$, is constant and equal to the demand value between C and D. So the system has two freedom degrees, i.e. one pair of independent variables is needed to describe completely the problem, for example $q=q_{AE}$ and $p=g_{1E}$.

Figure 2 shows in a plane (p,q) the contours of the system objective function Z(p,q), expressed by the total travel time (or any other system performance index). For a value q_0, the function $Z(p,q_0)$ takes the minimum for the value po. These pairs (p_0,q_0) define the curve $p=\Gamma(q)$ of the optimal value of the regulation parameter p for each flow value q, which expresses the solutions of a traffic signal setting model.

On the other hand, for a fixed traffic signal setting, link flow pattern is found through a traffic assignment model according to some user behaviour criterion. This relationship can be formally expressed as $q=\Phi(p)$. For example, in the case of symmetric deterministic equilibrium assignment with fixed demand, (assuming a user behaviour according to the first Wardrop's principle), this relationship, as noted above, corresponds to the solution of the following convex program:

$$\mathrm{Min}_{\mathbf{q}}\ I(\mathbf{p},\mathbf{q}) = \Sigma_{ij}\int_0^{q_{ij}} c_{ij}(\mathbf{p},t)dt \qquad \mathbf{q}\in Q$$

where

Q is the set of feasible flow patterns,

$c_{ij}(\mathbf{p},t)$ is the cost flow function for link ij.

Figure 3 shows the contours of the integral travel cost $I(\mathbf{p},\mathbf{q})$. For a value $\mathbf{p}_0$ the function $I(\mathbf{p}_0,\mathbf{q})$ takes the minimum for the value $\mathbf{q}_0$. These pairs $(\mathbf{p}_0,\mathbf{q}_0)$ define the curve $\mathbf{q}=\Phi(\mathbf{p})$ of the equilibrium flow pattern for each regulation parameter set, which expresses the solutions of a traffic assignment model.

When both signal setting parameters and link flows are not fixed, their combined evaluation is required, modelling the interaction between flow pattern and signal setting. This interaction can be seen as a game between two players: users and traffic agency [15]. User goal is the minimization of their individual costs, according to the first Wardrop's principle. Traffic agency chooses a control strategy to optimize a system performance index. This situation can be modelled by:

$$\mathrm{Min}_{\mathbf{p},\mathbf{q}} Z(\mathbf{p},\mathbf{q}) \qquad \mathbf{p}\in P,\ \mathbf{q}\in Q,\ \mathbf{q}=\Phi(\mathbf{p})$$

From this viewpoint the curve $\mathbf{p}=\Gamma(\mathbf{q})$ represents a relation between equilibrium flow pattern $\mathbf{q}$ and signal setting $\mathbf{p}$, and expresses the decision of traffic agency, without any knowledge of user behaviour. The curve $\mathbf{q}=\Phi(\mathbf{p})$, instead, represents a relation between the same variables but expresses the result of user choices. It is a constraint for the values of variables $\mathbf{p}$ and $\mathbf{q}$ and so a feasible pair $(\mathbf{p},\mathbf{q})$ should be on this curve.

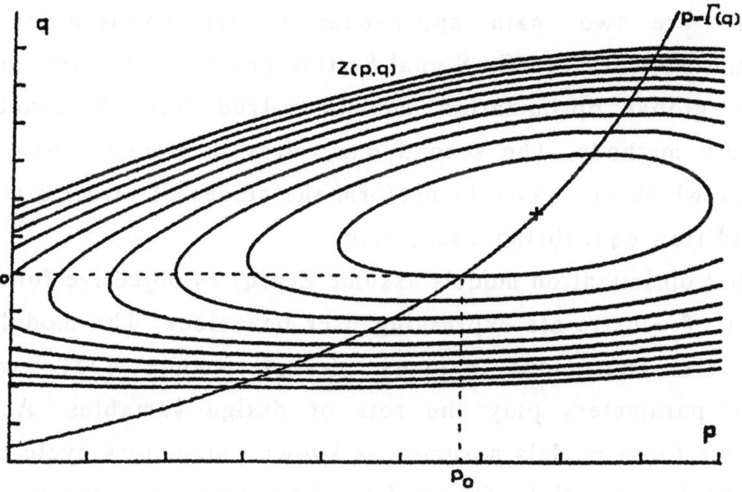

Fig.2. Signal setting:
contours of function Z(p,q) and curve p=Γ(q)

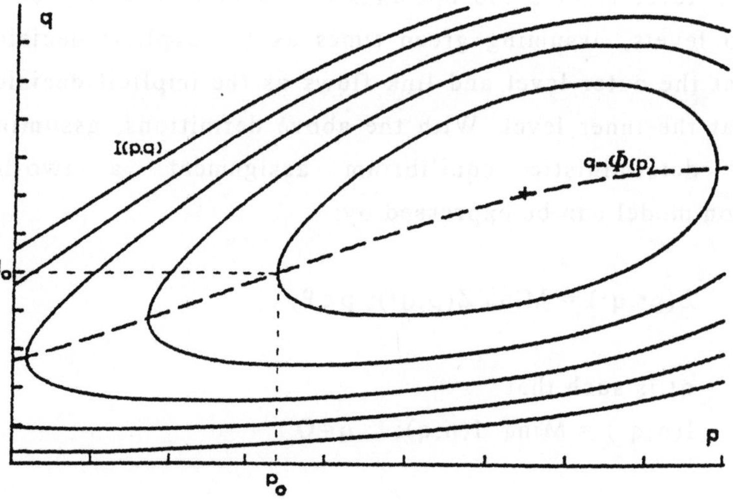

Fig.3. Traffic assignment:
contours of function I(p,q) and curve q=Φ(p)

There are two main approaches to the solution of the Equilibrium Network Traffic Signal Setting problem. The first one is based on global optimization models solved through exact or approximate methods. The second one refers to iterative heuristic procedures which successively perform the stages of traffic signal setting and flow equilibrium assignment.

Global optimization models assume **Z(p,q)** as objective function with a set of constraints expressing user behaviour. The models so obtained are continuous network design models [25], where regulation parameters play the role of design variables. At the present time these models assumes as known parameters cycle time, green scheduling and link offsets. Therefore green times are the only design variables. In conclusion, these models aim to the optimization of an objective function of flows and green times, where flow pattern must satisfy a user behaviour criterion like the widely used Wardrop's first principle. This requirement can be met in two different ways, leading to two level or one level optimization models.

In two level models the optimization is hierarchically carried out at two levels, assuming green times as the explicit decisional variables at the outer level and link flows as the implicit decisional variables at the inner level. With the above definitions, assuming a symmetric deterministic equilibrium assignment, a two-level optimization model can be expressed by:

$$Z(p^*,q^*) = \mathrm{Min}_p\ Z(p,q^*);\ p \in P,$$

where q^* is such that

$$I(p,q^*) = \mathrm{Min}_q\ I(p,q); \quad q \in Q$$

Hence for each objective function evaluation at the outer level for a fixed p, the associated flow pattern is to be computed by performing an equilibrium assignment at the inner level. This approach appeared in Tan et al.(1979)[32], who defined the model as an "hybrid optimization" model, solved via an augmented lagrangean method. More recently Leblanc and Boyce (1989)[24] proposed a solution method based on an heuristic generalization of techniques for bilevel linear programming to non linear problems.

In one level models both regulation parameters and link flows are explicit decisional variables. The user behaviour is expressed by a set of constraints for which some formulations were proposed ([8], [20], [26], [32]).

Iterative procedures can be expressed as the solution of the system of equations above introduced, $\mathbf{p}=\Gamma(\mathbf{q})$, $\mathbf{q}=\Phi(\mathbf{p})$. These procedures, starting with an initial assignment, try to reach a feasible solution of the problem solving sequentially equilibrium assignment and traffic signal setting until two successive flow patterns or signal settings are equal, within a specified tolerance. If convergence is reached in a finite number of iterations, the final solution can be said mutually consistent. This means that signal setting generates a set of link costs which determines a flow pattern such that signal setting is optimal for it.

Iterative procedures allow to assume all the regulation parameters (green times, green scheduling and offsets) as decision variables. They also generally allow to solve medium-scale problems with an acceptable computational effort. For these reasons at the present time it seems that real problems can be better approached via iterative procedures, even if they do not necessarily converge to the optimal solution. However, it should be stressed that some authors ([12], [26], [32]) noted that the final solution of an iterative procedure may depend on the initial assignment. Furthermore, Dickson (1981) [14] showed that the network total cost, assumed as a performance index, can increase during the procedure, clearly leading to a non optimal solution.

Cantarella et al. (1991) [7] described a brief state of the art of the solution approach proposed for the ENSS problem. Moreover an heuristic procedure, named ENETS, is proposed and a graphical representation of the solution generated by ENETS for a simple network with one isolated junction or two interacting junctions is shown. This representation is very helpful in understanding the behaviour of an iterative procedure.

2.3. Equilibrium Network Design and Signal Setting (ENDSS)

To take into account not only the continuous variables related to signal setting parameters (green times and offsets), but also the variables (continuous and discrete) related to the geometric characteristics of the road links, a new more complex model has to be formulated, from the combination of the END and ENSS models. This model, which can be defined Equilibrium Network Design and Signal Setting model assumes the following configuration:

$$\mathrm{Min}_{y,p}\ Z(y,p,q^*)$$

s.t.

$$y \in Y$$
$$q^* \in Q,\ q^* = \phi(y, p)$$
$$p \in P$$
$$h^T \cdot y \le B$$

where:

$q^*=\phi(y,p)$ is the equilibrium flow pattern for a given network y with a signal setting p, such that $I(y,p,q^*)=\mathrm{Min}_q I(y,p,q)$, with $q \in Q$.

A first solution approach to this problem has been proposed by Cantarella e Vitetta (1993) [10]. Another model could be formulated in the case of a network formed by a set of very close junctions, constituting a complex traffic node, for which the design refers to the building of by-pass-links for increasing the node capacity, for example overpass or underpass. The decisional variables are related to "node" topology, to by-pass-link capacities and also to signal setting parameters. The flow pattern cannot be assumed fixed, but for each solution y, each user has no path alternatives. This model presents the following configuration, with the known symbols:

$$\mathrm{Min}_{y,p}\ Z(y, p, q)$$

s.t.

$$y \in Y$$
$$p \in P$$
$$q = q(y)$$
$$h^T \cdot y \le B$$

3. Solution of the ENSS Problem Through an Iterative Procedure

In the following section the main features of the iterative procedure ENETS proposed in Cantarella et al. (1991) [7] are described. Moreover the results of new experimental tests on a small network are presented. These tests confirm already known results about convergences of iterative procedures and allow to better understand their behaviour.

3.1 Procedure ENETS

The procedure ENETS shows some differences from the other proposed iterative procedures. In particular, the traffic signal setting is performed in two successive steps: single junction signal setting and network coordination. In this way a more efficient method for network traffic signal setting is obtained, and the green time scheduling, that is stage composition and sequence, can be included in the decision process. Moreover, the single junction signal setting is approached using the capacity factor maximization as objective function, so assuring the lowest congestion level. In addition the adopted delay-flow functions for movement links at junctions can be applied both for interacting and non interacting junctions. In the latter case the well known Webster two term-formula is obtained.

ENETS is based on three optimization models, related to the isolated junctions signal setting, the signalized network coordination and the traffic assignment. Each step of the procedure uses an exact algorithm for a particular urban traffic optimization problem. Previous experiences with each algorithm show short running times for real-size problems. The network representation adopts a distinction between road links and movement links at the junctions.

3.2 Experimental Tests of ENETS on a Simple Network

ENETS was applied to the simple network shown in fig.4, with 6 nodes and 7 links, derived from the network of fig.1. Nodes E and B represent signalized junctions. There are 4 origin-destination pairs: (A-F) which can use two alternative paths, (A-D) which can use three alternative paths, (C-F) on a single path, (C-D) which can use two alternative paths. The test network, expanded with all the movement links, is shown in fig.5. Table 1 reports the fundamental data.

The cost functions proposed by Davidson (1966) [13] and generalized by Akcelik (1978) [1] are adopted for road links (see also [33]). The delay-flow functions briefly described in the previous section are adopted for movement links (for more details see [7]).

The experimentation was carried out varying cycle time, link lengths and o/d demand pattern and studying the effects on different parameters that may be observed, such as link flows, offsets and total costs.

Cycle time C assumes the values in the range 60÷120 sec, with step 10. The lengths of the links are modified with respect to an initial set of values, in the following named L1, according to a multiplying factor which varies in the range 2÷6, with step 1, so generating the sets of lengths L2÷L6. The o/d demand assumes two configurations, named D1 and D2 and reported in table 2.

The junctions E and B are assumed interacting and so the set of decisional variables contains the offset f_{EB} (link 5-8) between the start of green g_{1E} (link 6-14) and the start of green f_{1B} (link 8-11). The offset is assumed an integer variable in the range between 0÷C.

As above mentioned, the convergence test is performed comparing the flow values obtained in two successive iterations. If the difference between the two patterns satisfies a given approximation, the procedure stops, otherwise a new iteration of the procedure should be performed. In the tests performed the approximation parameter ε assumed the values 0.001 and 0.01.

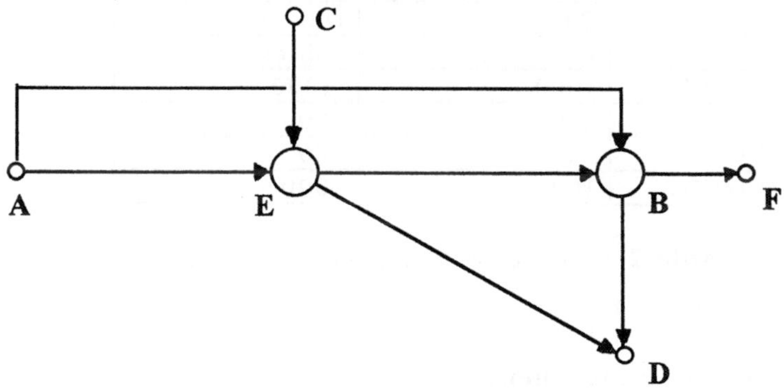

Fig.4. Test network

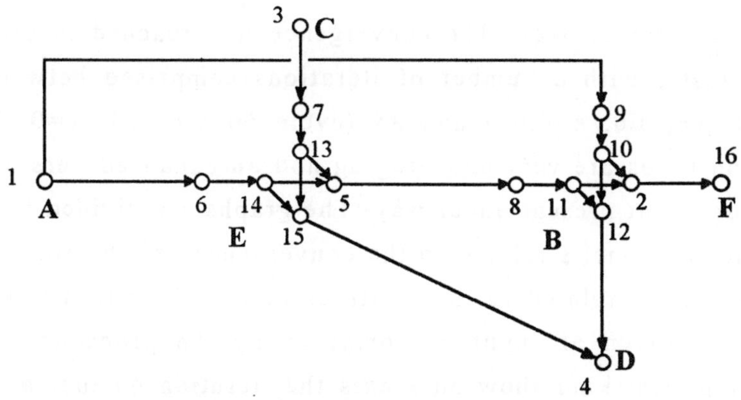

Fig.5. Expanded test network

(a)

	road links					i/o links	
link i-j	1-6	1-9	5-8	12-4	15-4	2-16	3-7
capacity	1500	1500	1500	1500	1500		
length	100	200	100	100	200		

(b)

	movement links					
link i-j	6-14	14-5	14-15	7-13	13-5	13-15
sat. flow	1500	1500	1500	1800	1800	1800
link i-j	8-11	11-2	11-12	9-10	10-2	10-12
sat. flow	1500	1500	1500	1800	1800	1800

Table 1. Test network data:
(a) road links, (b) movement links

	D1	D2
(O/D)	vehicles/h	vehicles/h
(A/F)	1000	1000
(A/D)	200	200
(C/F)	200	200
(C/D)	100	1000

Table 2. O/D pairs and related demands

3.3 Results of the Experimental Tests

The results of experimentation are summarized in table 3 and in the graphs of figures 6-21. The convergence was reached in all the performed tests, with a number of iterations comprised between 8 (cycle 120 sec, L6, $\varepsilon=0.01$) and 38 (cycle 60 sec, L1, $\varepsilon=0.001$). Computation times are very encouraging and an enhanced version of the procedure is at present under way. The graphs are divided in two groups. The first one is related to the convergence of the procedure. The second one is related to the relations among the parameters of the problems, given as input or computed by the procedure. The graphs of the first kind show on x-axis the iteration number and in y-axis one of the following parameters:

-flow on a link

-offset between adjacent junctions E and B

-total cost (travel cost+delay cost) on the network.

In particular in fig.6 the flow on the link 5-8 is reported for a given link length (L1, base length) and for a given demand pattern (D1), varying the cycle time in the range 60-120 sec. In fig.7 the same flow is reported, for a given cycle duration (100") and for a given demand pattern (D1), varying the link length until 6 times the base length (L1-L6). Varying the length of the links allows to highlight that when junctions are very close one to each other the relevance of coordination is higher, and in this case signal setting should be carefully considered in any decision concerning network topology and link capacities. As above mentioned, the convergence is reached in a number of iterations less than 40, but the graphs show that the solution reaches a good stability condition after 20÷25 iterations. These graphs also show that for cases L2-L6 the free link flow is significantly different from link flow for L1, since the effect of junction interactions reduces as length increases. In fig.8 the offset f_{EB} is reported for a given length (L1), and for a given demand pattern (D1), varying the cycle time in the range 60-120. Also in this case, after 15-20 iterations the offset is substantially constant, until the end of procedure, for all the cycle values. Figure 9 reports the values of total cost and delay cost during the procedure. The second group of graphs expresses the relation between the given parameters (link length, cycle time and demand pattern) and the computed parameters (flows, offset and total cost). In these graphs, each of the computed parameters is expressed in function of the length and the cycle, for the two demand pattern. In particular figure 10 reports the flow q_{58} in function of the length for demand D1, varying the cycle, and the figure 11 reports the same flow in function of the cycle, for demand D1, varying the length. As already noted, as the length increases (fig.10), the flow varies until the set of lengths L3, and then remains stable, meaning that for lengths over a threshold four times the base length, the effect of signal setting coordination is no longer relevant. The effect of cycle time on flows is relevant for any length (fig.11). Figures 12 and 13 are related to the offset between junctions E and B, expressed in function of the length varying the cycle (fig.12), and in function of the cycle varying the length (fig.13). From analysis of output it results that as cycle time is increased by 10 sec, the green period, g_{1B}, is increased by about 7 sec. Therefore the offset f_{EB} is decreased by 3 sec (remember that only integer values are considered for offset) to keep roughly constant the time between the end of up-stream green (link 6-14) and the beginning of down-stream green (link 8-11). So the offset is quite linearly decreasing with cycle time. Figures 14 and 15 show, for demand D1, the total cost and delay cost in function of the length, varying the cycle time, while figures 16 and 17 express the same cost values in function of the cycle time, varying the link length. The total cost in figure 14 is quite linear against length. This is mainly related to the fact that with demand pattern D1 the network

□		L 1		L 2		L 3		L 4		L 5		L 6	
		ε=0.001	ε = 0.01	ε=0.001	ε = 0.01	ε=0.001	ε = 0.01	ε=0.001	ε = 0.01	ε=0.001	ε = 0.01	ε=0.001	ε = 0.01
cycle 60 sec	iter. numb.	38	29	18	17	15	13	14	11	17	10	15	9
	total cost	153907	152931	238689	237343	315309	315268	398731	397544	477643	478272	555500	554308
cycle 70 sec	iter. numb.	31	28	20	15	25	15	14	11	17	10	18	9
	total cost	150321	150854	232895	232604	318493	318885	393984	393624	474220	474351	552804	554761
cycle 80 sec	iter. numb.	31	27	18	15	13	11	19	10	12	9	11	10
	total cost	149673	149010	231327	231779	307337	307000	392429	393138	472563	473682	552837	553374
cycle 90 sec	iter. numb.	28	24	17	14	37	18	15	10	14	9	16	8
	total cost	148645	149172	232516	231811	314944	315018	392505	392445	474639	473925	553946	554160
cycle 100 sec	iter. numb.	38	24	17	14	35	11	21	10	14	9	16	8
	total cost	151399	151087	232632	232868	314668	318301	390317	391637	474303	474022	554615	554781
cycle 120 sec	iter. numb.	33	26	19	14	38	21	13	10	11	10	19	8
	total cost	153394	154697	235369	235336	319116	319806	392032	392604	476873	476796	557094	559626

Table 3. Results of experimentation of ENETS on test networks

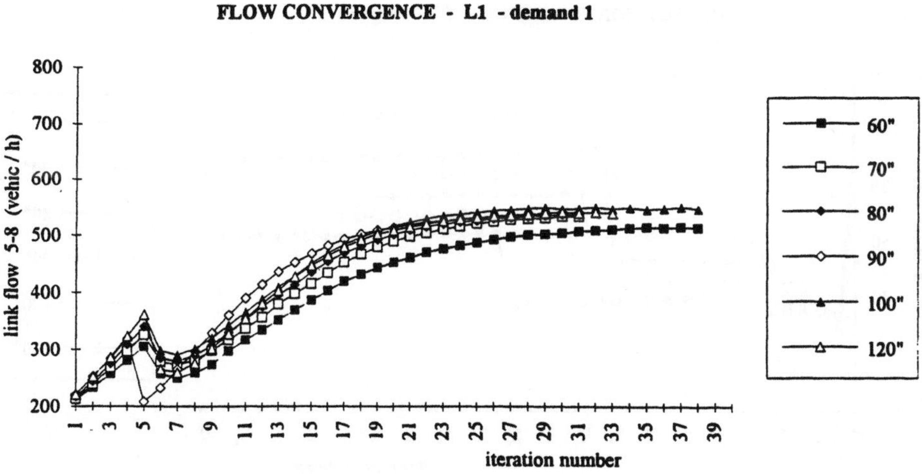

Fig. 6. Flow convergence for a fixed length, varying the cycle

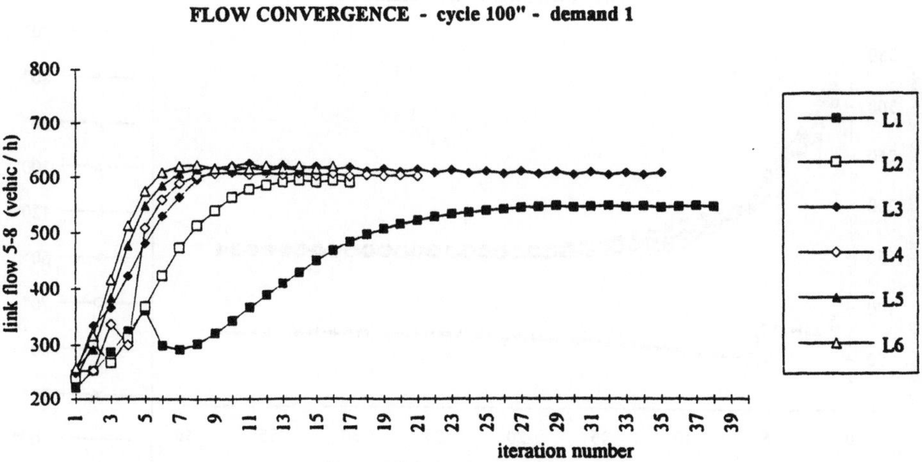

Fig. 7. Flow convergence for a fixed cycle, varying the length

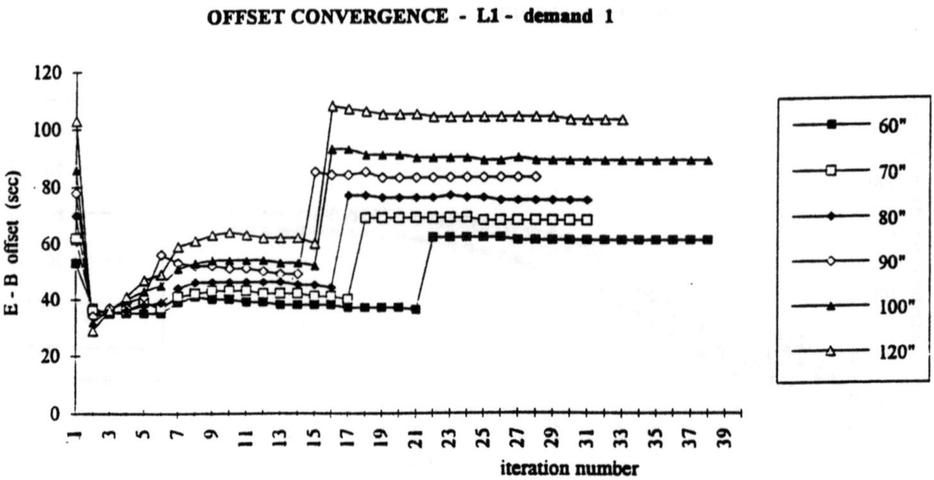

Fig. 8. Offset convergence for a fixed length, varying the cycle

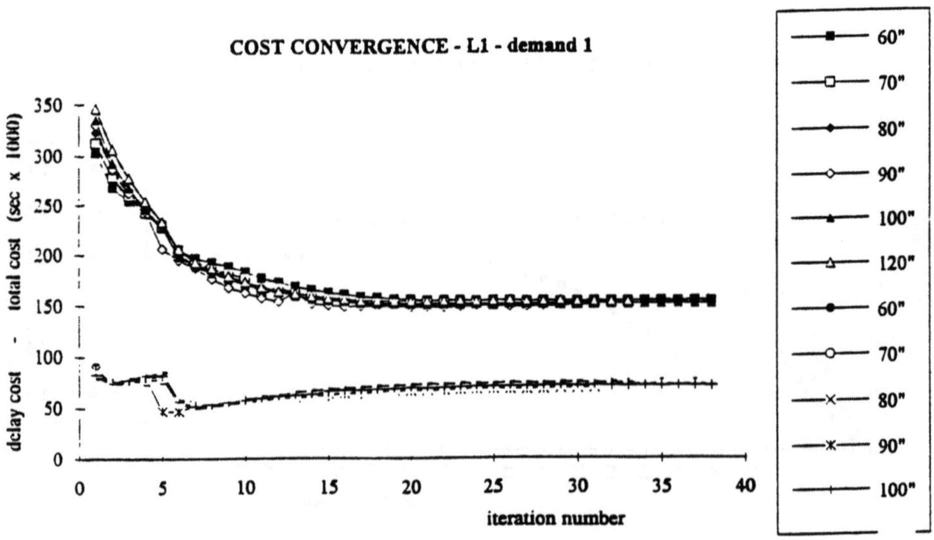

Fig. 9. Total and delay cost convergence for a fixed length, varying the cycle

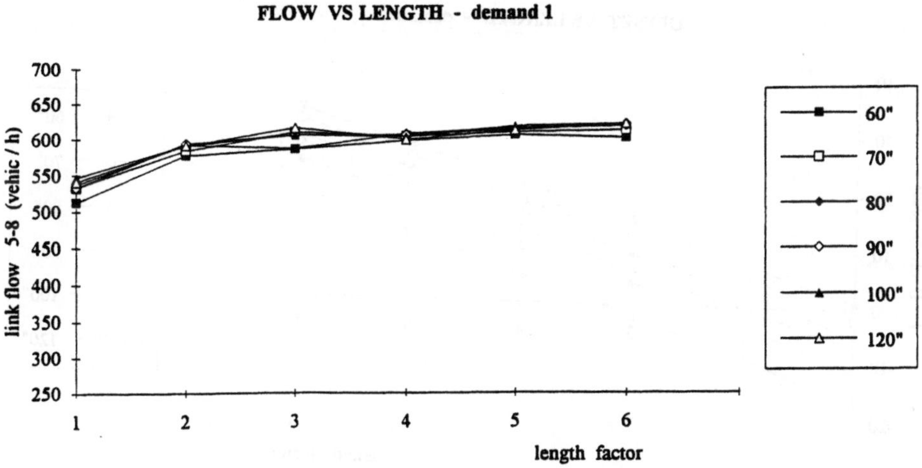

Fig. 10. Link flow 5-8 vs link length, varying the cycle

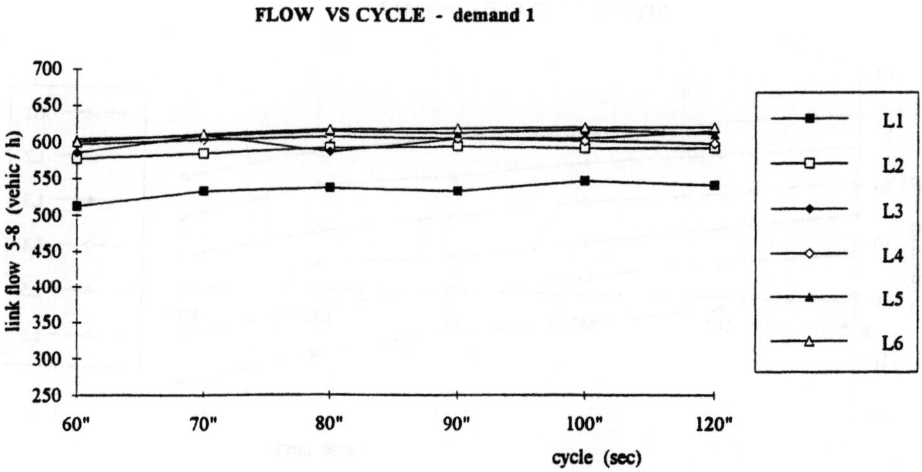

Fig. 11. Link flow 5-8 vs cycle, varying the length

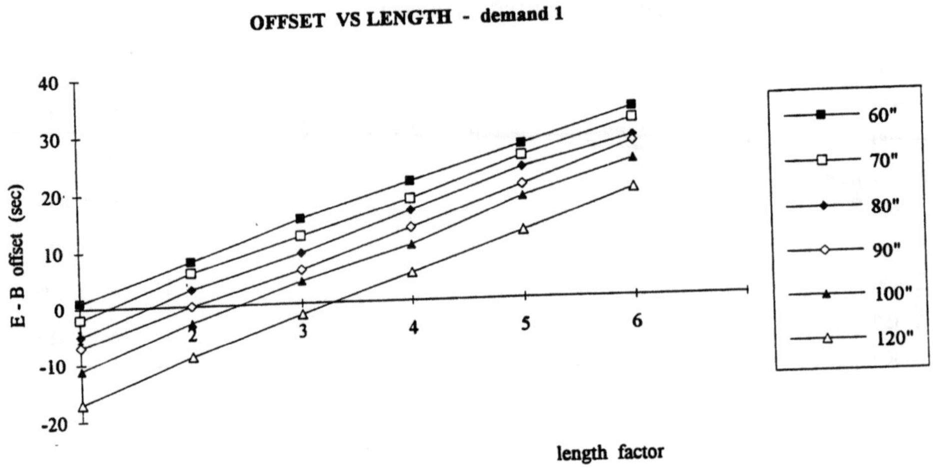

Fig. 12. **E-B** offset vs link length, varying the cycle

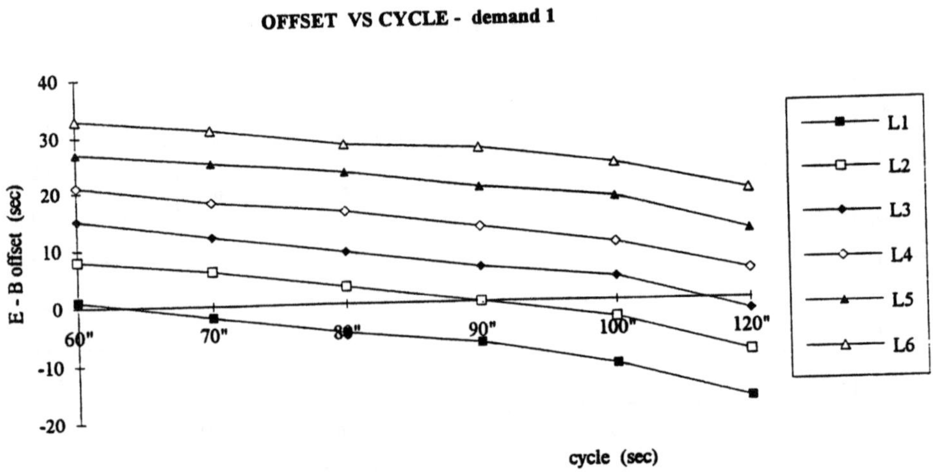

Fig. 13. E–B offset vs cycle, varying the length

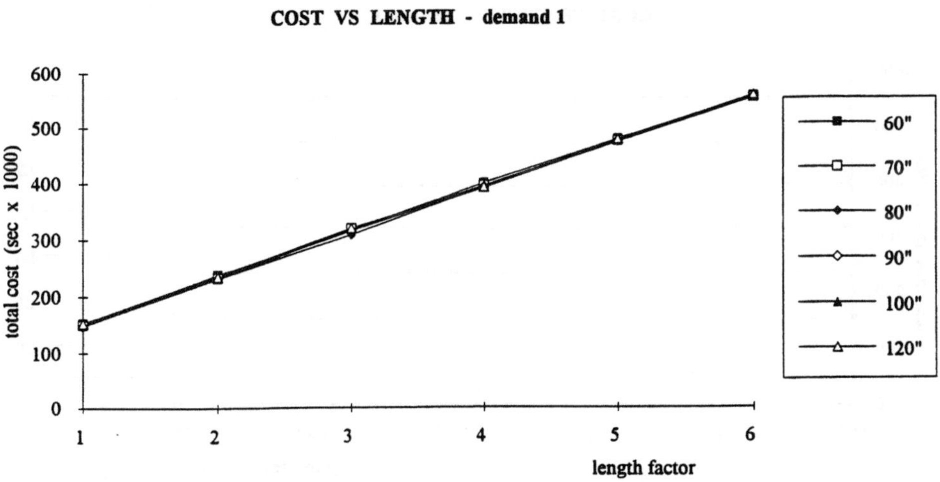

Fig. 14. Total cost vs link length, varying the cycle (demand D1)

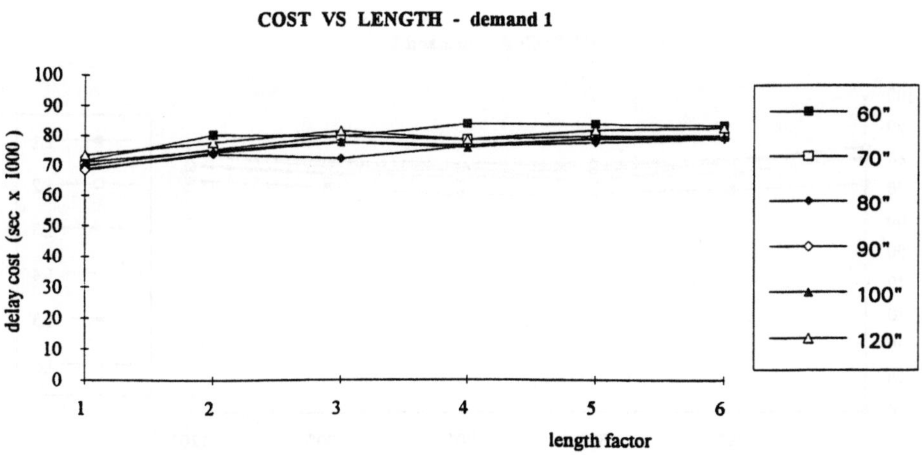

Fig. 15. Delay cost vs link length, varying the cycle (demand D1)

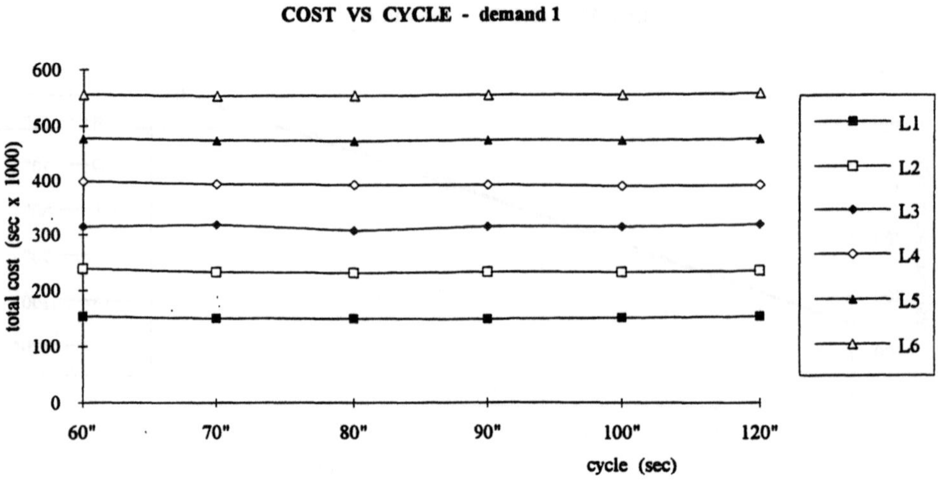

Fig. 16. Total cost vs cycle, varying the length (demand D1)

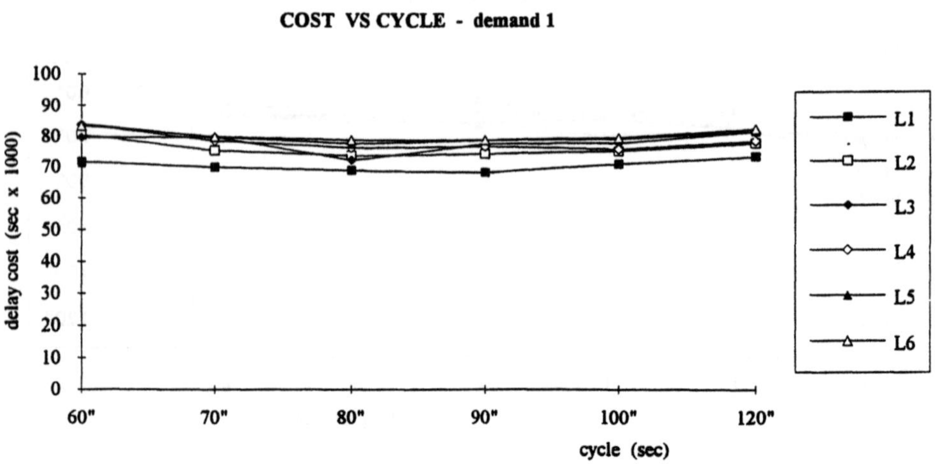

Fig. 17. Delay cost vs cycle, varying the length (demand D1)

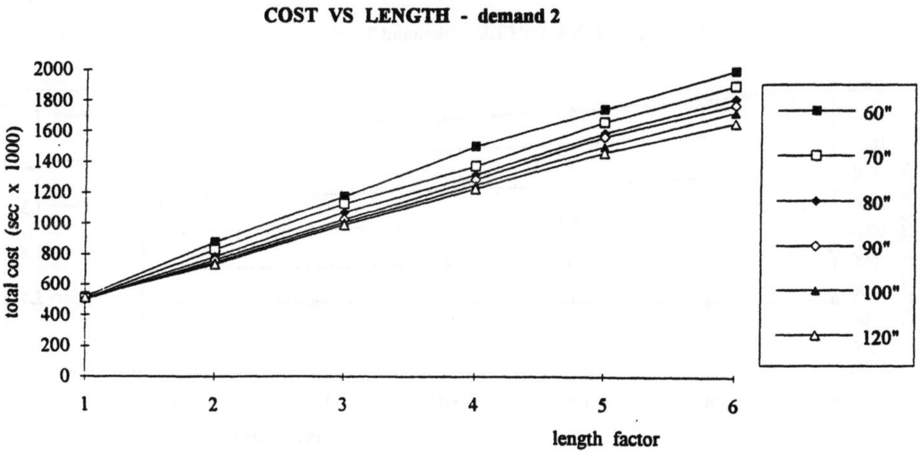

Fig. 18. Total cost vs length, varying the cycle (demand D2)

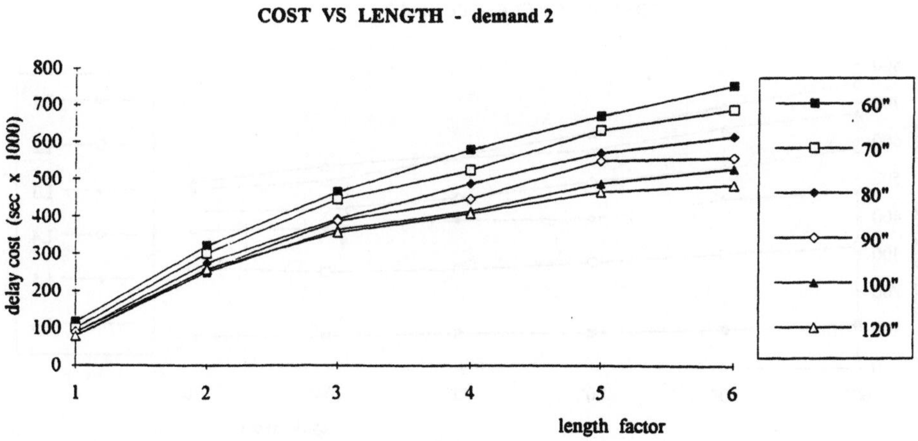

Fig. 19. Delay cost vs length, varying the cycle (demand D2)

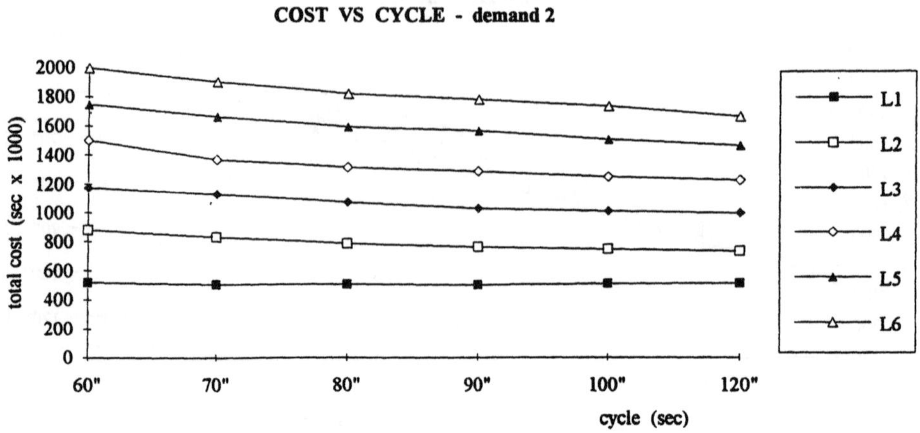

Fig. 20. Total cost vs cycle, varying the length (demand D2)

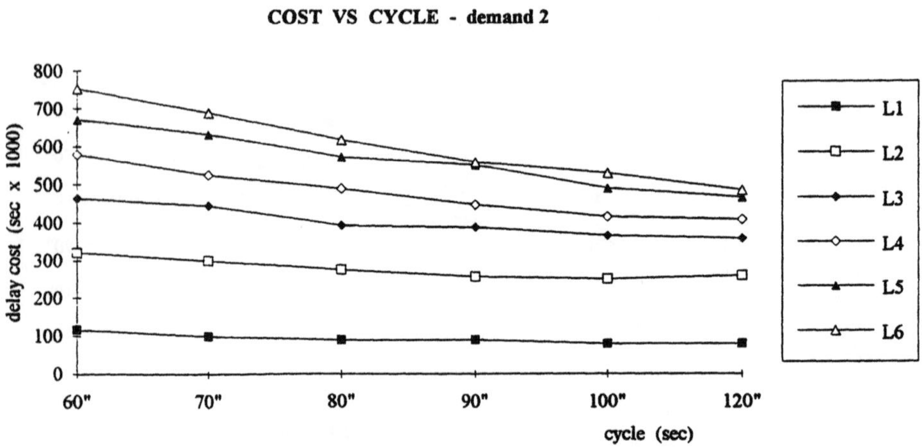

Fig. 21. Delay cost vs cycle, varying the length (demand D2)

is not very congested, so, simply increasing the length, the total time increases almost linearly and the relevant part of total cost is on running time rather more than on delay time (fig.15). It is also confirmed by figures 16 and 17, where the total cost is not much varying with cycle time, since there is no large effect of signal setting when the network is not very congested. The figures 18-21 show the same relations, for demand D2.

Conclusive Remarks

This paper mainly dealt with network design models which can be formulated for urban traffic management and planning. Equilibrium network design refers to a family of models for determining link capacity, taking into account also the user behaviour which defines the link flow pattern.

In extra-urban contexts the link capacity only depends on the geometric characteristics and two different model formulations are possible, a continuous or a discrete one. They are quite similar in the kind of provided results, and differ only with respect to the type of adopted solution techniques.

In urban context the link capacity depends also on the signal setting and therefore the set of variables have to be enlarged to comprehend green times and offsets. In this way a problem close to continuous Equilibrium Network Design is obtained, called Equilibrium Network Signal Setting, in which variables are the green times and offset, and their values are decided taking into account user behaviour. This task can be accomplished, of course, through a deterministic user equilibrium model, but, in principle, any assignement model can be embedded in this framework.

Trying to combine both the problems, that is design of network topology and signal setting, a new combined problem, the Equilibrium Network Design and Signal Setting, that is at the same time discrete and continuous, is obtained. Indeed, in the most general case in urban areas it is necessary to combine together both discrete variables to define the network topology (that is the circulation scheme), and continuous variables related to the signal setting, green times and offsets.

A common feature to the network design models is that the values of the decision variables are decided trying to optimize some general objective independent from the objective function adopted to model the user choice criterion (the latter function is embedded within the map through which the link flow pattern depends on the control strategy). In this case some budget constraints can also be included, as usual in network design problems, or constraints about the type of network topology, or for example about pollutant emission.

This framework allows to stress that network design problems are quite different in urban areas and in extra-urban areas. In the first case they are quite interesting but more complicated, because there is a large amount of different parameters to be taken into account. They should be formulated as mixed integer problems and it is relevant that Signal Setting should be taken into account even at the level in which the network topology is decided. It is also relevant that this kind of formulation allows a coherent framework to assess control strategies that operate at different levels, from demand management to signal setting.

In the second part of the paper the experimental tests of an iterative procedure for the equilibrium network traffic signal setting problem was described. The results of the application of the procedure to a test network are presented, with a large set of graphs. They show a good convergence to a feasible solution within an acceptable computation time. Procedure ENETS has been applied succesfully also to the network reported in Charlesworth (1977), with 6 signalized junctions and 22 o/d pairs demands, expanded to a network with 71 nodes and 83 links, already solved with a procedure based on TRANSYT method (Vincent et al., 1980). The computation times are also in this case very encouraging.

References

[1] Akcelik R., *A New Look at Davidson's Travel Time Function.* Traffic Engng. Control 19, pp. 459-463, 1978.

[2] Allsop R.E., *Some Possibilities for Using Traffic Control to Influence Trip Destination and Route Choice.* Proc. of the Sixth Int. Symp. on Transportation and Traffic Theory. Elsevier, Amsterdam, pp. 345-374, 1974.

[3] Allsop R.E. and Charlesworth J.A., *Traffic in a Signal Controlled Road Network: An Example of Different Signal Timing Inducing Different Routing.* Traffic Engng. Control, 18, pp. 262- 264, 1977.

[4] Boyce D.E. (Editor). *Transportation Network Design.* Special Issue of Transportation Research B, 13B, 1, 1979.

[5] Cantarella G.E. and Improta G., *Capacity Factor and Cycle Time Optimization: A Graph Theory Approach,* Transpn. Res. 22B, 1, pp. 1-23, 1988.

[6] Cantarella G.E. , Improta G. and Sforza A., *Road Network Signal Setting: Equilibrium Conditions.* Concyse Encyclopedia of Traffic and Transportation, M. Papageorgiou editor, pp.366-371. Pergamon Press, 1991.

[7] Cantarella G.E., Improta G. and Sforza A., *Iterative procedure for equilibrium network traffic signal setting.* Transportation Research A, 25A, 5, pp.241-249, 1991.

[8] Cantarella G.E. and Sforza A., *Methods for Equilibrium Network Traffic Signal Setting, in Flow Control on Congested networks,* Odoni A.R. et al. eds., Springer &Verlag, 1987.

[9] Cantarella G.E. and Sforza A., *Traffic Assignment.* Concyse Encyclopedia of Traffic and Transportation, M. Papageorgiou editor, pp. 513-520. Pergamon Press, 1991.

[10] Cantarella G.E. e Vitetta F., *Algoritmi genetici per il progetto delle reti di trasporto (in italian);* presented at Annual Conference of the Operational Research Society of Italy, Capri, 1993.

[11] Cascetta E., *Metodi quantitativi per la pianificazione dei sistemi di trasporto(in italian).* Cedam, Padova, 1990.

[12] Charlesworth J.A., *Mutually Consistent Traffic Assignment and Signal Timings for a Signal -Controlled Road Network.* Transport

Oper. Res. Group, Research Report n.24. Newcastle, England, 1977.

[13] Davidson K.B., *A Flow-Travel Time Relationship for Use in Transportation Planning*. Proc. 3rd Austr. Road Res. Board Conf. 3(1), pp. 183-194, 1966.

[14] Dickson T.J., *A Note on Traffic Assignment and Signal Timings in a Signal Controlled Road Network.* Transpn. Res., 15B, pp. 267-271, 1981.

[15] Fisk C.S., *Game Theory and Transportation Systems Modelling.* Transpn. Res.,18B, 4/5, pp. 301-313., 1984

[16] Gartner N.H., *Area Traffic Control and Network Equilibrium. Traffic Equilibrium Methods* (Edited by M.Florian) Springer-Verlag, Berlin, pp. 274-297, 1976.

[17] Gartner N.H., Gershwin S.B., Little J.D.C. and Ross P., *Pilot Study of Computer-Based Urban Traffic Management.* Transpn. Res., 14B, pp. 203-217, 1980.

[18] Gartner N.H., Little J.D.C., and Gabbay H., *Optimization of Traffic Signal Settings by Mixed Integer Linear Programming. Part I: The Network Coordination Problem; Part II: The Network Synchronization Problem,* Trans. Sci., 9,pp. 321-363, 1975.

[19] Harker P.T. and Friesz T.L., *Bounding the Solution of the Continous Equilibrium Network Design Problem.* Proc. of the Ninth Int. Symposium of Transportation and Traffic Theory, Delfth. VNU Science Press, pp. 233-252, 1984.

[20] Heydecker B.G., *Some Consequences of Detailed Junction Modelling in Road Traffic Assignment.* Transpn. Sci., 17, 263-281, 1983.

[21] Improta G., *Signal Control at Individual Junctions: Phase - Based Approach. Concyse Encyclopedia of Traffic and Transportation,* M. Papageorgiou editor, Pergamon Press, 1991.

[22] Improta G. and Cantarella G.E., *Control System Design for an Individual Signalized Junction.* Transpn. Res. 18B, pp 147-167, 1984.

[23] Improta G. and Sforza A., *Optimal Offsets forTraffic Signal Systems in Urban Networks.* Transpn. Res. 16B, pp. 143-161, 1982.

[24] Leblanc L.J. and Boyce D.E., *A Bilevel Programming Algorithm for Exact Solution of the Network Design Problem with User Optimal Flows*. Transpn. Res., 20B. 259-265, 1986.

[25] Magnanti T. and Wong R., *Network Design and Trasportation Planning Models and Algorithms*. Transp. Sci. 18, 1-55, 1984.

[26] Marcotte P., *Network Optimization with Continuous Control Parameters*. Transpn. Sci., 17, pp. 181-197, 1983.

[27] Sheffi Y., *Urban Transportation Networks*. Prentice Hall Inc. Englewood Cliffs, N.J., 1985.

[28] Smith M.J., *The Existence of an Equilibrium Solution of the Traffic Assignment Problem when there are Junction Interactions*.Transpn. Res., 15B, 443-451, 1981.

[29] Smith M.J., *Properties of a Traffic Control Policy which Ensure the Existence of a Traffic Equilibrium Consistent with the Policy*. Transpn. Res., 15B, 453-462, 1981.

[30] Smith M.J., *Traffic Signals in Assignment* Transpn. Res. 19B, pp.155-160, 1985.

[31] Smith M.J., Van Vuren T., Heydecker B.J. and Van Vliet D., *The Interaction between Signal Control Policies and Route Choice, in Transportation and Traffic Theory*, Gartner N.H. and Wilson N.H.M. Eds., Elsevier, 1987.

[32] Tan H., Gershwin S.B. and Athans M., *Hybrid Optimization in Urban Traffic Networks*. Report No. DOT-TSC-RSP- 79-7,MIT Cambridge, MA, 1979.

[33] Taylor M.A.P., *A Note on Using Davidson Function in Equilibrium Assignment*. Transpn. Res. 18B, pp. 181-199, 1984.

[34] Vincent R.A., Mitchell A.I. and Robertson D.I., *TRANSYT Version 8*. Road Res. Lab. Rep.,LR 80, Crothorne, Berkshire, 1980.

[35] Wardrop J.G., *Some Theoretical Aspects of Road Traffic* Research. Proc. Inst. Civ. Eng., Part. II, 1, pp. 325-378, 1952.

[36] Webster F.V., *Traffic Signal Setting. Road* Research Technical Paper n.39 HMSO, London, 1958.

[37] Webster F.V. and Cobbe B.M., *Traffic Signals*. Road Research Technical Paper n.56 HMSO, London, 1966.

DEVELOPMENT OF A COMBINED TRAFFIC SIGNAL CONTROL-TRAFFIC ASSIGNMENT MODEL

Mohammed Al-Malik

Department of Civil Engineering, King Saud University, Riyadh, Saudi Arabia

Nathan H. Gartner

Department of Civil Engineering, University of Massachusetts, Lowell, MA, USA

Having recognized the mutual interaction that exists between traffic signal control and route choice behavior, this paper considers the traffic assignment problem for a traffic signal controlled network. The interaction is characterized by (1) the effects of delay, due to traffic signals, on route choice, and (2) the influence of the resulting traffic assignment, due to the changes in route selection, on the timing of traffic signals. In this paper Webster's method for calculating green splits and the HCM delay model are used to develop a link-performance function that implicitly produces optimal signal settings for the arriving traffic volumes. This link-performance function is then used in an assignment procedure to develop a combined traffic signal control/ traffic assignment model. The model provides simultaneous solution of the traffic signal control problem and the traffic assignment problem. The result is a set of signal settings that optimally matches the assigned traffic flows. The combined model results in a non-convex problem which produces multiple equilibria. A modification of Wardrop's two principles is proposed to produce an assignment which converges to a single solution.

1. Introduction

The problems of traffic signal control and traffic routing, individually, are quite well understood in terms of their theoretical and computational aspects. A state of the art discussion for both was given by Cantarella and Sforza (1987) and by Improta (1987). Their separate treatment, however, avoids the interplay between the two, where a change in one can affect the optimality of the other. Thus, combining the two problems and exploring the degree of their interaction would contribute to measuring the stability of an equilibrium solution. Both problems can play very significant roles within the context of IVHS (Intelligent Vehicle Highway Systems). These systems are currently the focus of research for future traffic control.

It is well known that traffic signal control and route choice behavior mutually influence each other. This mutual influence reflects the way city traffic engineers react to changes in traffic flow patterns. A city traffic engineer, for example, will regularly readjust the traffic signal system settings according to actual traffic observations. This mutual influence also reflects the way drivers respond to changes in signal timings. A driver may switch to another route as a result of an implementation of a new signal timing plan. It is not known, however, if a mutually consistent solution between traffic signal control and route choice behavior can be obtained. A mutually consistent solution is the state at which a traffic signal control policy and the network link assignments match each other. This can be summarized as follows:

1) Signal settings match a given link flow pattern.
2) One of Wardrop's principles is satisfied.

This mutual influence between traffic signal control and route choice behavior is the main focus of this paper and is depicted in Fig. 1. The task here is to explore the two problems simultaneously and investigate the theoretical aspects of both, so that a natural, meaningful and practical relationship between them is developed. This problem, which is referred to as "Equilibrium Network Traffic Signal Setting" (Cantarella et al 1991), has been the focus of a number of studies in the past several years.

Allsop (1974) was the first to address the interaction between traffic signal control and traffic assignment by studying the possibility of using traffic signal control to influence route choice behavior. Gartner (1976 and 1977a) suggested a procedure to incorporate route choice into the traffic signal optimization process. The suggested procedure was the inclusion of traffic demands as decision variables in the objective function of the usual traffic signal timing problem. Thus, rather than assuming that the set of traffic demands is known and fixed, Gartner suggested that these variables be determined simultaneously with the signal traffic control variables. Charlesworth and Allsop (1977) showed that the

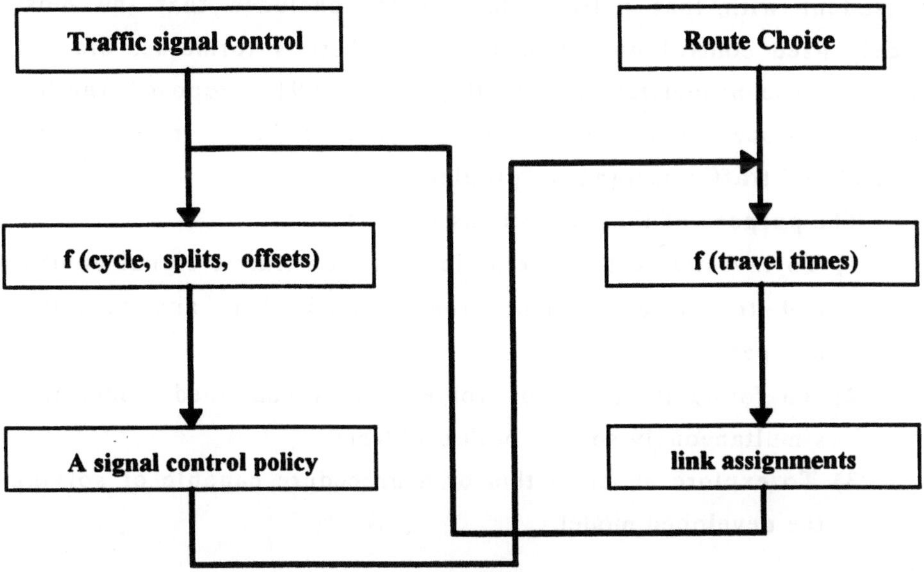

Figure 1. A Mutual Interaction Between Traffic Signal and Route Choice Behavior

same objective (combining route choice and traffic signal control) can be achieved in principle via combining a traffic signal optimization program with a traffic assignment program.

Gartner, et al (1980) proposed two iterative procedures for modeling the combined route choice and traffic signal setting problem. Both procedures were based on iterating between the traffic assignment problem and the traffic signal setting problem. Tan, et al (1979) combined the problems of traffic assignment and signal setting calculation by recognizing the role of drivers as independent decision makers in the traffic signal optimization process. They went on to formulate the "Hybrid-Optimization Problem", i.e., optimizing the system performance (system-optimization), subject to optimizing individual route choice (user-optimization).

Dickson (1981) discussed the problem of using signal timings to influence equilibrium flows such that a network performance index was optimized. Smith (1981) gave properties of a traffic control policy that encourages the existence of a traffic equilibrium consistent with that policy. He further concluded that the most widely used control policy, based on Webster's formulas, did not satisfy these properties. Cantarella, et al (1991) proposed another iterative procedure for the solution of the combined traffic signal control and traffic assignment problem.

The purpose of this paper is as follows:

1) To investigate the interaction between traffic signal control and traffic assignment from a theoretical and practical perspective.
2) To formulate the two problems as a combined model that simultaneously solves both problems.
3) To explore an algorithm or a procedure capable of solving the developed model.

2. A Flow-Dependent Traffic Signal Control Model

The interaction between traffic signal control and route choice behavior was explained and defined in the introduction. In this section, we develop a flow-dependent traffic signal control model. This model will be used to cast the interrelated problems of traffic signal control and route choice behavior as a combined optimization model. Gartner (1977b) and Gartner et al (1980) suggested that a delay function (Webster's in particular) be used in a procedure that optimizes both signal settings and traffic assignment. Smith (1985) showed that, with an appropriate choice of a link cost function, the problem of determining an equilibrium flow pattern that is consistent with a signal control policy is equivalent to a convex optimization problem.

2.1 *The Link Performance Function*

A number of functions have been developed to estimate average delay for signalized intersection links at under saturated conditions. The delay formulas that are predominant in practice are those developed by Webster (1958), Miller (1963), and Newell (1965). The TRANSYT 7-F program uses a modified version of Webster's delay function. The Highway Capacity Manual (HCM, 1985) proposed a new delay formula that can estimate delay for over saturated conditions as well.

The level of service associated with a given link in a transportation network may reflect many characteristics of that link. Travel time, safety and cost are some of these characteristics. The most important of these and the one which has been mostly used as a measure of the level of service is travel time. Sometimes we consider a combination of travel impedance other than travel time.

A typical link performance function is shown in Fig. 2. As shown, this function satisfies the monotonicity condition; that is, the greater the flow the higher the delay, or travel time. Much of the development of the theory of traffic assignment has concentrated on cases that satisfy this condition. The assumption of monotonicity is realistic as long as the interaction among different modes is relatively weak (Dafermos 1982).

Most delay formulas behave like the one shown in Fig. 2. Actual delays, however, may be much less, especially at or near capacity. The travel time on a signalized intersection link is the sum of two terms. The first term represents the time it takes a vehicle to travel on that link. It is simply the distance divided by the average travelling speed and, generally, can be assured to be little affected by the rate of flow. The second term represents the average delay a vehicle encounters at the intersection. It is the estimation of the second term that we are interested in.

Two models which have been extensively used for the estimation of average delay at signalized intersections will be mentioned here. The earliest, and probably the most used, is the one developed by Webster (1958). According to Webster, the average delay in seconds per vehicle on a signalized intersection link is calculated as

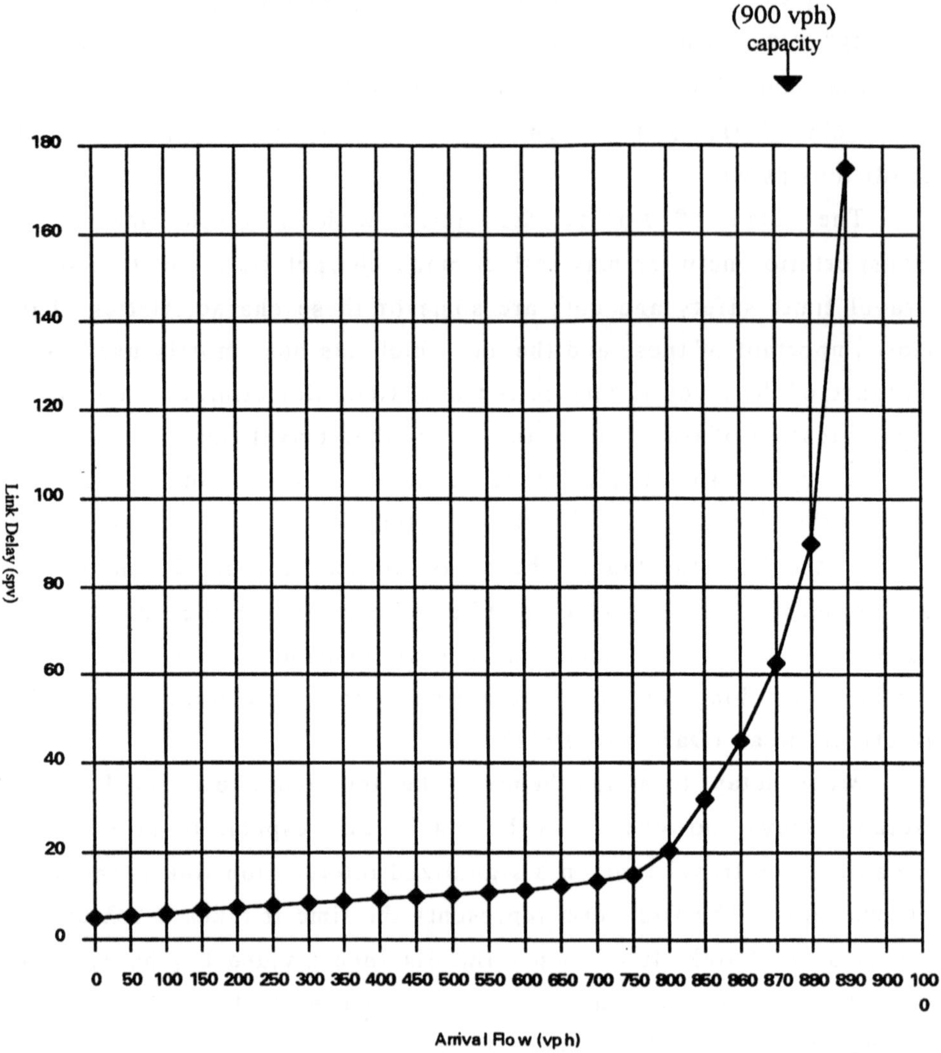

Figure 2. Behavior of Webster's Delay Function cycle = 60 sec, g = 0.5, s = 1800 vph

$$D = 0.45 \left[\frac{C(1-g)^2}{(1-q/s)} + \frac{3600x^2}{q(1-x)} \right] \tag{2.1}$$

where,

C = signal cycle time (sec)

g = proportion of the cycle which is effectively green

q = approach arrival flow rate (vph)

s = approach saturation flow rate (vph)

$x = q/gs$, degree of saturation

The second model is the Highway Capacity Manual (1985) model. This model seems to perform better than Webster's, especially when the degree of saturation approaches or exceeds 1.0. Both models will be considered in developing the flow-dependent signal control model.

2.2 *Expressing Signal Controls as Flow Variables*

The delay functions mentioned above are functions of several variables. Signal cycle, green splits, arrival and saturation flow rates are the most important. In this paper we will be dealing with an isolated intersection with the following simplifying characteristics:

(i) Only two conflicting streams;
(ii) Two phase operation;
(iii) Cycle length is fixed and given;
(iv) Lost time is assumed to be 10% of the cycle length;

The only remaining variables to be determined are the green splits. Gartner (1977b) has suggested that Webster's delay function be used in a procedure that optimizes both signal settings (specifically, green splits) and traffic assignments. The green time split, g, in Eq. (2.1) determines the capacity of the link. One can eliminate the green time split variables, the g's, from explicit consideration by expressing them in terms of the traffic flows on the conflicting links, as described below:

Consider an isolated signalized intersection with the characteristics described above. Then,

$$G_1 + G_2 + L = C \tag{2.2}$$

where,

G_1 = phase 1 green time (sec)

G_2 = phase 2 green time (sec)

L = intersection lost time (sec)

C = signal cycle time (sec)

Dividing by C, Eq. (2.2) can be rewritten as

$$g_1 + g_2 + \frac{L}{C} = 1 \tag{2.3}$$

Following Webster's method of timing traffic signals, we have

$$\frac{g_1}{g_2} = \frac{q_1 / s_1}{q_2 / s_2} \tag{2.4}$$

Eq. (2.4) is an approximate expression of the optimal green splits at an intersection.

Eq. (2.4) can be rewritten as

$$g_1 = g_2 \left(\frac{q_1 s_2}{q_2 s_1}\right) \tag{2.5}$$

Substituting in Eq. (2.3), we get

$$g_2 \left(1 + \frac{q_1 s_2}{q_2 s_1} \right) = \lambda \tag{2.6}$$

where, $\lambda = 1 - L/C$. The green split variables g_1 and g_2 can be expressed by their conflicting links flows

$$g_2 = \frac{\lambda q_2 s_1}{q_2 s_1 + q_1 s_2} \tag{2.7}$$

$$g_1 = \frac{\lambda q_1 s_2}{q_1 s_2 + q_2 s_1} \tag{2.8}$$

If the green splits are calculated according to these equations, they will be optimally matched (according to Webster's method) to the traffic flows. Next, we will study the behavior of both Webster's and the HCM delay functions when the above equations for calculating green splits are incorporated into them.

2.3 *Webster's Delay Function*

Webster's delay function is given in Eq. (2.1). The behavior of this function when the arrival flow rate, q, varies is the one actually shown in Fig. 2. The capacity for this approach is 900 vph. As the arrival flow rate approaches capacity the function tends to overestimate delay under practical circumstances (it goes to infinity). This occurs because the function calculates average delay for a steady-state queueing process in which the load factor (flow/capacity ratio) approaches 1. In practice, due to the very high delays (approaching infinity), traffic will be discouraged and the near-saturation condition will be short-lived. In other words, the steady-state condition will not be realized. Therefore, Webster's delay function is adequate only for undersaturated conditions.

Expressing the green split, g, as shown in Eq. (2.8) and substituting it in Eq. (2.1), yields the following delay function:

$$D_1=0.45\left[\frac{C\left(1-\left(\frac{\lambda q_1 s_2}{q_1 s_2+q_2 s_1}\right)\right)^2}{\left(1-\frac{q_1}{s_1}\right)}+\frac{3600\left(\frac{q_1 s_2+q_2 s_1}{\lambda s_1 s_2}\right)^2}{q_1\left(1-\left(\frac{q_1 s_2+q_2 s_1}{\lambda s_1 s_2}\right)\right)}\right] \tag{2.9}$$

The resulting delay model given by Eq. (2.9) is no longer an explicit function of green splits. From now on, these splits will be expressed by their conflicting link flows. This model states that the delay on a certain link is a function not only of the flow on that link, but also on the flow on the conflicting link.

Using the above model we can estimate the average delay on a particular link, call it link 1. By varying the arrival flow rate on that link, while keeping the flow on the conflicting link (call it link 2) fixed, we obtain the relationship shown in Fig. 3. Fig. 3 represents a link performance function for link 1. It is obvious that this function does not satisfy the monotonicity condition defined earlier. The delay curve shown in Fig. 3 has a section where delay decreases as flow increases. This occurs due to the increase in green time allocated to this link according to the Webster optimality ratio. Thus, the higher the flow on a particular link, the greater the split

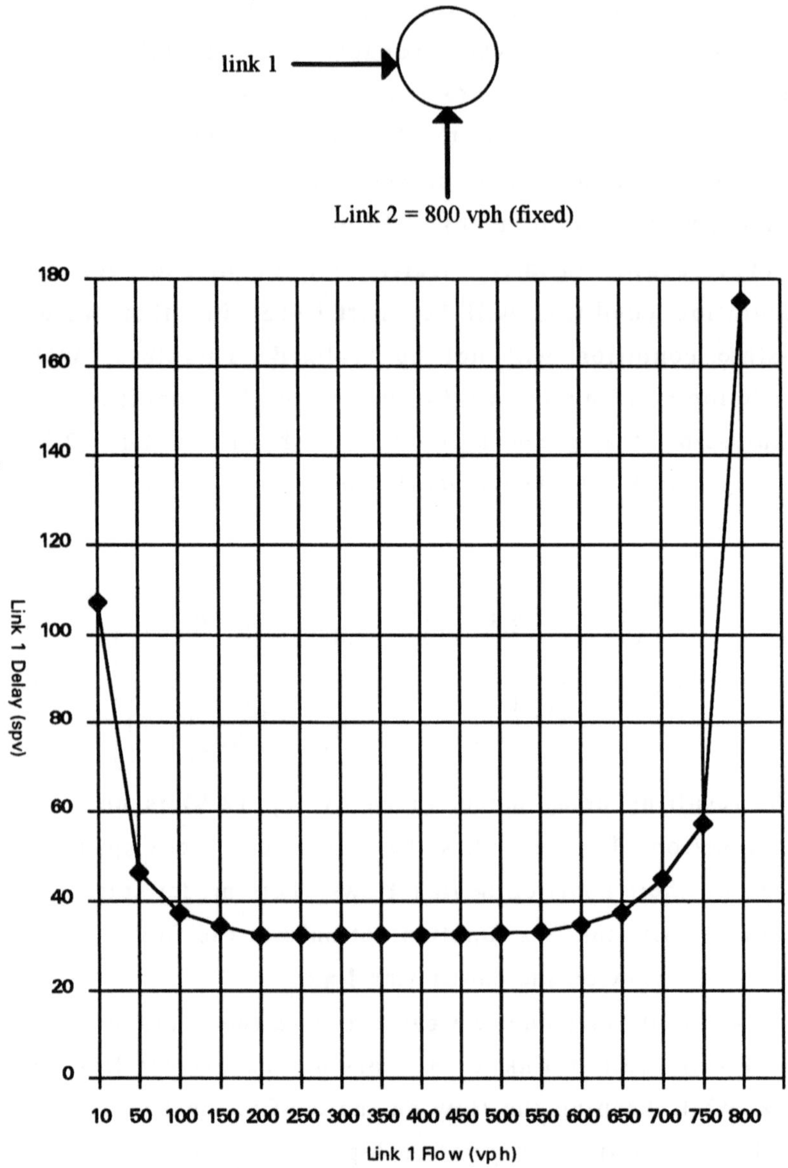

Figure 3. Volume-Delay Relationship
Link 2 Flow Rate = 800 vph

is for that link. If we use the same model to estimate the delay on link 1 when the flow on that link is kept fixed, while the flow on link 2 is varied, the curve shown in Fig. 4 results. This curve satisfies the monotonicity condition.

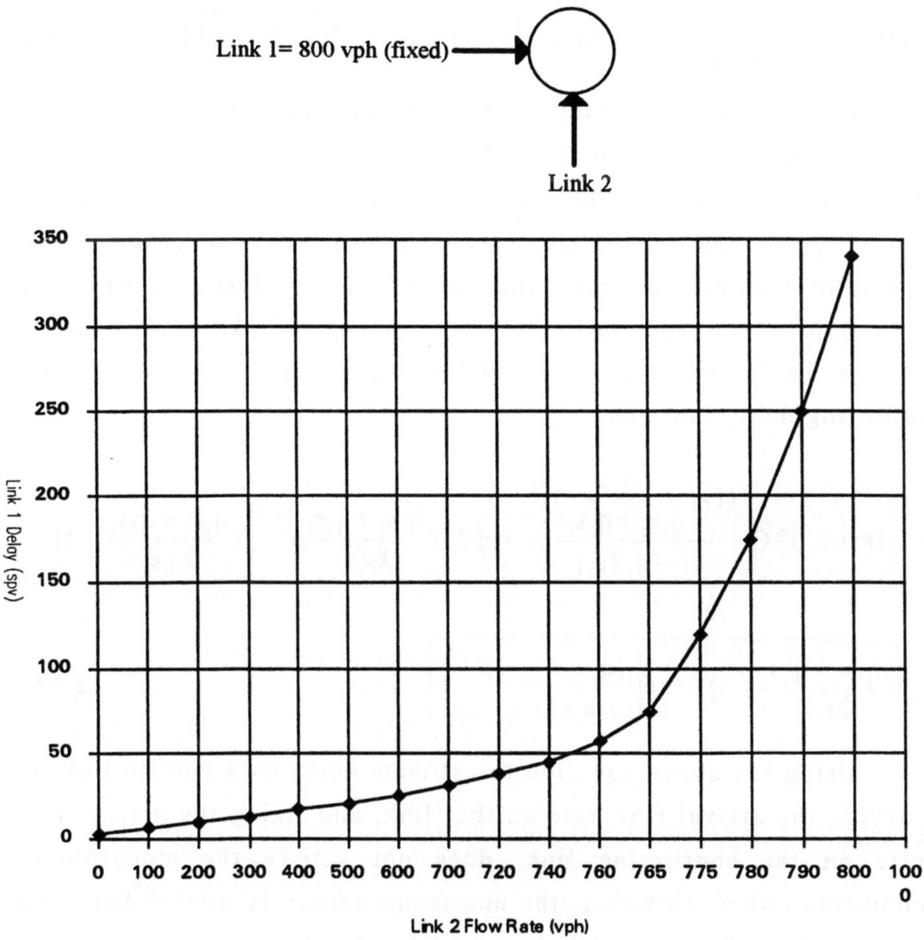

Figure 4. Volume-Delay Delay Relationship
Link 1 Flow Rate = 800 vph

2.4 *The HCM Delay Function*

The Highway Capacity Manual (1985) offered a new function for estimating average stopped delay at signalized intersections. Like Webster's function, this model also has two delay terms. The first term is based on uniform arrivals, while the second term is based on random arrivals. According to the HCM function, the average stopped delay per vehicle an a signalized intersection approach can be estimated by

$$D = 0.38\ C\ \frac{(1-g)^2}{(1-q/s)} + 173\ x^2 \left[(x-1)+\sqrt{(x-1)^2+(16x/gs)}\right] \qquad \textbf{(2.10)}$$

There is a significant difference between the Webster delay function and the HCM function. The HCM delay function estimates a finite average delay at both saturated and oversaturated conditions. This is predicated on the more realistic assumption of a short-lived oversaturation period, rather than an infinte (steady-state) high flow rate. Expressing green splits by their conflicting links flows as given by Eqs. (2.7) and (2.8), and substituting in Eq. (2.10), we get the following delay function:

$$D_1 = 0.38C\ \frac{\left(1-\left(\frac{\lambda q_1 s_2}{q_1 s_2 + q_2 s_1}\right)\right)^2}{(1-q_1/s_1)} + 173\left(\frac{q_1 s_2 + q_2 s_1}{\lambda s_1 s_2}\right)^2 \left[\left(\frac{q_1 s_2 + q_2 s_1}{\lambda s_1 s_2} - 1\right)\right.$$

$$\left.\sqrt{\left(\frac{q_1 s_2 + q_2 s_1}{\lambda s_1 s_2} - 1\right)^2 + 16\left(\frac{(q_1 s_2 + q_2 s_1)^{2}}{q_1(\lambda s_1 s_2}\right)^{2\ 2}}\ \right] \qquad \textbf{(2.11)}$$

Using the above equation to estimate delay on a certain link, by varying the arrival flow rate on that link, and fixing the arrival flow rate on the conflicting link, does not satisfy the monotonicity condition either. However, the non-monotonicity is weaker here than that resulting from using Webster's delay function.

2.5 *Total Intersection Delay and Optimal Splits*

The total intersection delay is calculated by summing the products of the link volumes and their average delay for all the intersection links. All delay models assume that the greater the flow on any link the higher the total intersection delay. Our combined model is no exception. An increase of the flow on any link will result in an increase in the total intersection delay. Fig. 5 depicts this relationship for a single intersection with two links. The average delay on each link is estimated using the HCM model.

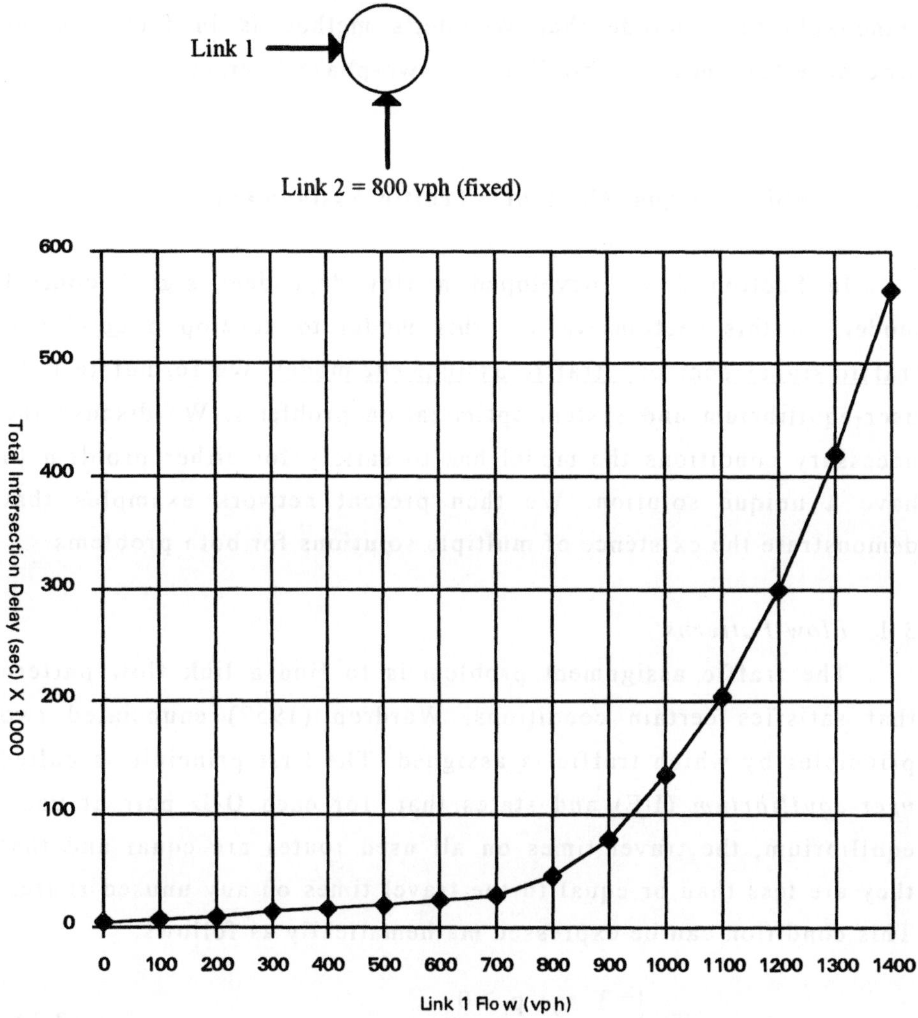

Figure 5. Flow-Total Intersection Delay Relationship

We also checked the optimality characteristics of Webster's method for setting the green splits, according to Eq. (2.4). The total intersection delay was calculated using the delay function given by Eq. (2.9), for different g_1/g_2 ratios. Webster's method does not always agree with the optimal g_1/g_2 ratio as far as minimizing the average intersection delay. It comes, however, close to the optimal ratio in most cases. The higher the arrival flow rate, the closer Webster's method is to the optimal g_1/g_2 ratio. Therefore, it is reasonable to conclude that Webster's method is in fact a good procedure for timing traffic lights at two-phase intersections.

3. Combined Signal Control - Traffic Assignment

In Section 2 we developed a flow-dependent signal control model. In this section we use this model to develop a combined traffic signal control - traffic assignment model. We formulate both user-equilibrium and system-optimization problems. We discuss the necessary conditions the model has to satisfy for either problem to have a unique solution. We then present network examples that demonstrate the existence of multiple solutions for both problems.

3.1 *Flow Patterns*

The traffic assignment problem is to find a link flow pattern that satisfies certain conditions. Wardrop (1952) enunciated two principles by which traffic is assigned. The first principle is called *user-equilibrium* (UE) and states that, for each O-D pair at user-equilibrium, the travel times on all used routes are equal and that they are less than or equal to the travel times on any unused routes. This condition can be expressed mathematically as follows,

$$T_i \begin{cases} = T_e & , \quad p_i > 0 \\ \geq T_e & , \quad p_i = 0 \end{cases} \tag{3.1}$$

Where,

T_i = travel time on route i

T_e = travel time at equilibrium

p = the amount of traffic flow using route i

The above principle suggests that each user chooses the least travel time route. This principle may be reasonable in some cases, but it depends on several assumptions that cannot always be assumed to hold true (Sheffi, 1985). One assumption is that users have complete information on the characteristics of their respected alternative routes. A second assumption is that users always make the correct selection. A third assumption is that all users are identical in their travel choice behavior. These assumptions can be relaxed by introducing a stochastic behavior to the UE definition given above. In this paper we limit our discussion to the deterministic UE condition given by Eq. (3.1).

The second principle stated by Wardrop leads to a *system-optimized* (SO) assignment. The system-optimization link flow pattern is achieved when users act together so as to minimize the total travel cost incurred by the system. This principle states that for each O-D pair, at optimum, the marginal contribution to travel time on all used routes is equal and that it is less than or equal to the marginal contribution on any unused route. The marginal contribution on a particular route is defined as the additional system delay that an additional traveller on that route will cause to all users. This condition is expressed mathematically as

$$MC_i \begin{cases} = MC_e \quad , \quad p_i > 0 \\ \geq MC_e \quad , \quad p_i = 0 \end{cases} \tag{3.2}$$

where MC_i denotes the marginal contribution to travel time (or, marginal cost) of route i. The marginal contribution of a particular route is the summation of the marginal contributions of the links that form that route.

3.2 *Link Interaction*

Before the basic formulations of the UE and SO problems are presented, we discuss the type of link interaction. Link interactions can be symmetric or asymmetric. Interaction is symmetric when the effect of the flow of one link on the delay on a second link equals the effect of the flow of the second link on the delay of the first link. Mathematically, the symmetry condition is expressed as follows.

$$\frac{\partial t_i(q_i, q_j)}{\partial q_j} = \frac{\partial t_j(q_j, q_i)}{\partial q_i} \tag{3.3}$$

When the interactions are symmetric, a flow pattern that satisfies Wardrop's first principle can be found by solving an equivalent optimization program. When the interactions are asymmetric, there is no known equivalent optimization program that can be used to find a user-equilibrium flow pattern (Sheffi 1985). A system-optimization flow pattern on the other hand, can always be found by solving an equivalent optimization program, irrespective of the existance of the symmetry condition. The following section presents a formulation of the user-equilibrium and system-optimization problems when interaction is symmetric. In this study we assume that link interaction is symmetric. When applied to our problem, symmetric interaction is obtained when intersecting links have the same physical characteristics.

3.3 *Optimization Formulation*

A traffic flow pattern that satisfies one of Wardrop's principles can be obtained by solving the following problem,

$$\min Z(Q) \tag{3.4}$$

subject to the following conservation and non-negativity constraints,

$$\Sigma k \; p_k^{mn} = f_{mn} \qquad \forall \; k,m,n$$

$$k \; p_k^{mn} \geq 0 \qquad \forall k, \; m,n \tag{3.5}$$

The flow on a single link can be calculated by

$$q_a = \Sigma_m \Sigma_n \Sigma_k \; p_k^{mn} \; \delta_{a,k}^{mn} \qquad \forall a \tag{3.6}$$

$$\delta_{a,k}^{mn} = \begin{cases} 1, & \text{if link a is on path k} \\ 0, & \text{otherwise} \end{cases}$$

Where, p_{kmn} denotes the flow on route k connecting O-D pair m-n, and f_{mn} denotes the trip demand rate between origin m and destination n. If $t_a(q_a, q_b)$ denotes the average travel time on link a (q_b denotes the flow on the conflicting link b), the user-equilibrium objective function (3.4) becomes

$$Z(Q)=\frac{1}{2}\sum_a\left[\int_0^{q_a} t_a(\omega,q_b)d\omega+\int_0^{q_a} t_a(\omega,0)d\omega\right] \quad (3.7)$$

The system-optimization objective function is, correspondingly

$$Z(Q) = \Sigma_a\, q_a\, t_a(q_a,q_b). \quad (3.8)$$

3.4 *Uniqueness Conditions*

Sheffi (1985) gave the following conditions for the UE problem to have a unique solution:

1. The travel time on a link is a strictly increasing function of the flow on that link, that is

$$\frac{\partial t_a(q_a,q_b)}{\partial q_a}>0 \qquad \forall\ a \quad (3.9)$$

2. The main dependence of a link's travel time is on its own flow. In other words, the derivative of t_a with respect to q_a is larger than the derivative of t_a with respect to the flow on any other link.This condition is stated mathematically as

$$\frac{\partial t_a(q_a,q_b)}{\partial q_a}>\frac{\partial t_a(q_a,q_b)}{\partial q_b} \qquad \forall\ a \quad (3.10)$$

We have shown in Section 2 that the delay function that resulted from combining the traffic signal control and route choice behavior problems violated the first condition given above (namely, the monotonicity condition). We have shown that the link delay curve has a section where delay decreases as flow increases (Fig. 3). The delay function does not satisfy the second condition either. We can prove this by showing that the value of the derivative of the delay equation with respect to its respective link flow is less than the value of the derivative of the delay function with respect to the flow on the conflicting link. Evaluating these derivatives is a very tedious task. Therefore, we prove this point by a simple example.

Consider a signalized intersection with two identical conflicting links. Each link has a saturation flow rate of 1800 vph. The average delay each link encounters at the intersection is calculated by Eq. (2.11), based on a cycle length of 60 seconds. Now, assume that each link carries a flow of 500 vph. The estimated delay on each link based on that flow is 10.58 spv. Assume that we add 100 vph on link 1, i.e., link 1 carries 600 vph, and link 2 still carries 500 vph. The average delays based on these flows are 10.35 spv for link 1 and 12.79 spv for link 2, respectively. These delay figures contradict both necessary conditions stated above. The first condition states that the greater the flow on a link, the higher the delay on that link. The average delay on link 1 dropped from 10.58 spv to 10.35 spv. The second condition states that additional flow on a link should affect delay on that link more than it would affect delay on the conflicting link. Increasing the flow on link 1 by 100 vph has caused the average delay on link 2 to rise to 12.79 spv and the average delay on link 1 to drop to 10.35 spv.

Summing up, we conclude that the delay model developed in Section 2 violates both conditions. Consequently, we should expect that the objective functions given by Eqs. (3.7) and (3.8) to be nonconvex. To demonstrate the non-convexity of the UE and SO objective functions we analyze the two networks shown in Figs. 6 and 7. Both examples are calculated by the well-known Frank-Wolfe algorithm (Gartner, 1977b).

3.5 *Network Example 1*

Consider the network depicted in Fig. 6. The network consists of three intersections and ten links. The free travel time and the saturation flow rate of each link are shown in the figure. The average delay experienced at every signalized intersection link is calculated by Eq. (2.11), based on a 60 second cycle length. Thus, the travel time on a signalized intersection link is the summation of the free travel time and the average delay encountered at the intersection on that link. The network has two pairs of origin-destination trip demands. The first O-D is from origin A to destination D. The trip demand on this O-D is 1400 vph.

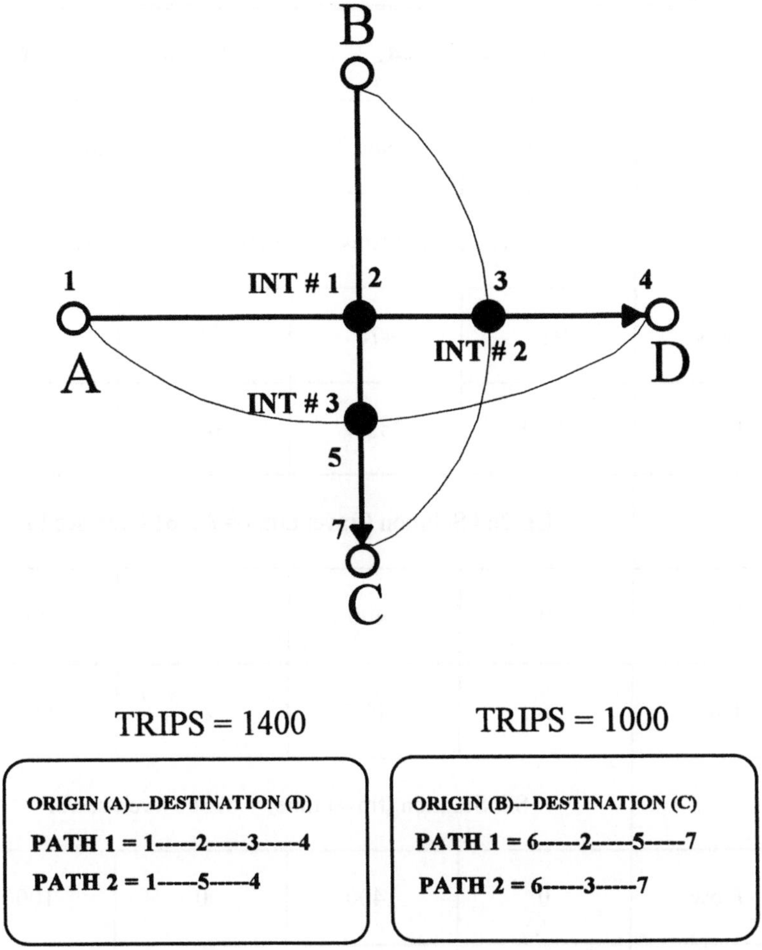

Figure 6. Network Example With Link Interaction

The second O-D is from origin B to destination C. The trip demand on that O-D is 1000 vph. There are two route choices for each pair.

This network problem is solved for UE and SO flow patterns using the Frank-Wolfe algorithm. Two different initial link flow patterns are used for the solution of the UE and SO problems. The reason for using different starting points is to see if the same link flow pattern is to be found. The results are summarized in Table 3.1. We can see that there are at least two link flow patterns that satisfy the UE condition for that network. The first flow pattern seems to be

	A---------------D		B-------------------C	
	Route 1	Route 2	Route 1	Route 2
	UE 1st Solution (travel 1 time) = 856674 veh-sec/hr			
Flow	922	478	626	374
Time	353.1	353.4	362.1	362.4
	UE 2nd Solution (travel time) = 823614 veh-sec/hr			
Flow	0	1400	0	1000
Time	∞	344.7	∞	779.1
	SO Solution (travel time) = 823614 veh-sec/hr			
Flow	0	1400	0	1000
Marginal	973041	378.3	3158517	344.1

Table 3.1 Frank - Wolfe Algorithm Applied to the Network Shown in Figure 6

a stable solution. It can be reached from different directions. The second flow pattern, on the other hand, can be attained only by starting from a point close to it. Solving for SO produces the same link flow pattern regardless of the starting point. This, however, does not mean that the SO problem is a convex one. The SO link flow pattern is similar to the second link flow pattern found for UE. This flow pattern is simply an all-or-nothing type of assignment.

An interesting point that is worth mentioning is the nature of the all-or nothing solution found above. Route 1 for each O-D pair was assigned no flow. This means that had these routes not existed, the same total network delay would have resulted; i.e., we can close these routes and cause the network no more delay. Actually, we can argue that by closing these routes, or by not building them to begin with, we may decrease the total network delay because we will eliminate the occurrence of UE-1.

This example has demonstrated the non-convexity of the UE problem by showing multiple UE equilibria solutions. The following example illustrates the existence of multiple UE and SO solutions.

3.6 *Network Example 2*

Consider the network depicted in Fig. 7. The network consists of four links. Links 1 and 2 are identical and cross each other at a signalized intersection. Through traffic is the only movement allowed at the intersection. Links 3 and 4 are also identical. The free travel time on each link is shown in the figure. The average delay encountered on links 1 and 2 is calculated by Eq. (2.11). Now assume that a flow of 1600 vph is to travel from point A to point B. Two route choices are available. Route 1 goes through links 1 and 3. Route 2 goes through links 2 and 4. The Frank-Wolfe algorithm is again used to find the link flow patterns.

An obvious flow pattern that satisfies the UE condition is to assign half the traffic to each route, because of network symmetry. However, the only way to attain this solution is by forcing the traffic to divide evenly between the two routes. If route 1 is assigned more traffic than route 2, then route 1 will encounter less delay at the intersection than that encountered by route 2. This means less travel time on route 1, and consequently more traffic is attracted to this route. The more traffic route 1 attracts, the higher the delay route 2 encounters at the intersection. This cycle goes on until all traffic is assigned to route 1. If, on the other hand, we assign more traffic on route 2 than route 1, the travel time on route 2 will be less than that of route 1. As a result, route 2 will attract all the traffic.

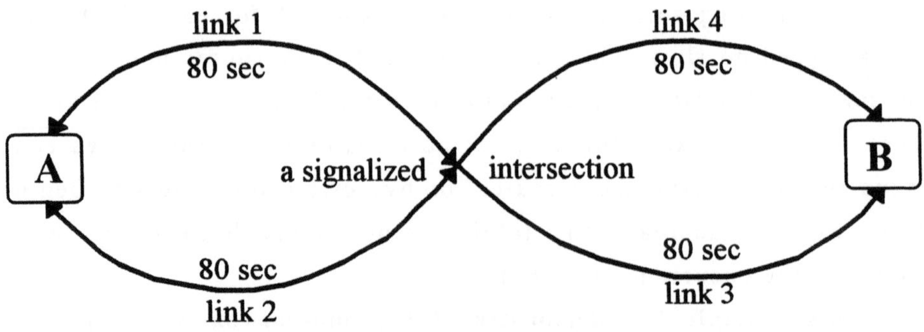

Figure 7. Network Example with Link Intersection

Table 3.2 summarizes the results. We obtain three solutions, each of which satisfies both the UE and SO conditions.

3.7 *Multiple vs. Single Solutions*

In view of the fact that the combined problem has multiple solutions we can pursue two options: (1) accept the fact that there are multiple solutions and search for the best among them; or, (2) influence the network to obtain a single solution. We shall analyze these options based on the examples, described above and others presented in Al-Malik (1991).

All patterns that were calculated include at least one link with a zero flow assignment. Such flow patterns result in zero green times for some links, since the green splits are automatically calculated to match the corresponding traffic assignments. Obviously, such settings are not realistic because an intersection approach should have a minimum of green time on each phase due to safety and other considerations. Furthermore, it is reasonable to assume that some drivers will select a route irrespective of the amount of delay that they encounter. This leads us to the idea of imposing lower limits which will not only meet the technical requirements but will also satisfy the requirements for a unique solution. This will be achieved by avoiding the downward portion of the link delay function (see Fig. 3). This is, of course, a departure from our initial assumption that green splits should always match traffic flows.

	Route 1	Route 2
1 st solution	Total Trave Time = 310380 veh-sec/hr	
Flow	800	800
Travel Time	194	194
Marginal	410.8	410.8
2ed solution	Total Travel Time = 282825 veh-sec/hr	
Flow	1600	0
Travel Time	176.7	∞
Marginal	376.9	∞
3ed solution	Total Travel Time = 282825 veh-sec/hr	
Flow	0	1600
Travel Time	∞	176.7
Marginal	∞	376.9

Table 3.2 Frank-Wolfe Algorithm Applied to the Network in Figure 7

At this point, the green split is not being calculated anymore based on the conflicting flows. If the assignment procedure assigns traffic in such a way that a particular green split is less than the specified minimum, the minimum is enforced. To implement this approach, we propose in the next section a more realistic method for the computation of the combined signal control-traffic assignment problem. This method is manifested by a modification of the statement of the two Wardrop principles.

4. A Modified Formulation Of The Combined Problem

We have demonstrated in Section 3 the existence of multiple solutions for both the user-equilibrium and the system-optimization problems. We have shown that without imposing some limits on green splits, these solutions tend to be of the 'all-or-nothing' type. A solution of this type results in zero green splits at some intersections, which is unrealistic. We can avoid zero splits by imposing minimum limits. However, imposing minimum limits on green splits changes the structure of both problems. This change contradicts our main argument that optimal signal settings should match the existing link flows. In this section we describe an approach that overcomes this problem.

4.1 *Modifying Wardrop's Two Principles*

According to Wardrop's first principle, routes connecting an O-D pair may be divided into two subsets. The first subset includes routes that carry flow, while the second subset includes routes with no flow. The travel times on the routes in the first subset are equal and less than the travel times on the routes in the second subset. According to Wardrop's second principle, routes between an O-D pair may also be divided into two subsets: one that carries flows and the second with no flows. The marginal costs in the routes in the first subset are equal and less than those in the second subset. In summary, the application of the above two principles may result in a traffic assignment in which some routes carry no flow. Consequently, some links in the network may not be assigned any flow. In most practical circumstances, such a situation is unrealistic.

In practice, a route with a higher travel time is often used anyway because of various reasons. First, the assumption that drivers are fully aware of the characteristics of all alternative routes may very well be unreasonable, especially at high traffic flows. Second, some drivers tend to stick to their usual driving routes regardless of the amount of delay they encounter. Third, there are many local trips made within smaller areas of the network, where drivers do not have as many routes to choose from. Fourth, and most importantly, is that in any transportation network, there is a background (ambient) traffic flow as a result of external-to-external O-D flows. These flows are assumed to use the network links irrespective of the signal control policy. Therefore, it is plausible to assume that each link in the network carries at least a minimum amount of traffic. This minimum will be determined later.

Henceforth, we will divide the trip matrix into two parts: fixed and variable. For each O-D pair, the fixed part is divided apriori among all alternative routes. The variable part, on the other hand, is determined by the assignment procedure. For example, assume a one origin-destination pair with a trip demand of 1000 vph. We divide the trip demand into a fixed part of 600 vph, and a variable part of 400 vph. Assume that there are four alternative routes between the origin and the destination. According to our proposed strategy, each route will carry a fixed flow of 150 vph. The 400 trips are to be assigned according to one of Wardrop;s principles.

The definitions of Wardrop's two principles have to be modified to accommodate the above mentioned changes. Assume that for each O-D pair, the trip demand is divided into a fixed part, p_{if}, and a variable part, p_{iv}. Then we obtain the following definitions:

(a) Wardrop's First Principle

We modify the definition such that at equilibrium all routes are used. The modified definition states that for each O-D pair, at equilibrium, the travel times on all routes with flow greater than the fixed minimum are equal and that they are less than or equal to the travel timea on any route with the fixed minimum flow.

Mathematically, this is written as

$$T_i \begin{cases} = T_e, & p_i > p_{if} \\ \geq T_e, & p_i = p_{if} \end{cases} \qquad (4.1)$$

where, T_i denotes the travel time on route i. T_e denotes travel time at equilibrium.

(b) Wardrop's Second Principle

We modify this definition such that at equilibrium, all routes carry at least the fixed amount of flow. The modified definition states that for each O-D pair, at equilibrium, the marginal costs on all routes with flow greater than the fixed minimum are equal, and they are less than or equal to the marginal costs on any route with the fixed minimum flow. This new definition can be expressed mathematically as

$$MC_i \begin{cases} = MC_E, & p_i > p_{if} \\ \geq MC_E, & p_i = p_{if} \end{cases} \qquad (4.2)$$

where, MC_i denotes the marginal costs on route i. The marginal costs on a particular route equals the summation of the marginal costs on the links that form that route.

The following section presents the formulation of the user-equilibrium and the system-optimization problems based on the modified Wardrop's principles.

4.2 *User-equilibrium*

A link flow pattern that satisfies the user-equilibrium condition given by Eq. (4.1) can be obtained by solving the following mathematical program. Let b denote the link that intersects link a.

$$\min Z(Q) = \frac{1}{2}\sum_a \left(\int_0^{q_{av}} t_a\left(\left[q_{af}+\omega\right],\left[q_{bf}+q_{bv}\right]\right)d\omega + \int_0^{q_{av}} t_a\left(\left[q_{af}+\omega\right],q_{bf}\right)d\omega \right) \qquad (4.3)$$

Subject to the flow conservation and the non-negativity constraints. These constraints are given by the following equations.

$$\begin{aligned} &\textstyle\sum_k p_k^{mn} = f_{mn} && \forall m,n \\ &p_k^{mn} \geq 0 && \forall k,m,n \end{aligned} \qquad (4.4)$$

The above objective function includes two terms for each link in the network. The first is the integral of the link's delay function with respect to its own flow, when the flow on the conflicting link is held constant. The second term is the integral of the link's delay function with respect to its own flow, when the flow on the conflicting link is held at its fixed minimum. To show that the solution of the mathematical program given in Eq. (4.3) results in a user-equilibrium link flow pattern, we have to prove that the structure of this program is similar to the UE program presented in Section 3. This is accomplished by showing that the derivative of this program with respect to the variable flow on a particular link equals the actual travel time on that link. To show this, let us evaluate the first derivative of the objective function with respect to the variable flow on a link, say link a.

The derivative of the first sum in Eq. (4.3) with respect to q_{av} is given by

$$\frac{\partial}{\partial q_{av}} \int_0^{q_{av}} t_a\left(\left[q_{af}+\omega\right],\left[q_{bf}+q_{bv}\right]\right) d\omega + \frac{\partial}{\partial q_{av}} \int_0^{q_{bv}} t_b\left(\left[q_{bf}+\omega\right],\left[q_{af}+q_{av}\right]\right) d\omega \qquad (4.5)$$

The above derivative includes two terms. The first one equals the actual travel time on link a.This is given by $t_a([q_{af}+q_{av}],[q_{bf}+q_{bv}])$ The second term can be shown to reduce to

$$t_a([q_{af}+q_{av}],[q_{bf}+q_{bv}]) - t_a([q_{af}+q_{av}],q_{bf}).$$

The derivative of the second sum in Eq. (4.3) with respect the variable flow on link a is given by

$$\frac{\partial}{\partial q_{av}} \int_0^{q_{av}} t_a\ ([q_{af}+\omega]\ ,\ q_{bf})\ d\omega \qquad (4.6)$$

This is equal to $t_a([q_{af}+q_{av}],q_{bf})$. We can now calculate the derivative of the objective function, Eq. (4.3), with respect to the variable flow on link a, by adding the derivatives obtained above. We obtain

$$1/2\ (2t_a([q_{af}+q_{av}],[q_{bf},q_{bv}]) + t_a([q_{af}+q_{av}],q_{bf}) - t_a([q_{af}+q_{av}],q_{bf})\ ) = \quad t_a([q_{af}+q_{av}],[q_{bf}+q_{bv}]) \qquad (4.7)$$

Thus, the derivative of the objective function with respect to the variable flow on a particular link is the travel time on that link.

4.3 *System-optimization*

A link flow pattern that satisfies the system-optimization condition given in Eq. (4.2) can be found by solving the following mathematical program.

$$\min Z(Q) = \Sigma_a (q_{af}+q_{av})\, t_a([q_{af}+q_{av}], [q_{bf}+q_{bv}]) \qquad \textbf{(4.8)}$$

subject to the same constraints given for user-equilibrium. The objective function of this program is the total travel time spent in the network. The flow pattern that minimizes this program represents an equilibrium situation characterized by Eq. (4.2). Eq. (4.2) states that the marginal costs on all used routes are equal and are less than or equal to those of unused routes. The marginal costs of a particular link is obtained by calculating the first derivative of the above objective function with respect to the variable flow on that link. The derivative with respect to the variable flow on a particular link, say link a, is given by

$$\frac{\partial}{\partial q_{a_{l_v}}} [\, \Sigma_a (q_{af}+q_{av})\, t_a([q_{af}+q_{av}], [q_{bf}+q_{bv}])\,]$$

This equals to: $$t_{al}+q_{alf} \frac{\partial t_{a_l}}{\partial q_{a_l v}}+q_{alv} \frac{\partial t_{a_l}}{\partial q_{a_l v}}+q_{blf} \frac{\partial t_{b_l}}{\partial q_{a_l v}}+q_{blv} \frac{\partial t_{b_l}}{\partial q_{a_l v}}$$

which can be expressed as

$$t_{al}+q_{al} \frac{\partial t_{a_l}}{\partial q_{a_l v}} + q_{bl} \frac{\partial t_{b_l}}{\partial q_{a_l v}}.$$

By definition, the marginal contribution of a variable flow on a particular link is equal to the marginal contribution of the flow on that link. Thus, the derivative given in (4.17) can be expressed as

$$t_{al} + q_{al} \frac{\partial t_{a_l}}{\partial q_{a_l}} + q_{bl} \frac{\partial t_{b_l}}{\partial q_{a_l}}.$$

We conclude, therefore, that the structure of the system-optimization problem remains the same, despite the fixed flows imposed on the network.

4.4 *Computational Considerations*

The objective of dividing the trip matrix into fixed and variable matrices is twofold. First, to force the assignment procedure to converge to a single solution. Second, to avoid the all-or-nothing type assignments that result in zero green splits for some links. The question that arises at this point is what percentage of the trip matrix should be fixed. Following are some considerations. We want the percentage of the trip matrix that is fixed to be large enough such that the above objectives are achieved. We also want that percentage to be small enough such that the assignment procedure models the travel choice behavior as good as possible. Furthermore, we need the percentage to be small enough such that when the fixed-trip matrix is distributed onto the network, no link will have a flow that exceeds its capacity.

Computational experiments done by Al-Malik (1991) have shown that imposing even a small percentage of the trip matrix onto the network as fixed flows has resulted in forcing the assignment procedure to converge to the same solution. Thus, by imposing fixed flows, we can force the solution domain of the problem to be shaped such that there is a single link flow pattern with the least objective function value for both UE and SO. We have also succeeded in avoiding the all-or-nothing type solutions that we encountered earlier. As a result, green splits at all intersections are adjusted to match link assignments.

It is important to emphasize the fact that the fixed-trip matrix represents background or intrinsic traffic. This traffic exists and varies according to the specific circumstances and attributes of the transportation network. Therefore, the percentage of the trip matrix that is fixed is not necessarily a controllable parameter, rather, it should be an observed parameter.

5. Conclusion

The modeling and optimization of the interaction between the assignment of traffic flows in an urban network and the signal controls at the intersections of the network has long been considered an important and challenging problem. The conventional approach has been to treat each problem independently, assuming it is unaffected by the other: traffic signals are set assuming given constant volumes while demands are assigned to routes assuming fixed signal settings. This approach is obviously unsatisfactory since traffic volumes and signal settings are mutually interdependent. Attempts to solve this problem by iteration have also been unsatisfactory: either the problem does not converge or, if it converges, it does not attain an optimal solution.

In this paper, for the first time, an optimization model is developed that *simultaneously* accounts for both route choice behavior and optimal signal controls. In fact, since the signal timings at the intersections are directly determined by the approach volumes, the signal control variables can be expressed as equivalent flow variables. Thus is obtained an optimization model formulation that inherently reflects the mutual consistency between traffic volumes and signal settings.

In the problems discussed in this paper, only signal splits are being optimized while the cycle time is kept fixed. Performance functions are constructed on the basis of both Webster's queueing delay theory and the Highway Capacity Manual formula. The resulting Combined Model formulation is a non-convex optimization problem that produces multiple equilibria. Under practical conditions, this problem can be alleviated by a suitable adaptation of the two Wardropian principles. The demand flows are divided into those that are unaffected by the signal controls (the *ambient* flows) and into those that are subject to influence by the signal controls (i.e., a change in signal timings will cause a change in route choice).

Computational results are given for sample networks with both user optimization and system-optimization model formulations.

Acknowledgment

The material in this paper is based on a Ph.D. thesis prepared by M.S. Al-Malik while being a doctoral candidate at the Georgia Institute of Technology, under the guidance and supervision of Prof. N.H. Gartner of the University of Massachusetts, Lowell.

References

[1] Al-Malik, M.S. (1991). *"An Investigation and Development of a Combined Traffic Signal Control - Traffic Assignment Model."* Ph.D. Thesis, Georgia Institute of Technology.

[2] Allsop R.E.(1974).*"Some Possibilities for Using Traffic Control to Influence Trip Distribution and Route Choice."* Proceedings 6th International Symposium on Transportation and Traffic Theory, Sydney.

[3] Allsop R.E. and J.A. Charlesworth (1977). *"Traffic in a Signal-Controlled Road Network: An example of Different Signal Timings Inducing Different Routing."* Traffic Engineering and Control, 18(5), 262-264.

[4] Cantarella G.E. and A. Sforza (1987). *"Methods for Equilibrium Network Traffic Signal Setting."* Flow Control of Congested Networks, Springer-Verlag, Berlin.

[5] Cantarella G.E., G. Improta and A. Sforza (1991). *"An Iterative Procedure for Equilibrium Network Traffic Signal Settings."* Transportation Research 25B(4).

[6] Dafermos S. (1972).*"The Traffic Assignment Problem for Multiclass-User Transportation Networks."* Transportation Science 6, 73-87.

[7] Dickson, T.J. (1981) *"A Note on Traffic Assignment and Signal Timings in a Signal Controlled Road Network."* Transportation Research 15B, 267-271.

[8] Gartner N.H. (1976). *"Area Traffic Control and Network Equilibrium."* Proceedings International Symposium on Traffic Equilibrium Methods (Edited by Florian, M. A.), Springer Verlag.

[9] Gartner N.H. (1977a). *"Influencing Traffic Equilibrium in a Network by Control Measures."* Transportation Research Record 644, 125-129.

[10] Gartner N.H. (1977b). *"Analysis and Control of Transportation Networks by Frank-Wolfe Decomposition."* Proceedings 7th International Symposium in Transportation and Traffic Theory (Edited by T. Sasaki and T.Y. Yamaoka) Tokyo, Japan.

[11] Highway Capacity Manual. *"Special Report 209"*, TRB, National Research Council, Washington, D.C.

[12] Improta G. (1987). *"Mathematical Programming Methods for Urban Network Control."* Flow Control of Congested Networks, Springer-Verlag, Berlin.

[13] Miller A.J. (1963). *"Settings for fixed-cycle traffic signals."* Oper. Res. Q., 14: 373-386.

[14] Newell G.F. (1965).*"Approximation methods for queues with applications to the fixed-cycle traffic light."* SIAM Rev.,7(2):223-240.

[15] Sheffi Y. (1979). *"Urban Transportation Networks."* Prentice Hall, Englewood Cliffs, N.J.

[16] Smith M.J. (1981). *"Properties of a Traffic Control Policy which Ensure the Existence of a Traffic Equilibrium Consistent with the Policy."* Transportation Research 18B(1), 63-65.

[17] Smith M.J. (1985). *"Traffic Signals in Assignments."* Transportation Research 19B(2), 155-160.

[18] Tan H., S.B. Gershwin and M. Athans (1979). *"Hybrid Optimization in Urban Traffic Networks"*. Report No. DOT-TSC-RSP-79-7, MIT Cambridge, MA.

[19] Wardrop J.G.(1952).*"Some Theoretical Aspects of Road Traffic Research."* Proceedings, Institution of Civil Engineers II(1),325-378.

[20] Webster F.V. *"Traffic Signal Settings."* Road Research Laboratory Report 39, Crowthorne, Berkshire, England.

TRAFFIC CONTROL UNDER POLLUTANT EMISSIONS CONSTRAINTS

Giuseppe Bruno, Gennaro Improta

Dipartimento di Informatica e Sistemistica
Università di Napoli Federico II
Via Claudio 21 - 80125 Napoli - Italy

The industrial development of the last years and its interaction with the environment is one of the most important problems of the next future. While the pollution due to industrial production is actually strictly regulated by laws that force companies to observe precise constraints, pollution due to vehicle emissions, because of pollutants sources general diffusion, is a problem for whose solution technological and political measures conditioning collective customs and manners are needed. The adoption of these measures, especially for emergency situations, is performed on the base of data provided by special detectors capable of measuring concentrations of some substances. This methodology can cause two great problems: (i) because of required equipment high costs, it is not always possible, on the ground of fixed detectors location in a certain area, to obtain reliable information on the distribution of the pollutant substances on the whole area; (ii) mistakes strongly depending on receptor localization can arise, for instance when detectors are located in proximity of pollutant sources. In order to solve these problems, general models for the evaluation of pollutants levels which take into account the different complex factors (i.e. traffic data, meterological conditions, topological characteristics of the observed area) influencing the phenomena of the emission and of the diffusion of the pollutants are needed.
In this paper a general description of the problems related to the air pollution due to vehicle emissions is firstly provided, pointing out several aspects which affect the phenomenon. A first idea on the characteristics of a general assignment model with the minimization of the maximum concentration of the air pollution is proposed. An example is shown in order to show the differences with classical assignment models.

1. Air pollution sources

Motor vehicles produce both chemical (toxic substances emission) and physical (noise) air pollution. Chemical pollution is mainly due to gasoline and gas oils combustion, as well as tires movement on asphalt. Fuels are made of hydrocarbons mixings of paraffinic, olefinic, naphtenic and aromatic series, and of antiknock additives lead grounded; car tires hold valuable amounts of amianthus. Many cancerous substances are produced by asphalt pulverization, car tires, brakes and clutch wear, as well as by engine exhaust pipe. Moreover, due to high temperature in combustion chamber, there can be chemical reaction between oxygen and air nitrogen which produces sulphur oxides.

The pollutant substances in atmosphere can be drawn together in two groups:

- **primary pollutants**, directly emitted in atmosphere by various sources;
- **secondary pollutants**, obtained by reactions in the atmosphere.

As regards motor exhaust gases, the fuel should give rise, as reaction products, only to water and carbon dioxide. The presence of impurities and additives in fuels and the not perfect combustion conditions lead to the formation of other tossic substances such as:

- carbon monoxide
- nitrogen oxides (nitrogen monoxide and dioxide)
- aliphatic hydrocarbons
- aromatic hydrocarbons (benzene,...)
- polycyclic aromatic hydrocarbons (benzoalphapyrene,...)
- oxygenated compounds (aldehydes, ketones,...)
- metal traces (lead, brake linings coming amianthus,...)
- sulphur dioxide
- suspended particulate

These substances are in each vehicle exhaust (except Pb which is only in gasolines); usually their rates and their relative emission ratios depend on different elements.

Except CO and NO_X, typical of all exhausts, there are very negligible amounts of the other pollutants in methane alimented vehicles; their concentration progressively increase for gasoline and GPL engines and gets to highest values in diesel vehicles.

In the last years it has been confirmed a general trend to a **decrease in emissions in the industrial sector**, and on the other side, an **increase in emissions in transportation system** ([4], [8]). The progressive **decrease of sulphur oxides (SO_x) emissions** is due to various reasons, like the gradual substitution of gas oil and combustible oil with natural gas, the lowering of sulphur contents in gas oil for heating systems and motor transport (the legal limit has been lowered from 1980's 1,1% to actual 0,3%), etc.. Most of the emissions of sulphur oxides (about 60%) is to be ascribed to **high altitude** emitting sources, whose influence on air quality is felt more at medium and large than at short distances from emitting point. This is the case of great plants (thermoelectric stations, refineries) for which there are generally measures and technologies for pollutant emissions control.

For all the other pollutants classes, most of the emissions (about 60% for nitrogen oxides and almost 100% for carbon monoxide) is due to **low altitude** emitting sources (motor vehicles, heating plants, small industries), whose great effect is felt in proximity of emitting points, expecially if weather conditions are against a quick dispersion of pollutants. For example, more than 20% of nitrogen oxides totally emitted by transport system in Italy is released in urban areas just like almost 80% of carbon monoxide and about 40% of particulate.

2. The main factors influencing emissions

Motor vehicle emissions are connected to various factors, particularly to *vehicle features* and *motion conditions*. In this section, the dependence on these different factors is illustrated.

Vehicle type.

Vehicle technical features significatively affect the emissions of pollutants. In the last years a great alarmism has spread about diesel motor vehicles, so producing a persistent uncertainty in the market and a dizzy falling down of sales. Although its superior economy in comparison to gasoline motor vehicles has always been recognized (about 20-30% saving in daily consumption), diesel motor has been defamed because of not exact definitions and wrong descriptions. In Table 1 a classification of the vehicles on the base of the technical features is provided. In Figures 1÷4 emissions for each considered vehicle class are shown.

CL1	Gasoline 2 stroke < 1.400 cm^3
CL2	" 4 stroke > 2.000 cm^3
CL3	" " 1.400÷2.000 cm^3
CL4	" " < 1.400 cm^3
CL5	Gasoline catàlytic without Ox sensor
CL6	" " with Ox sensor
CL7	Diesel without catalyst
CL8	" with catalyst

Table 1 - Vehicle classes.

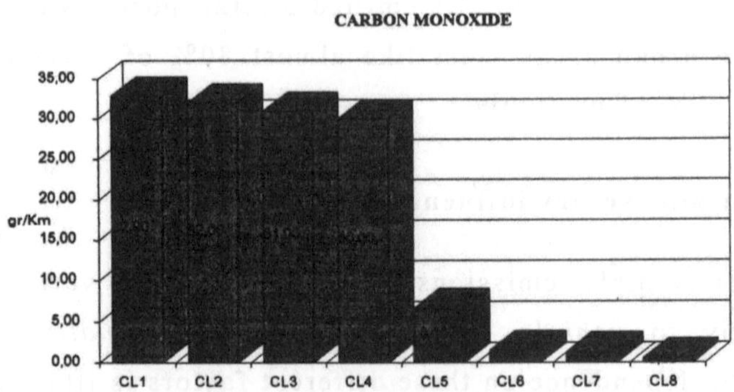

Figure 1

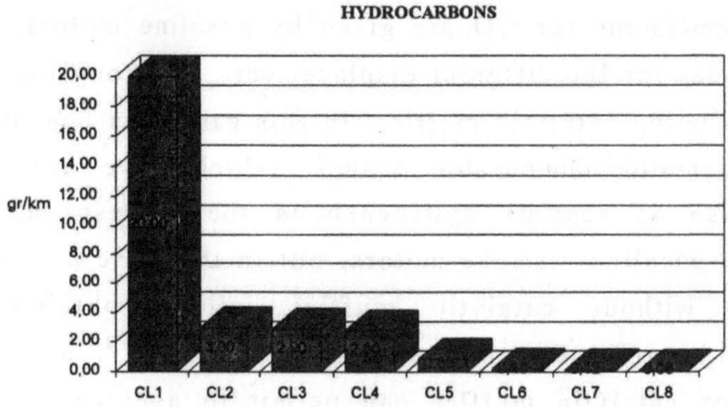

Figure 2

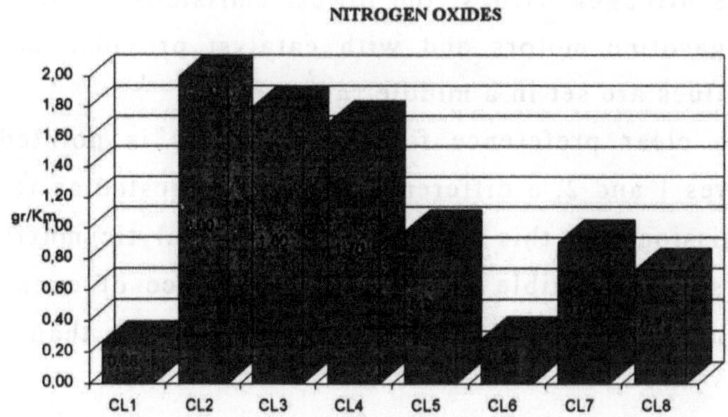

Figure 3

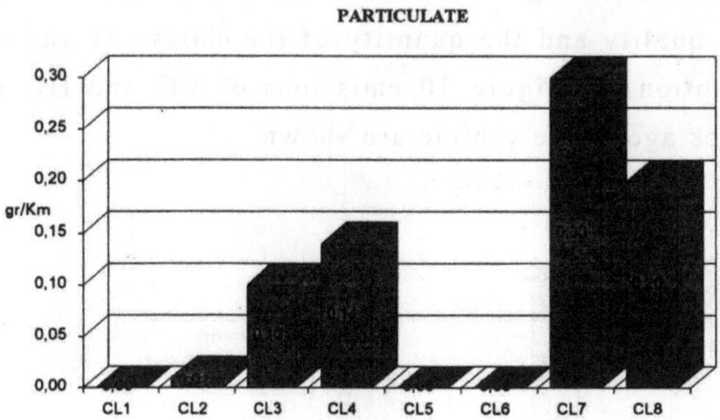

Figure 4

Highest emissions for **CO** are given by gasoline motors, with slight differences for the different displacements. A strong lowering is found introducing catalytic muffler. In this case it is possible to obtain diesel motor comparable values, which have very low emissions. Also as regards **hydrocarbons** the highest emission values are for gasoline 2-stroke motors, but in this case a 4-stroke motor, even without catalytic muffler, allows obtaining a significative improvement in emissions. Anyway, only the introduction of catalytic muffler can permit to achieve emission values similar to diesel motor ones.

As regards **nitrogen oxides**, the lowest emissions are obtained with 2-stroke gasoline motors and with catalyst provided motors. Diesel motor values are set in a middle range.

Whereas a clear preference for diesel motors is pointed out analyzing Figures 1 and 2, a different choice is suggested as regards **particulate** emissions. In this case applicating catalytic muffler to gasoline motors it is possible to cancel the presence of emissions. Anyway gasoline motors emissions are sensitively lower than diesel motors ones.

Age of circulating lot.

The age of a vehicle strongly influences the performances with reference to the quality and the quantity of the emissions and of the fuel consumption. In Figure 10 emissions of CO and HC in dependence on the age of the vehicle are shown.

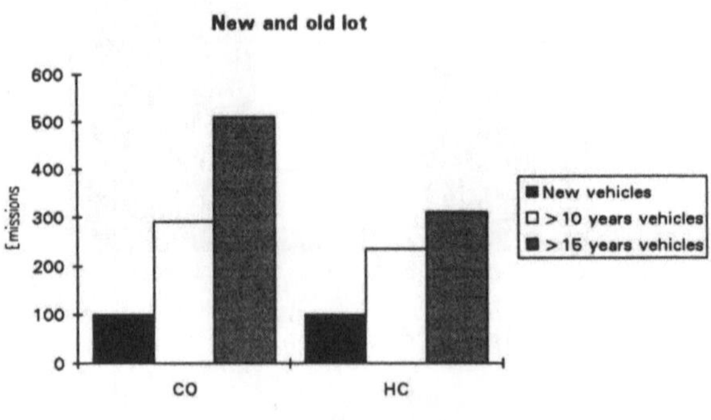

Figure 5

In order to reduce the effects of the age of the lot, controls on the emissions of the single vehicles should be adopted by authorities. Figure 6 indicates the average number of controls per vehicle in 15 years for different countries.

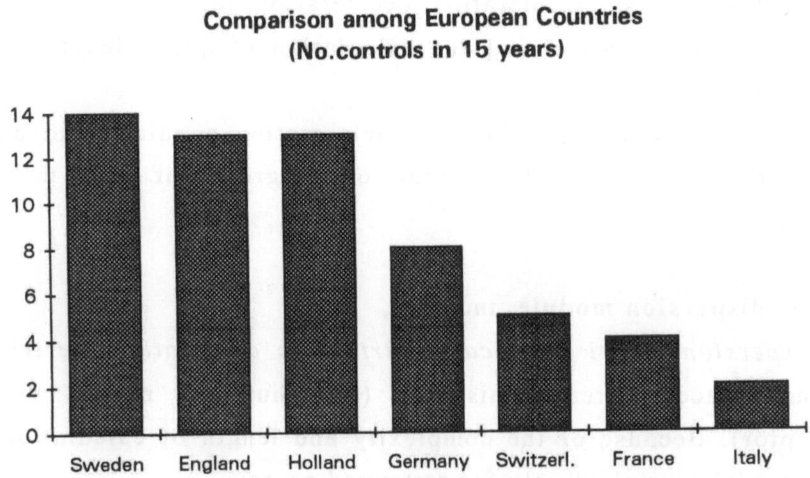

Figure 6

3. Air pollution models for vehicular traffic

The models for the calculation of vehicular traffic concentrations of gaseous pollutants (carbon monoxide CO, nitrogen oxides NO_x, hydrocarbons HC, sulphur dioxide SO_2, etc.) in urban areas are considerably complex essentially depending on the phenomena of gas **emissions** from vehicles and their **dispersion** in the atmosphere.

The concentration in a certain point of an inspected area is calculated as a sum of a local contribution, coming from immediate vicinity placed sources (local network), and of a background contribution, originated from all the sources located in the area (extended network).

The **emission module**, in the most general case, includes three sub-models:

1. an *emission model*, providing emission rates, i.e. pollutant amount emitted by each vehicle class, determined on the ground of type and age, in time and space unit depending on motion conditions (cruise, acceleration, deceleration, idle);
2. a *queue model*, permitting the calculation of queue length and waiting times;
3. a *downflow model*, providing vehicle motion conditions in any point or line of each network road, on the ground of queue model results.

The **dispersion module** includes:

1. a *dispersion model for local contribution calculation*, deriving from adjacent areas emissions (few hundred meters from receptor). Because of the complexity and length of calculations, this model is applied only for restricted areas;
2. a *dispersion model for background contribution calculation*, deriving from remote areas emissions: this contribution becomes considerable in middle and extended urban areas;
3. a *model for metereological data calculation*, required by dispersion models, obtained by the metereological data.

3.1 Emission models

Various elements influence the formulation of a general model for the calculation of emission rates of gaseous pollutants from motor vehicle traffic. As mentioned above, it is possible to group together the different elements in:

a) vehicle features:

- *vehicle type* (gasoline, diesel, GPL);
- *displacement*;
- *vehicle weight*;
- *age*;

- *maintenance state*;
- *fuel type*;

b) motion conditions:

- *motor temperature* (cold or warm engine);
- *external temperature*;
- *motion stroke* (acceleration, deceleration, cruise and idle).

At the present state, emission models can be subdivided in two classes. **The first class** describes the emissions only in dependence on the average speed of a flow on a link. In the following, some models are presented referring to CO, HC and NO_X emissions in the case of urban areas.

A model developped for the Australian case (PERTH) proposes the following analytical expression for the calculation of the emission rates ER [g/Km]:

$$ER_{CO} = 465 \cdot v^{-0.97}$$

$$ER_{HC} = 21.5 \cdot v^{-0.73}$$

$$ER_{NO_x} = 2.2 + 0.008 \cdot v$$

if v is the average vehicle speed in Km/h.

Kent and Mudford [9] have modified these expressions:

$$ER = a \times \frac{b}{v^c} \qquad \text{for CO and HC;}$$

$$ER = a + \frac{b}{v^c} \qquad \text{for } NO_X\text{;}$$

with a, b, c parameters of the model. This model well agrees with others based on United States experiences and it is very similar to the one developped by Rogers [14] for English urban areas.

In Figure 7 the emission vs speed diagrams for the three considered pollutants are sketched comparing Kent and Mudford's results with Rogers's data. The results are very similar for CO and HC while they present some differences for NO_X.

The second class is oriented to describe the vehicle emission phenomenon in a more complex way taking into account the numerous factors influencing the emission. Annual emission factors for the European countries for various pollutants have been estimated by the methodology CORINAIR [5]. The proposed methodology considers a classification of the emissions into 10 different major categories of traffic and in three different cases (urban road, rural road and highway).

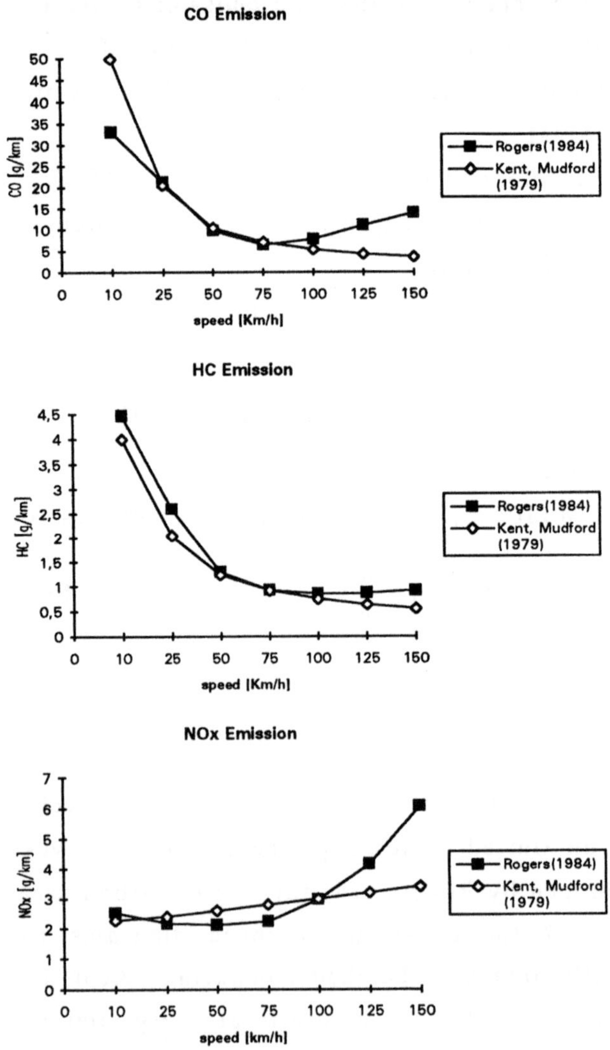

Figure 7

In general, consider a flow of vehicles q moving with an average speed v(q), an emission model can be represented in the following way:

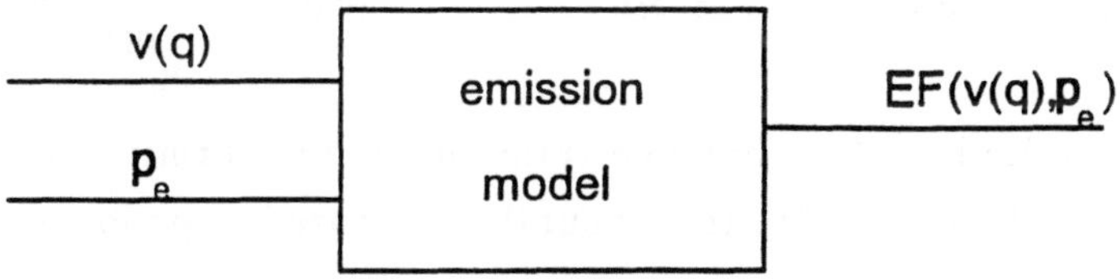

Figure 8

in which the dependence on the average speed of the flow q is pointed out, while p_e is a vector of parameters that can influence the phenomenon.

3.2 Dispersion models.

The dispersion of the produced emissions in the atmosphere is very complex because a lot of factors must be considered like:

- **metereological factors,** such as *wind direction and speed, atmospheric turbolence degree, mixing height;*
- **vehicular flow turbulence;**
- **inspected area features.**

In the following a brief description of the principal adopted models is provided.

Eulerian models

Eulerian models (K-models) are oriented to the calculation of pollutant concentration (C) in each tridimensional space position at subsequent time periods. Given initial conditions and tridimensional wind (u_i) and turbulence (K_i) components, diffusion equation terms are to be solved on a fixed grid in order to determine the concentration values $C^n_{j,k,l}$ for any point (j,k,l) at instant n.

$$\underbrace{\frac{\partial C}{\partial t}}_{\substack{\text{TIME}\\\text{DERIVATIVE}}} + \underbrace{u_i \frac{\partial C}{\partial x_i}}_{\substack{\text{TRANSPORT}\\\text{TERM}}} = \underbrace{\frac{\partial}{\partial x_i}\left(K_i \frac{\partial C}{\partial x_i}\right)}_{\substack{\text{DIFFUSION}\\\text{TERM}}} + \underbrace{S}_{\substack{\text{SOURCE}\\\text{TERM}}} + \underbrace{R}_{\substack{\text{TRANSFORMATION}\\\text{AND REMOTION}\\\text{TERM}}}$$

The complexity of turbolence terms does not permit a practical use of this kind of models. It is usually supposed a proportionality dependence on the space variations which is expressed in mathematical terms by K_i parameters; depending on the complexity of the expressions of K_i different kinds of models are defined. K-models are very suitable for simulations in which transformation and remotion due term (R) is not neglectable. Analytical expressions of the diffusion equation in such cases in which wind and turbulence fields are particularly simple are provided.

Lagrangian models

In these models pollutant dispersion is simulated through the motion of some particles whose trajectories allow the calculation of emitted substance concentration field. Basis hypothesis is that the trajectories of these particles simulate, on average, the movement in the atmosphere of air masses located in the same position at starting instant. The resulting equation system is:

$$\begin{cases} \dfrac{\partial U_i'}{\partial t} = -\beta U_i'(t) + \gamma & (1) \\[2ex] \dfrac{\partial X_i(t)}{\partial t} = U_i(t) + U_i'(t) & (2) \end{cases}$$

where "i" represents the three directions of a Cartesian system; $X_i(t)$ is the i-th component of particle position at time t, $U_i(t)$ is the average wind vector component in i-th direction at instant t, $U'_i(t)$ is the speed fluctuation component due to the atmospheric turbulence in the same direction.

Numerically integrating the equation (1), it is possible to obtain speed values $U'_i(t)$ as inputs of (2) which, integrating again, provides particle positions due to average wind and turbulence matched effects. Dispersion calculation is done casually generating a great number of particles, moving according to atmospheric parameters describing transportation and turbulence; average concentration estimate in a point is obtained counting the number of particles in a certain volume including that point.

Gaussian models

Because of their simplicity, Gaussian models are the most used dispersion models. The relationship between emission rate and the concentration at a certain point of the downwind source space, indeed, is analytic and it doesn't need great calculation resources.

Gaussian models are recommended for environment impact evaluation by various government corporations like E.P.A. (Environmental Protection Agency - USA).

These kind of models assume that an ideal point source produce emissions which are dispersed in the atmosphere in each direction with a Gaussian profile whose orientation is strongly influenced by the wind direction.

Given a Cartesian system with x-axis coinciding with the wind direction, the general equation used in Gaussian models for evaluating the concentration C in any point (x,y,z) of a substance emmitted from a source point at (0,0,H) is the following:

$$C(x,y,z)=\frac{Q}{2\pi\sigma_y\sigma_z V}exp\left\{-\frac{y^2}{2\sigma_y^2}\right\}\cdot\left\{exp\left(-\frac{(z-H_e)^2}{2\sigma_z^2}\right)+exp\left(-\frac{(z+H_e)^2}{2\sigma_z^2}\right)\right\} \quad (3)$$

where Q is the pollutant emission rate, V is the wind speed along the direction x, σ_y and σ_z are respectively the horizontal and vertical dispersion parameter and H_e is the effective plume height. The parameters in the mathematical formulation are easy obtainable also when a limited number of measurements is available.

Plume horizontal and vertical dispersion parameters are connected with turbulent diffusion coefficients used in K-models, and they can be empirically obtained by simple metereological measurements. Effective plume height represents smokes levelling height obtained as sum of emitting chimney geometrical height and plume rise due to temperature difference with air and output speed; various formulas have been developped for the calculation of this parameter on the base of numerous experimental measurements.

In Figure 8 some tridimensional concentration profiles are sketched referring to a point source at height H and effective plume height H_e.

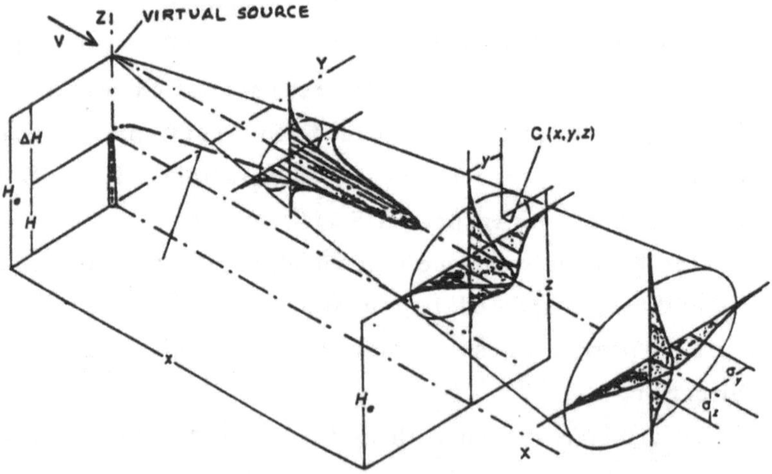

Figure 8

In dependence on the limits of integration of (3), it is possible to extend the point source model in order to define finite line models, indefinite line models, area source models and volume source models.

The state of the art offers various kinds of Gaussian models of different complexity and characteristics. In Table 2 Gaussian models actually presented in literature are listed together with the indication of the kind of considered source and the area for which the model can be used. The difference between "open air" model and urban model is in the fact that urban models have the possibility of describing the phenomena related to the presence of obstacles which can used to represent buildings or any kind of obstacles of a typical urban area.

MODEL	SOURCE	AREA	REFERENCES
EPA - CALINE 4	finite line	open / urban	Benson (1989)
G.M.	finite line	open	Chock (1977)
EPA - HIWAY2	finite line / area	open / urban	Petersen (1980)
GFLSM	finite line	open	Luhar and Patil (1988)
OMG	volume	urban	Kono and Ito (1990a,b)

Table 2 - Gaussian models.

In general, given an emission factor ER of a source, a diffusion model can be represented like in (Figure 9) in which the concentration of a substance in a certain point x is supposed as a function of the emission factor of a source and of a vector of parameters in dependence on the kind of the adopted model.

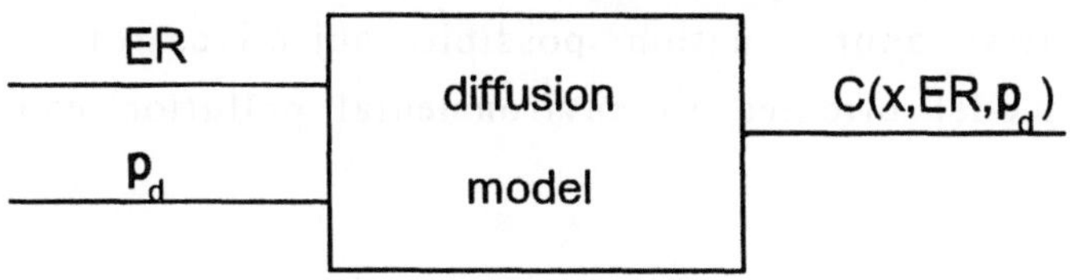

Figure 9

4. Optimising models for the reduction of the environment pollution

The Wardrop's principles [16] are the basis for the traffic assignment models.

The first principle defines the so-called user-optimising assignment for which, under equilibrium conditions, no single driver can reduce his own travel time by choosing alternative routes. In this way the travel times on all used routes must be equal and less than the travel time on any unused route. The hypothesys assumed by the model is that users choice their routes according to perceived travel times instead of real travel times.

System optimising assignment is oriented to the minimization of the total travel time on the entire network. The optimal conditions provide that the marginal travel times on all used routes are equal or less than on any unused routes.

For both user-optimising and system optimising the choice of the travel time function for the link is fundamental. User-optimising and system-optimising models with travel time functions are not capable to describe equilibrium situations in which the level of the air pollution is assumed as objective.

In the field of the enviromental pollution due to vehicular traffic different objectives can be assumed. It is clear that user optimising objectives have no sense because the users have not a perception of the produced emissions function to minimize and, hence, they have no interest to reduce the emissions during their trips. Consequently the only assignment models which can be defined is a system optimising model.

In a first approximation possible objectives of a system optimising model oriented to environmental pollution can be the following:

- the minimization of the total emissions in the network;
- the minimization of the maximum concentration in a certain area.

In both the cases the results depend on the kind of the adopted models to describe the emission and the diffusion models.

According to the adopted schemes for the emission and the diffusion models, a general system optimising assignment models with an air pollution oriented objective can be described by a three stage model (Figure 10).

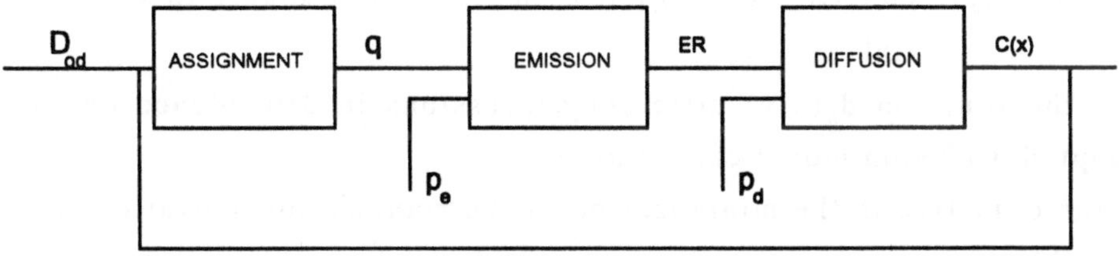

Figure 10

where $\mathbf{D_{od}}$ is the vector of the o-d pairs demand, **q** is the vector of the flows on the network, **ER** is the vector of the emission rates of the sources related to the flows and **C(x)** is the concentration of a considered pollutant at the point x of the network and p_e and p_d are respectively the vectors of the parameters of the emission and diffusion models.

The principal characteristic of the described model is that the concentration at a certain point x can be expressed by a function:

$$C(x) = C(\mathbf{D_{od}}, \mathbf{q}, \mathbf{p_e}, \mathbf{p_d})$$

In a first approximation, if the function C(x) is supposed constant for each point in proximity of the link (i,j), it is possible to define a cost function foe each link. The expression Cij(x) of this function can be given by:

$$C_{ij}(x) = \sum_{1}^{n} I_k(x)\, q_k \tag{4}$$

where n is the number of links of the network, q_k is the flow on the link k and $I_k(x)$ is the concentration at point x in proximity of the link (i,j) due to a vehicle belonging to the flow q_k. Suppose that each link can be considered a linear source, the concentration $I_k(x)$ can be calculated in the following way:

$$I_k(x) = \frac{ER_k}{w} \int_0^{l_k} d_k(x,y)\,dy$$

with:

w: wind speed (m/sec);

ER_k: emission rate (g/km·sec) of a vehicle on the link k;

$d_k(x,y)$: concentration (g) at x due to a point source in y on link k;

For the function $d_k(x,y)$ different expressions in dependence on the adopted diffusion model can be used.

If the objective is the minimization of the maxima concentration it is possible to use a user equilibrium model in which the cost of a link is given by $C_{ij}(x)$. In fact, according to the first Wardrop's principle, as the equilibrium condition coincides with a situation in which it is not possible to reduce the cost of the single driver, it means that, using (4) as cost flow function to determine the equilibrium, no single driver can change own route in order to reduce the marginal pollution on the links.

According to the definition of Dafermos ([6], [7]), the assignment model oriented to the calculation is an asymmetric model, as the objective function C_{ij} depends on the flows on the entire network through the terms introduced by the diffusion model.

The model can be solved using the Streamlined Diagonalization Algorithm for the solution of an asymmetric equilibrium model [15].

In order to point out the differences of the results obtainable compared to the classical assignment model, a numerical example is shown.

A comparison between the user equilibrium with cost-flow function and system equilibrium for the minimization of the maximum concentration may be furnished using a network test of Figure 11 similar to the one used by Akcelik [1]. Consider a single origin-destination network with three routes having the same characteristics and an O-D demand Q.

An assignment to the network consists of the definition of a flow pattern q_1, q_2 such that

$$q_1 = f\,Q \qquad q_2 = (1-f)\,Q \qquad 0 \leq f \leq 1$$

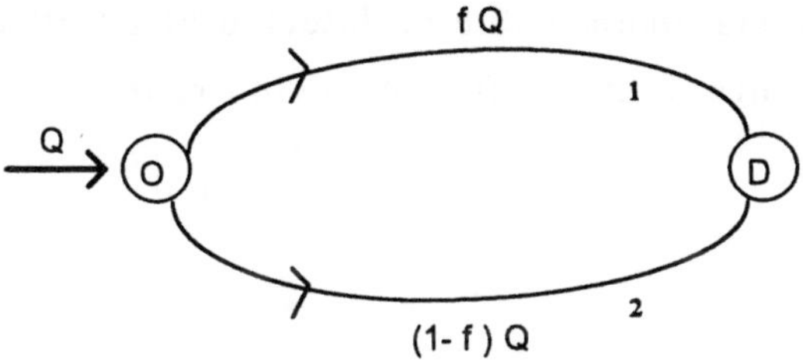

Route	lenght	free flow speed	saturation flow
1-2	4000 m	120 km/h	3600 veh/hr

Figure 11

As the routes have the same characteristics the user-equilibrium assigns the same flow to the routes (f = 0.5).
Consider now the situation represented in Figure 12 showing two parallel routes with the characteristics of the network test of Figure 11.

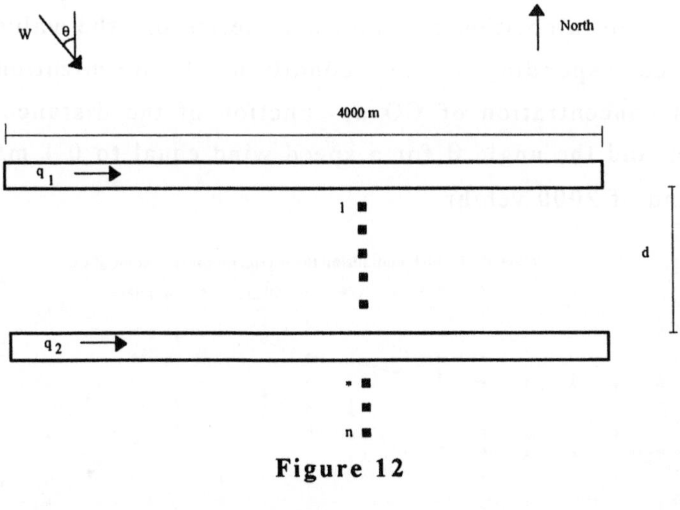

Figure 12

The points 1...n represent the positions of concentration detectors (i.e. CO detectors) and W is the vector of the wind speed. If an assignment q1=q2 is considered, using a dispersion module, a tipical pattern of the CO concentration at points 1,..,n is shown in

Figure 13. The maximum value is detected at point labelled with "*" in Figure 12 in proximity of the downwind route.

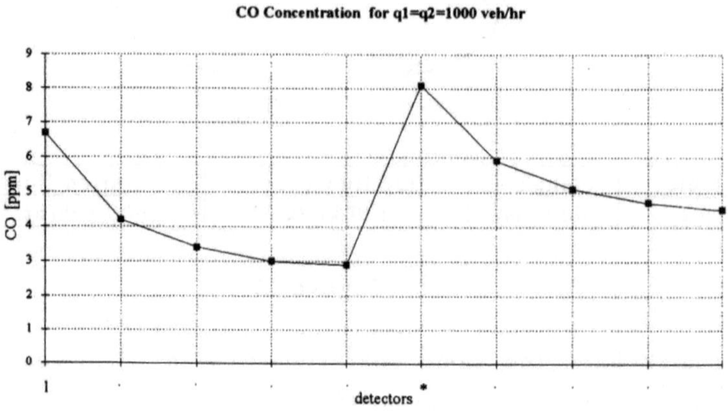

Figure 13

If the objective function is the minimization of the maximum concentration, the assignment must be such that q1>q2 i.e. f > 0.5. In Figure 14 the results obtained using the streamlined diagonalization algorithm are shown in terms of the values of the flow q_1 corresponding to the condition of minimization of the maximum concentration of CO as function of the distance between the routes and the angle θ for a speed wind equal to 0.1 m/s and an o/d demand of 2000 veh/hr.

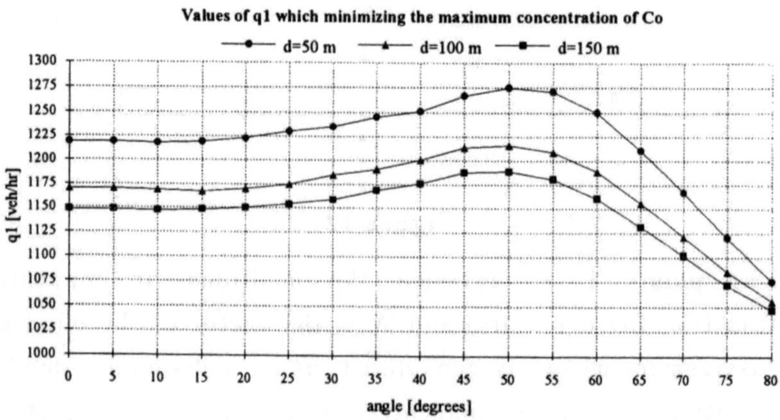

Figure 14

The obtained results are quite different compared to the results of the user-equilibrium assignment. On average the differences are about of 15÷27 percent for values of $\theta \in [0,50]$ and distances between 50 and 100 meters.

Conclusions

At the present state the adoption of measures to reduce the effects of the emissions due to the vehicular traffic, especially for emergency situations, is performed on the base of data provided by special detectors capable of measuring concentrations of some substances. This methodology can cause different kind of mistakes which may be solved through the use of general models for the evaluation of pollutants levels which take into account the different complex factors (i.e. traffic data, meterological conditions, topological characteristics of the observed area) influencing the phenomena of the emission and the diffusion of the pollutants.

It has been shown that a model for the minimization of the maximum concentration of emissions can be classified as an asymmetric equilibrium model. The streamlined diagonalisation algorithm can be used to determine the distribution of the flows which minimise the maximum concentration. The algorithm has been applied to a simple network test in order to show the results obtainable using the dispersion model CALINE 4.

The accuracy of the results are strongly dependent on the used dispersion model; it should be interesting to verify how the equilibrium flows are dependent on some parameters like metherological characteristics, mutual distance between the routes and so on. These aspects should suggest a development of the research in order to point out assignment models more appropriate to represent the phenomena of emissions and dispersion of pollutants on a certain area. In order to apply this approach to an urban area, the dispersion models should be calibrated to take into account the presence of obstacles.

References

[1] R. Akcelik. A Graphical Explanation of the two Principles and two Techniques of Traffic Assignment.Transpn. Res. Vol. 13A, pp 179-184, 1979.

[2] P. Benson. Caline 4 - A dispersion model for predicting air pollution concentrations near roadways. State of California, Department of Transportation - Division of New Technology and Research, 1989.

[3] D. Chock. A Simple Line Source Model for Dispersion Near Roadways. Atmospheric Environment 12, pp. 823-829, 1977.

[4] M. Cirillo. L'inquinamento dell'aria in Italia. Impresa ambiente, Gennaio 1991, pp. 13-23, 1991.

[5] Egglestone H.S., Gaudioso D., Gorissen N., Joumard R., Rijkeboer R.C., Samaras Z. and K.H. Zierock. CORINAIR methodology. Final Report Volume 1, 1991.

[6] S. Dafermos. Relaxation Algorithms for the General Asymmetric Traffic Equilibrium Problem. Transportation Science Vol.16, No. 2, pp 231-240, 1982.

[7] S. Dafermos. Sensitivity Analysis for the Asymmetric Network Equilibrium Problem. Mathematical Programming 28, pp 174-184, 1984.

[8] D. Howard. Energy, Transport and the Environment. Transnet, 1990.

[9] Kent J.H. and N.R. Mudford. Motor Vehicle Emissions and Fuel Consuption Modelling. Transpn. Res. 13A, pp 395-406, 1979.

[10] Kono H. and S. Ito. A Microscale Dispersion Model for Motor Vehicle Exhaust Gas in Urban Areas -OMG Volume Source Model. Atmospheric Environment 24B, pp. 243-251, 1990a.

[11] Kono H. and S. Ito. A Comparison of Concentration Estimates by OMG Volume Source Dispersion Model with Three Line Source Dispersion Models. Atmospheric Environment 24B, pp. 253-260, 1990b.

[12] Luhar K. and R. Patil. A General Finite Line Source Model for Vehicular Pollution Prediction. Atmospheric Environment 23, pp. 555-562, 1988.

[13] W.B. Petersen. User's Guide for HIGHWAY-2, a Highway Air Pollution Model. EPA-600/8-80-018, Office of Research and Development U.S. Environmental Protection Agency, Research Triangle Park, North Carolina, 1980.

[14] F.S.M. Rogers. A Revised Calculation of Gaseous Emission from UK Motor Vehicles. Warren Spring, LR 508 (AP) M, 1984.

[15] Y. Sheffi. Urban Transportation Networks. Prentice-Hall, Englewood Cliff, NJ, 1985.

[16] J.G. Wardrop. Some Theoretical Aspects of Road Traffic Research. Proc. Instn. Civ. Engrs. Pt II, 1, pp 325-378, 1952.

Dynamic Network Equilibrium

TRAFFIC MODELS FOR DYNAMIC ASSIGNMENT

J. D. Addison

Transport Studies Group, University College Lòndon

B. G. Heydecker

Transport Studies Group, University College London

1. Introduction

The purpose of this paper is to illustrate the effects on estimated traffic flow of different traffic models in solving the dynamic traffic assignment problem. We will compare two distinct traffic models, the kinematic wave model and the flow model. We will show that the model's resulting estimates differ substantially. The kinematic wave model is regarded as the more realistic of the two but it is difficult to work with, requiring large computational effort, while the flow model is less realistic but is computationally more tractable. The flow model has been adopted widely in the development of existing dynamic traffic assignment techniques (Wie, Friesz and Tobin, 1990; Ho, 1990, Vythoulkas, 1990a; 1990b; Carey, 1986; 1987; 1988). According to this model, the outflow from each link is a function of the mean density on the whole link.

The kinematic wave model (Lighthill and Whitham,1955) has seen much less use although Newell(1988) has use it in studying the morning commute and Makigami et al(1981) have used it to study the spread of congestion. In this paper we will consider in detail the case of a known time varying demand. Thus we will allow route choice but not departure time choice. The parameters we use for traffic description are the flow and the density. These are assumed to take real values and are piecewise continuous functions of time and space.

2. Equilibrium

For origin-destination pairs a, b, c, ... let Q_a (s) etc. denote the demand at time s. For a fixed o-d pair a with routes i, j, k,... let T_i (s), T_j (s) denote the arrival times at the end of the routes for traffic leaving the origin at time s. The travel time is then given by T_i (s) $-s$.

For equilibrium we require:
for any route i in use ($q_i > 0$) at departure time s

$$T_i\ (s) = T_a^*\ (s)$$

and for any route j not in use ($q_i = 0$) at this time

$$T_j\ (s) \geq T_a^*\ (s)$$

so that T_a^* (s) denotes the least cost of travel time for traffic between the o-d pair a and entering the network at time s. Let q_i denote the flow of traffic entering route i at time s, then we must have $\Sigma\, q_i = Q_a\ (s)$.

The dynamic equilibrium assignment problem is to find the functions T_a^* (s) and the associated assignments $q_{b,j}$. For given assignments, the travel times T_i depend on the assignment functions. i.e. They are of the form T_i $(s, q_{b,i}, \ldots)$ depending on the functions $q_{b,j}$ and not just on their values at time s.

Because there are only a finite number of routes the time interval of interest is divided into subintervals during each of which the number of routes in use is fixed. For such an interval, if i and j are routes in use, between the same o-d pair then we must have

$$T_i\ (s) = T_j\ (s).$$

Differentiating with respect to s gives

$$\frac{dT_i}{ds}(s) = \frac{dT_j}{ds}(s)$$

In intervals where a fixed number of routes is in use the solution of the dynamic assignment problem is equivalent to solving the system of ordinary differential equations. In most cases, however, writing down the required equations is a non-trivial exercise and depends on the traffic model that is adopted.

3. The Kinematic Wave Model

Here we will recall the elements of the kinematic wave model of traffic flow. (Lighthill and Whitham, 1955). The kinematic wave model provides a good description of the small scale phenomena in traffic on the link. What follows is not an exhaustive discussion of the model but concentrates on those results that are of immediate relevance to this paper. The density and flow are taken to be functions of x, the distance along the link and t, the time. At any point on a link the density k and the flow q are related by $q = vk$, where v is the space mean speed, and satisfy the conservation equation

$$\frac{\partial k}{\partial t} + \frac{\partial q}{\partial x} = 0$$

It is further assumed that the speed of a vehicle at any point on the link is a function of the density alone, given by $v = v(k)$. Differentiating the relation $q = vk$ and using the conservation equation we obtain the partial differential equation

$$\frac{\partial k}{\partial t} + \omega \frac{\partial k}{\partial x} = 0$$

for k, where $\omega = v'(k)\ k + v(k)$. This partial differential equation describes the evolution of the density. The characteristics of this equation are the lines of constant density. They describe the propagation of a wave of constant density along the link.

The speed of propagation of this wave is given by ω. The general solution of the equation is given by $\phi(x - \omega t,\ k) = 0$.

Usually we are interested in solutions that satisfy a given boundary condition such as the known density at the entrance to the link. Finding the appropriate function ϕ is not always easy. However we can describe a patch of such a solution surface by

$$\Sigma(t,s) = (t,\ \omega(g(s))(t-s),\ g(s))$$

where the density at the entrance is given by $k = g(t)$.

In what follows we shall assume the Greenshields relation $v(k)=V-\alpha k$ between v and k,where $0 \le k \le V/\alpha$ and V is the free flow speed. For the Greenshields relation k and q are related by the quadratic equation

$$\alpha k^2 - V + \overline{q} = 0 \qquad (3.1)$$

The wave velocity is given by $\omega = V - 2\alpha k$. If $k > V/2\alpha$ the density wave will travel upstream along the link. This corresponds to the link being congested.

3.1 Some solutions

In this section we will describe the solution for k in two cases; when there is a discontinuous increase in demand at the start of the link and when the demand flow is linear in time.

For the first case we assume that at $x = 0$ and $t = 0$ there is an increase in q from q_o to q_1 with k_o and k_1 the corresponding densities. Then

$$k(x,t) = \begin{cases} k_o & x \ge \omega_o t \\ \dfrac{V}{2\alpha} - \dfrac{x}{2\alpha t} & \omega_o t \ge x \ge \omega_1 t \\ k_1 & x \le \omega_1 t \end{cases} \qquad (3.2)$$

where ω_0 and ω_1 are the wave speeds associated with k_0 and k_1 respectively. This is the surface swept out by a line rising on the k-axis from k_0 to k_1 and rotating so that the slope of its projection decreases from $\omega(k_0)$ to $\omega(k_1)$.

We have a parametric representation $\Sigma(t,s) = (t, \omega(s)t, s)$. In the time-distance diagram in the (t,x)-plane there is a fan of lines of constant k radiating from the origin. Figure 1 shows a fan of such lines followed by lines from an increasing linear demand.

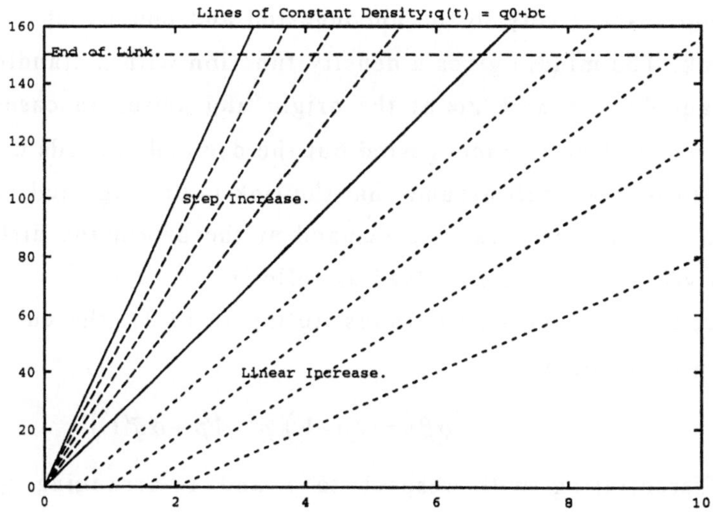

Figure 1. - Waves of constant density

For the second case the demand is given by $\overline{q}(t) = q_o + \beta t$. Then at $x=0$ the density k satisfies $\alpha k^2 - Vk + \overline{q}(t) = 0$. There are two solutions of k, corresponding to uncongested and congested conditions on the link. The lines of constant k are given by

$$x = (V - 2\alpha k)(t - \tau)$$

$$\alpha k^2 - Vk + \overline{q}(\tau) = 0 \qquad (3.2)$$

Here τ is the time at which the wave of density k leaves the origin. Solving for τ and substituting gives

$$\beta x = (V - 2\alpha k)\left(\alpha k^2 - Vk + \bar{q}(t)\right) \quad (3.3)$$

or

$$2\alpha^2 k^3 - 3\alpha V k^2 + \left(V^2 + 2\alpha\bar{q}(t)\right)k - V\bar{q}(t) + \beta x = 0 \quad (3.4)$$

The surface patch we are interested in lies on this cubic surface. This equation has three positive solutions for k, at least for the values that arise in modeling traffic. The smallest and largest correspond respectively to the uncongested and congested conditions on the link. The middle gives a density function with a standing wave of constant density $k=V/2\alpha$ at the origin and arises in cases where the flow on the link is uncongested but the demand exceeds capacity. In what follows we will assume that the link is uncongested, so that $k \leq V/2\alpha$. For linearly varying demand at the origin the differential equation $v(k)=V-\alpha k$ can be solved as follows.
Let $p=\dot{x}$, then $k = (V - p)/\alpha$. Substituting in (3.3), the cubic for k, gives x in terms of p.

$$\alpha\beta x = (2p - V)\left(p^2 - Vp + \alpha\bar{q}(t)\right) \quad (3.5)$$

Differentiating with respect to t and re-arranging gives an equation of the form $\dot{p}=F(p,t)$. Inverting this gives a linear differential equation for t in terms of p. Solving this we arrive at

$$2\alpha\beta t = -3p^2 + 2Vp - 2\alpha q_o + \frac{C}{(V-p)^2} \quad (3.6)$$

Combining (3.5) and (3.6) gives a pair of parametric equations for the trajectory in terms of the speed:

$$\alpha\beta x = (2p - V)\left(p^2 - Vp + \alpha\bar{q}(t)\right)$$

$$\alpha\beta t = -3p^2 + 2Vp - 2\alpha q_o + \frac{C}{(V-p)^2}$$

We can find C, the constant of integration, in terms of the initial speed. For a trajectory leaving the origin at time s, p has the value $p_o = V - \alpha k$ where $\alpha k^2 - V + \overline{q}(s) = 0 = 0$. We observe that p_o satisfies the equation $p_o^2 - V + \alpha\overline{q}(t)$. Substituting in the equation for t shows that $C = (p_o(V - p_o))^2$.

3.2 Shock Waves

The kinematic wave model gives rise to shock waves. These are points on the link where there is a discontinuous increase in density in the direction of travel. The presence of shock waves is the major source of difficulty when working with the kinematic wave model.

Shock waves can arise on an uncongested link when the demand is continuous but diminishing. We consider the case when the demand falls linearly after time $t = 0$ i.e. $\overline{q}(t) = q_o + \beta t, \quad t \geq 0, \beta < 0$. For this falling linear demand a shock wave will arise on the link, if it is sufficiently long. The shock wave is formed at the point where the line of density k_o, associated with the demand $\overline{q}(0)$, meets the envelope of the lines $\beta x = (V - 2\alpha k)(\alpha k^2 - Vk - \overline{q}(t))$. The time and location of the formation are given by

$$2\alpha\beta x = -(V - 2\alpha k_o)^3$$
$$2\alpha\beta t = -(V - 2\alpha k_o)^2$$

After this time the shock wave propagates with the velocity

$$u = V - \alpha(k_l + k_r)$$

where k_l and k_r are the densities on either side of the the shock wave. Because the traffic on the link is uncongested these shock waves always travel forward along the link but at a slower speed than the traffic in the regions on either side of the shock wave. If one of the densities is constant then it is possible to find a parametric representation of the trajectory of the shock wave.

4. The Flow Model

The flow model assumes that conditions along the link are uniform. The equation for the density is given by

$$l\frac{dk}{dt} = a(t) - b(t) \tag{4.1}$$

where $a(t)$ is the flow into the link and $b(t)$ is the flow out of the link. In what follows we will assume that there is no bottleneck at the end of the link and so $b(t) = Vk - \alpha k^2$. If the flow into the link is constant so that $a(t) = q_o,\ t \geq 0$, then the differential equation can be solved. For an initially empty link ($k(0) = 0$)

$$k(t) = \frac{2q_o \tanh\left(\frac{\sqrt{d}t}{2l}\right)}{\sqrt{d} + V\tanh\left(\frac{\sqrt{d}t}{2l}\right)} \tag{4.2}$$

and for a link with $k(0) = k_o$ and $q_o = 0$

$$k(t) = \frac{k_o V \exp(-Vt/l)}{V + \alpha k_o(\exp(-Vt/l) - 1)} \tag{4.3}$$

5. A Comparison of the Two Models

Before applying these two models to the dynamic assignment problem it is instructive to compare the results from the two models when applied to a single link. We will examine the density at the downstream end of the link for a brief pulse of traffic, with the inflow $\bar{q}(s)$ given by

$$\bar{q}(s) = \begin{cases} 0 & t < 0 \\ q_o & 0 \leq s \leq t_o \\ 0 & t > t_o \end{cases}$$

The link we will use has the following characteristics: a free flow speed, V, of 16 m/s (58 kph), a jam density, V/α, of 1/7 vehicles per metre (142.8 vehicles per kilometre) and a length of 300 m.

Figures 2 and 3 show the density at the downstream end of the link and travel times respectively. The travel time for the flow model is calculated using the density on the link at the time of entry.

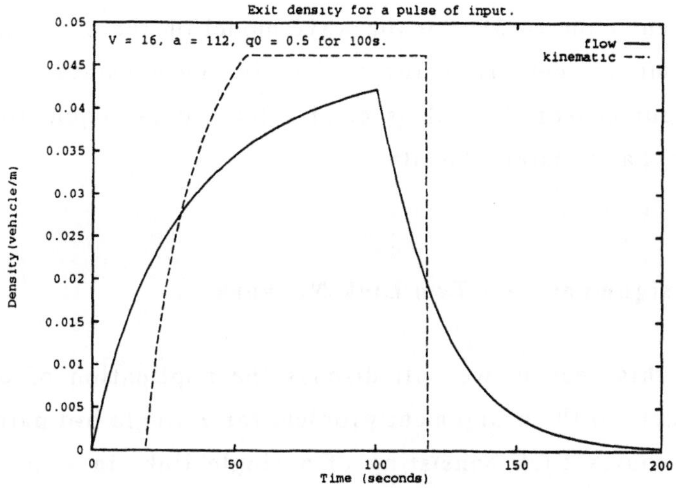

Figure 2. - Downstream Density

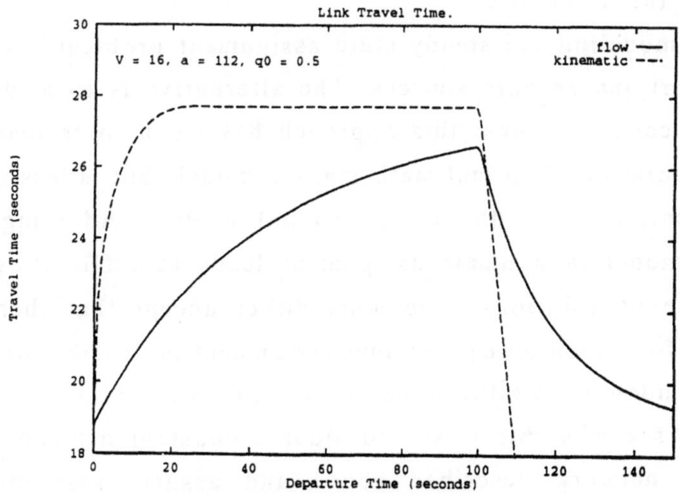

Figure 3. - Travel Time

From these figures we see that the results are qualitatively different. The flow model has traffic appearing at the end of the link immediately traffic enters the link while for the wave model output does not appear until after a delay equal to the free flow travel time along the link. For the flow model the density and the travel time rise slowly and never reaches a constant value. After traffic stops entering the link the decay is very slow and the link never clears, as is apparent from (4.3). For the wave model on the other hand the the travel time and density increases quickly to a constant value and after input ceases the link is clear after a delay equal to the travel time for that constant density.

6. Assignment on a Two Link Network

In this section we will discuss the application of each of the two models to the assignment problem for a single o-d pair connected by two routes each consisting of a single link. In what follows we will assume that link 1 has the smaller free flow travel time. We will denote by $\mu(s)$ the proportion of traffic that is entering the network at time s that is assigned to link 1. The arrival times $T_i(s)$, $i = 1,2$, are then functions of μ.

In modeling the steady state assignment problem it is usual to regard origins as pure sources. The alternative is to model sources as high capacity links; this approach has much to recommend it in the formulation of formal mathematical models but is less used in the implementation of the computational models. Adopting the pure source model in dynamic assignment leads to difficulties with the existence of solutions. One must either accept that there may be periods for which no equilibrium assignment is possible or the model must be adapted to allow queuing to occur at the origins.

To see why this is so, consider a constant demand q_0 arising for the network described above and assume that initially the network is empty. Suppose further for c_1 and c_2, the capacities of the two links, that $c_1 < q_0 < c_1 + c_2$. The demand at the origin is less than the steady state capacity of the network and there is no difficulty doing steady state assignment. However to achieve dynamic equilibrium initially we can assign traffic only to the faster link, link 1, but then we have a demand $q_0 - c_1$ that is unsatisfied and assigning this balance to link 2 will result in a non-equilibrium assignment. To take account of this we can introduce a queue at the origin. Traffic will wait in the queue until the travel time on link 1 has risen to equal the free flow travel time on link 2 at which time some traffic will use link 2. We note that this example shows that the dynamic capacity of a network may be smaller than its steady state capacity.

An examination of Figure 3 above shows that the predicted time at which the second route will come into use can different greatly between for the two traffic models. The kinematic wave model predicts that the second link will come into use very quickly while the flow model predicts a considerable delay.

6.1 The Differential Equations

In Section 2 we showed that when a fixed number of routes are in use the solution to the dynamic assignment problem is given by the solution of a system of differential equations. We will derive the relevant equations for our example network assuming a constant demand q_0 from time $t = 0$.

6.1.1. The Kinematic Wave model

First of all we give the general form of $\frac{dT}{ds}$ for an inflow $\bar{q}(t)$.

Let $x(t, s)$ be the trajectory with starting time s, i.e. $x(s, s) = 0$. The time at which the end of the link is reached is given by

$$l = x(T(s), s) \qquad (6.1)$$

From this, using standard results for the derivative of the

solution of a differential equation with respect to its boundary conditions that (See the appendix for the derivation.), we find

$$\frac{dT}{ds}(s)=\frac{V-\alpha k(s,0)}{V-\alpha k(T(s),l)}\exp\int_s^{T(s)}\frac{\alpha\bar{q}'(r)dt}{2\alpha(t-r)\bar{q}'(r)+\left(V-2\alpha k(t,x(t,s))\right)^2} \tag{6.2}$$

where

$x(t,\ s)\ =\ (V\ -\ 2\alpha k\ )(t-r)$, $\alpha k^2-Vk+\bar{q}(r)=0$ define $k(t,\ x)$ and r.

On link 1 we have $\bar{q}_1(s)=\mu(s)q_o$ and on link 2 $\bar{q}_2(s)=(1-\mu(s))q_o$. Substituting for $\bar{q}_i',i=1,2,$ in the expression above we obtain $\frac{dT_1}{ds}$ and $\frac{dT_2}{ds}$. Equating them gives the required differential equation.

$$\frac{V_1-\alpha_1 k_1(s,0)}{V_1-\alpha_1 k_1(T_1(s),l_1)}\exp\int_s^{T_1(s)}\frac{\alpha_1\mu'(r_1)q_o dt}{2(t-r_1)\alpha_1\mu'(r_1)q_o+\left(V_1-2\alpha_1 k_1(t,x_1(t;s))\right)^2}=$$

$$\frac{V_2-\alpha_2 k_2(s,0)}{V_2-\alpha_2 k_2(T_2(s),l_2)}\exp\int_s^{T_2(s)}\frac{\alpha_2\mu'(r_2)q_o dt}{2(t-r_2)\alpha_2\mu'(r_2)q_o+\left(V_2-2\alpha_2 k_2(t,x_2(t;s))\right)^2}$$

where

$$x_i(t,s)=(V_i-2\alpha k_i)(t-r_i) \tag{6.3}$$

$$\alpha_i k_i^2-V_i k_i+\mu(r_i)q_o=0 \tag{6.4}$$

define $k_i(t,\ x)$ and r_i, $i\ =\ 1,2$.

While this equation is of theoretical interest, it does not seem to offer a practical approach to solving the assignment problem in this case. It may be possible to obtain a more tractable equation by differentiating twice and using the resulting equations to eliminate the exponential terms. This would give a third order equation for μ.

6.1.2. The Flow Model

The differential equation for k using the flow model with no bottleneck at the downstream end is

$$l\frac{dk}{dt} = q(t) - kv(k) \tag{6.5}$$

where $v(k)$ is the speed when the density on the link is k. The travel time is given by $tt(s)=l/v(k(s))$. It follows that

$$\frac{dtt}{ds} = \frac{\alpha\left(q(s) - kv(k)\right)}{v(k)^2} \tag{6.6}$$

For our network we must solve the differential equation

$$\frac{dtt_1}{ds} = \frac{dtt_2}{ds}$$

which is

$$\frac{\alpha_1\left(q_1(s) - k_1 v_1(k_1)\right)}{v_1^2(k_1)} = \frac{\alpha_2\left(q_2(s) - k_2 v_2(k_2)\right)}{v_2^2(k_2)}$$

We are considering constant demand q_o after time $t = 0$, so $q_1=\mu(s)q_o$ and $q_2=(1-\mu(s))q_o$. Substituting and solving for μ we get

$$\mu q_o = \frac{B(k_1,k_2) + C(k_1)q_o}{A(k_1,k_2)} \tag{6.7}$$

where

$$\begin{aligned} A(k_1,k_2) &= \alpha_1 v_2^2 + \alpha_2 v_1^2 \\ B(k_1,k_2) &= v_1 v_2\left(\alpha_1 v_2 k_1 - \alpha_2 v_1 k_2\right) \\ C(k_1) &= \alpha_2 v_1^2 \end{aligned}$$

This expression for μ as a function of k_1 and k_2 means that in this case we have a pair of coupled differential equations

$$l_1 \frac{dk_1}{ds} = \mu(s) q_o - k_1 v_1$$
$$l_2 \frac{dk_2}{ds} = (1-\mu(s)) q_o - k_2 v_2$$

This coupled system is easily integrated numerically using standard techniques. In order to perform numerical integration we need to know the value of μ at the start of the interval. Let s_o be the splitting time i.e. the time at which the travel time on link 1 becomes equal to the free flow travel time on link 2. We need to know $\mu(s_o^+)$, thelimit of $\mu(s)$ as s approaches s_o from above. Because of k_i must be continuous $k_2\ (s0)\ =\ 0$. Letting $k_{1,0}$ be the density on link 1 at this time. Substituting in the expression for μ we find

$$\mu(s_o^+) = \frac{q_o}{1+b}\left(1 + b\frac{k_{1,0}\left(V_1 - \alpha_1 k_{1,0}\right)}{q_o}\right)$$

where

$$b = \frac{\alpha_1 V_2^2}{\alpha_2 \left(V_1 - \alpha_1 k_{1,0}\right)^2}$$

The flow $k_{1,0}\ (V_1\ -\ \alpha_1\ k_{1,0}\ )$ is the flow out of the link at $s0$ and is less than q_o. From this we see that $\mu(s_o^+)$ is less than 1. Before the splitting time s_o, μ had the constant value 1. It follows that at s_o, the time when the second route comes into use, the flows into links 1 and 2 are discontinuous.

6.2 The Continuity of Kinematic Wave Assignments

Having observed the discontinuity in the assigned flows when using the flow model it is natural to ask if corresponding discontinuities occur for the kinematic wave model. We shall show below that at the instant of a discontinuous increase in flow into a link, $\frac{dT}{ds}$ is infinite. It follows that a discontinuity in assignment at splitting would result in a sharp increase in the travel time on link 2, to match this on link 1 would require a corresponding increase in assigned flow. Clearly this is not possible so the assignment must be continuous in time.

For simplicity in demonstrating the behaviour of $\frac{dT}{ds}$ we will restrict ourselves to the case of constant flow q_1 into an initially empty link. Let k_1 denote the associated density. We know from (3.1) that when a discontinuous increase in flow occurs a fan of density waves arises. The trajectories of traffic in this region are given by

$$x^2 - (2Vx + c^2)t + V^2 t^2 = 0 \tag{6.8}$$

If a trajectory enters this region at (t_o, x_o) then

$$c^2 = \frac{(x_o - Vt_o)^2}{t_o}$$

But $x_o = (V - 2\alpha k_1)t_o$ giving $c^2 = 4\alpha^2 k_1{}^2 t_o$. The time T at which the trajectory leaves the link is given by

$$l^2 - (2Vl + c^2)T + V^2 T^2 = 0.$$

We differentiate this with respect to s to get

$$\left(2V^2T - 2Vl - c^2\right)\frac{dT}{ds} = 2cT\frac{dc}{ds} \tag{6.9}$$

The expression for c^2 shows that

$$2c\frac{dc}{ds} = 4\alpha^2 k_1^2 \frac{dt_o}{ds}$$

Under our assumption of constant input $t_o = \dfrac{(V - \alpha k_1)}{\alpha k_1}s.$

Letting s go to 0 we see that the right hand side of (6.9) is non zero while the coefficient of $dTds$ becomes zero since $T(0) = l / V$ and c goes to 0 as s goes to 0. This argument is easily adapted to deal with the general case.

7. Conclusion

In the course of this paper we have discussed a number of phenomena that arise in the study of the dynamic equilibrium assignment problem. We have shown that careful modelling of origins is needed and that the dynamic equilibrium capacity may be less than the steady state capacity of a network. By contrasting two distinct traffic models we have shown that they can make substantially different predictions. The flow model does not behave realistically for short term transient phenomena and it will under estimate the rate at which congestion spreads through the network.

On the other hand, the flow model is relatively easy to use and may still be of use in examining slowly varying phenomena on congested networks. We have seen that the more realistic kinematic wave model can be difficult to work with. This is particularly true of the trajectory based approach used in this paper. The existence of shock waves makes the development of suitable software difficult. Recent work by Newell(1991) holds out the prospect of using accumulated traffic flows at the end of the links. His results for the partial differential equation for the accumulated flow suggest the possibility of developing computationally more tractable formulations using the kinematic wave model.

Acknowledgements

This work was supported by a grant from the Science and Engineering Research Council. We would also like to thank Professor Richard Allsop for his interest and encouragement.

References

[1] CAREY, M. (1986) *"A constraint qualification for a dynamic traffic assignment model"*, Transportation Science, 20, 55-8.

[2] CAREY, M. (1987) *"Optimal time-varying flows on congested networks"*, Operations Research, 35, 58-69.

[3] CAREY, M. (1988) *"Some problems in modelling dynamic traffic assignment"*, Paper presented to the ORSA/TIMS Joint Meeting, Washington DC.

[4] HO, J.K. (1990) *"Solving the dynamic traffic assignment problem on a hypercube multicomputer"*, Transportation Research, 24B, 443-51.

[5] HUREWICZ, W. (1958) *"Lectures on Ordinary Differential Equations"*, The M.I.T. Press.

[6] MAKIGAMI, Y., NAKANISHI, T., TOYAMA, M. and MIZOTE, R. (1981) *"On a Simulation for the Traffic Stream on the Freeway Merging Area"*, Eighth International Symposium on Transportation and Traffic Theory, 163-173.

[7] NEWELL, G.F. (1988) *"Traffic Flow for the morning commute"*, Transportation Science, 22, 47-58.

[8] NEWELL, G.F. (1991) *"A simplified Theory of Kinematic Waves. I General Theory"*, Research Report UCB-ITS-RR-91-12. Institute of Transportation Studies, University of California at Berkley.

[9] VYTHOULKAS, P.C. (1990a) *"A dynamic stochastic assignment model for the analysis of general networks"*, Transportation Research, 24B, 453-69.

[10] VYTHOULKAS, P.C. (1990b) *"Two models for predicting dynamic stochastic equilibria in urban transportation networks"*, In: Transportation and Traffic Theory (ed M. Koshi). New York: Elsevier, 253-72.

[11] WIE, B.-K., FRIESZ, T.L. and TOBIN, R.L. (1990) *"Dynamic user optimal traffic assignment on congested multidestination networks"*, Transportation Research, 24B, 431-42.

A. Appendix

Let $x(t,s)$ be a parametrized family of solutions of the differential equation $\dot{x}=f(t,x)$ such that $x(s,s) = 0$. Let $\bar{x}(t)$ denote the solution for $s=\bar{s}$. Then the partial derivative of x with respect to s at $(t,\bar{s})$ is

$$\frac{\partial x}{\partial s}=-f(\bar{s},0)\exp\int_{\bar{s}}^{t}\frac{\partial f}{\partial x}(t,\bar{x}(t))dt \tag{A.1}$$

This is a special case of a more general result. Hurewicz(1958) proves a different special case but this is easily adapted. We want to find $\frac{dT}{ds}$ where $T(s)$ is determined by

$$l = x(T(s),\ s) \tag{A.2}$$

and $x(t,s)$ is the solution of $\dot{x}=F(k,(t,x))$. Differentiating equation (A.2) with respect to s gives

$$\begin{aligned}0&=\frac{\partial x}{\partial t}(T(s),s)\frac{dT}{ds}+\frac{\partial x}{\partial s}(T(s),s)\\&=\dot{x}(T(s),s)\frac{dT}{ds}+\frac{\partial x}{\partial s}(T(s),s)\end{aligned} \tag{A.3}$$

$$=F(k(T(s),l))\frac{dT}{ds}+\frac{\partial x}{\partial s}(T(s),s) \tag{A.4}$$

Solving for $\frac{dT}{ds}$ we obtain

$$\begin{aligned}\frac{dT}{ds}(s)&=-\frac{1}{F(k(T(s),l))}\frac{\partial x}{\partial s}(T(s),s)\\&=\frac{F(s,0)}{F(k(T(s),l))}\exp\int_{s}^{T}(s)\frac{\partial F}{\partial x}(t,x(t;s))dt\end{aligned} \tag{A.5}$$

using (A.1).

In our case $F(k(t,x)) = V - \alpha k(t,x)$, so $\frac{\partial F}{\partial x} = -\alpha \frac{\partial k}{\partial x}$. The two equations

$$x = (V - 2\alpha k)(t - r), \quad \alpha k^2 - Vk + \bar{q}(r) = 0$$

give k as a function of t and x, from which we find

$$\frac{\partial k}{\partial x} = \frac{-\mu'(r) q_o}{2(t-r)\alpha\mu'(r) q_o + (V - 2\alpha k(t,x))^2}$$

Substituting in (A.5) we arrive at (6.2).

A DYNAMIC USER EQUILIBRIUM ASSIGNMENT MODEL - A RIGOROUS FORMULATION

Helena B. B. Cybis

Institute for Transport Studies, The University of Leeds Leeds, U.K.

This paper proposes a time dependent assignment model based on a rigorous formulation of dynamic equilibrium that aims to equilibrate real travel times. The paper also describes the algorithms devised to solve this problem and their tests on simplified networks

1. Introduction

Dynamic assignment models have been appointed as a necessary tool to cope with increasing congested roads and to adequately represent the evolution of peak period traffic along the network.

This problem has been tackled basically by two approaches:

(1) variable demand models - that are especially concerned with drivers reaction to congestion modelling demand variability over time.

(2) predefined demand models - they are dynamic versions of classic network transportation models concerned with route the choice process and estimation of travel times.

In this paper we propose an assignment model for a previously determined time varying demand based on a rigorous formulation of dynamic user equilibrium. In contrast to the concept of instantaneous equilibrium it aims to equilibrate the actual times vehicles face when travelling between any origin-destination pair. Equilibrium is defined as the state when no driver departing at any time interval, travelling between any origin-destination pair can change to a route that presents cheaper average cost for departure within that interval.

The model's mathematic formulation is presented in section 3. The algorithm devised to solve this problem is based on an iterative procedure that, similarly to Frank-Wolfe, combines current flows with the current all-or-nothing minimum cost flows. Its basic version is described in section 4. This algorithm has been tested on a simplified network. Section 5 describes the tests, and the variations that have been introduced to the algorithm's basic structure to overcome convergence problems.

2. Previous studies on dynamic traffic assignment

A traffic assignment model can be viewed as the interaction between two aspects:

(1) demand - characterized by the principles that are assumed to represent user's choice of route.

(2) supply - represented by the network reaction to these choices, the network performance.

The supply characterization of dynamic assignment models has been represented in the most variable ways. Some models assume simple link impedance functions like Janson (1989). The most used formulation, however, assumes that travel time is the sum of a constant running time and queuing delay. More detailed formulations include concepts like maximum link capacity (Driss-Kaitouni and Hameda-Benchekro, 1989) and link blocking back effect (Leonard et al.,1989).

In a general way, however the model's supply characterization, defined by the trip loading procedure and travel cost estimation, is highly limited by the model's structure and demand assumptions.

One interesting feature of most dynamic models is the fact that, although quite different in essence, their demand characterization is based on dynamic generalizations of classic steady-state equilibrium principles. In the literature we can find models like Merchand and Nemhauser (1978), Carey (1986), Wie (1989) that pursue dynamic system equilibrium. The majority of the works, however, are based on dynamic user equilibrium principles. In the effort of developing time varying generalizations for equilibrium, authors usually introduce some simplifications that deserve further analysis.

Independently of their approach (1) simulation or (2) optimization, the majority of the proposed dynamic assignment models present incremental structures, meaning that the route choice process and network loading procedures progress in successive time increments.

One incremental formulation that has been adopted in several studies (e.g. Wie et al., 1990) is the optimal control theory. It leads to the concept of instantaneous equilibrium where:

"At each instant in time, for each origin-destination pair, the instantaneous expected travel cost for all paths that are being used are identical and equal to the minimum instantaneous path travel cost."

This formulation has been indicated as appropriate to the complex task of guiding vehicles along networks. However, from an intuitive point of view, it would be more reasonable to assume that drivers are looking for the minimization of their total cost of travel, and the sum of instantaneous minimum costs is not necessarily a minimum cost route for the whole trip.

One important problem of pure incremental models is the fact that they are not able to reevaluate past decisions or indicate routes based on the congestion vehicles would face ahead in the network. It seems that whenever possible, for instance for strategic modelling purposes, iterative procedures constitute a more adequate structure to represent the dynamic integration of traffic in networks.

Another problem that derives from the dynamic generalization of static equilibrium is the flow continuity problem. It arises when authors define equilibrium as a link cost minimization problem in a dynamic environment. This problem has been solved either by structuring the model in terms of time increments, as discussed above, or by introducing limitations in the time period length. Time period length limitations, however, restrict demand characterization.

This paper presents a dynamic user equilibrium model that aims to equilibrate the average total travel time of all origin-destination pairs for departures at each time period. The proposed algorithms follow an iterative procedure and are formulated in terms of path flows.

3. Problem formulation

3.1 Notation

Time:

t - time interval

τ - continuous time

$\Delta\tau$ - length of time interval

0-T - total modelled period

```
                Δτ
|--------|--------|--------|----->τ
0   t=1      t=2      t=3   T
```

Network:

ij - origin-destination pair

pij - path linking i to j

Demand:

Demand is defined in terms of number of vehicles /unit of time.

f^t_{ij} - flow rate entering the network during time interval t from origin i to destination j

$f_{ij}(\tau)$ - flow rate from i to j entering the network at time τ

Flow split:

$f_{pij}(\tau)$ - flow rate departing at time τ assigned to pij

f^t_{pij} - flow rate departing at time interval t assigned to path pij

g^t_{pij} - all-or-nothing flows assigned to path pij at time interval t

h^{t}_{pij} - modified flows resultant from the combination of f^{t}_{pij} and g^{t}_{pij}

Number of vehicles:

V^{t}_{pij} - number of vehicles departing at time interval t using pij

Link characteristics:

$f_a(\tau)$ - total flow entering link a at time τ

$Q_a(\tau)$ - queue size faced by vehicles entering link a at time τ

$D_a(\tau)$ - queuing delay for vehicles entering link a at time τ

S_a - saturation flow

Travel costs:

$C_a(\tau)$ - cost of travel along link a having entered the link at time τ

$C_{pij}(\tau)$ - cost of travel from i to j along pij for cars departing at τ

C^{t}_{pij} - average cost of travel along pij for car departing during time interval t

C^{t*}_{ij} - minimum average cost of travel from i to j for cars departing during time interval t

3.2 Model formulation

Using the above notation we can define the proposed model according to the following assumptions and relationships.

The demand is assumed to be constant during a time interval.

$$f^{t}_{ij} = f_{ij}(\tau) \qquad \text{for} \quad (t-1)\Delta\tau \leq \tau \lesssim t\Delta\tau$$

The proportion of flow assigned to each path is constant within each time interval.

$$f^{t}_{pij} = f_{pij}(\tau) \qquad \text{for} \quad (t-1)\Delta\tau \leq \tau \leq t\Delta\tau$$

The total number of vehicles departing during time interval t travelling along pij is therefore:

$$V^t_{pij} = f^t_{pij} \cdot \Delta\tau$$

The process of loading trips to the network respects a "first in - first out" principle allowing the estimation of $f_a(\tau)$ and hence $Q_a(\tau)$.

For the purposes of this model, cost is equal to travel time. The cost of travelling along link a, at any point in time (τ) is given by ($C_a(\tau)$), that represents the time taken to transverse link a, for a vehicle entering the link at time τ.

Link costs depend only on the vehicles transversing and queuing on that link:

$$C_a(\tau) = C_a(f_a(\tau), Q_a(\tau))$$

The cost of travel along pij having departed at τ_i is the sum of successive link travel times for all links along pij.

$$C_{pij}(\tau_i) = \sum_{a=i}^{j} (C_a(\tau_a))$$

The average cost of travel along pij for departures at time interval t is:

$$C^t_{pij} = \frac{\int_{(t-1)\Delta\tau}^{t\Delta\tau} C_{pij}(\tau)d\tau}{\Delta\tau} \qquad (1)$$

A feasible solution satisfies equilibrium when:

$$\forall\, f^t_{pij} > 0 \qquad C^t_{pij} = C^{t*}_{ij}$$
$$\forall\, f^t_{pij} = 0 \qquad C^t_{pij} \geq C^{t*}_{ij}$$

Equilibrium is achieved when no driver departing at any time interval, travelling between any O-D pair can change to a route that presents cheaper average cost for departure within that time interval.

In analogy to the static case, it is possible to define a parameter (δ) to measure the relative success in achieving equilibrium, where:

$$\delta = \frac{\sum_{t}\sum_{pij}\left[V_{pij}^{t}\left(C_{pij}^{t} - C_{ij}^{t*}\right)\right]}{\sum_{t}\sum_{ij} V_{ij}^{t} \times C_{ij}^{t*}} \qquad (2)$$

According to the equilibrium conditions, for all time intervals and paths, either V^{t}_{pij} or $(C^{t}_{pij} - C^{t*}_{ij})$ is equal to zero. Therefore following this formulation we can define dynamic user equilibrium as the state where δ is equal to zero.

It is not considered here the possibility of changing departure time to decrease travel time/costs.

4. Basic algorithm

The basic algorithm consists of a sequence of iterations where the current flows allocated to each route at every time interval are combined with current minimum cost route flows for departures at that interval. It can be described by the following steps:

1. Set iteration number n=1
2. Calculate all-or-nothing flows for unloaded network (g^{t}_{pij})
3. Set current flows $f^{t}_{pij}=g^{t}_{pij}$
4. Load all f^{t}_{pij} to the network to obtain $f_a(\tau)$ and $Q_a(\tau)$
5. Calculate average cost of travel for all used paths at each time interval (C^{t}_{pij})
6. Calculate convergence parameter δ
7. Terminate if flows have reached an acceptable level of convergence ($\delta \leq \delta_{min}$); otherwise:

8. Calculate new all-or-nothing flows following the current minimum cost routes for all O-D pairs at each time interval (g^t_{pij})
9. Calculate modified path flows (h^t_{pij}) using a combination of the current and all-or-nothing flows:

$$h^t_{pij} = (1-\lambda)\, f^t_{pij} + \lambda\, g^t_{pij} \qquad (3)$$

where the calculation of the step length λ and alternative equations are discussed in the next section.
10. Set current flows $f^t_{pij} = h^t_{pij}$
11. Increment iteration number ($n=n+1$)
12. Return to step 4

Note that due to the dynamic evolution of traffic in the network, equation 3 is only applied to path flows. We cannot build a link equivalent equation.

One interesting feature of this model's structure is the multipath assignment for different O-D pairs and time intervals. It allows a sensible representation of the route choice process under varying traffic conditions. The same O-D pair may present a different set of routes or flow split for each time interval, based on the real travel costs vehicles face when transversing each link of the network.

Up to this stage, the effort of developing this proposition has concentrated on testing and improving the demand side of the model, especially on the calculation of new flows (h^t_{pij})and the criterion for choice of λ, described in step 9.

Different variations of this basic algorithm have been developed and tested on a simplified network.

5. Tests and alternatives to the basic algorithm

The algorithms have been tested on a simplified network that consists of one origin destination pair and two alternative routes, a shorter route with smaller capacity and a longer one with larger saturation flow.(figure 1).

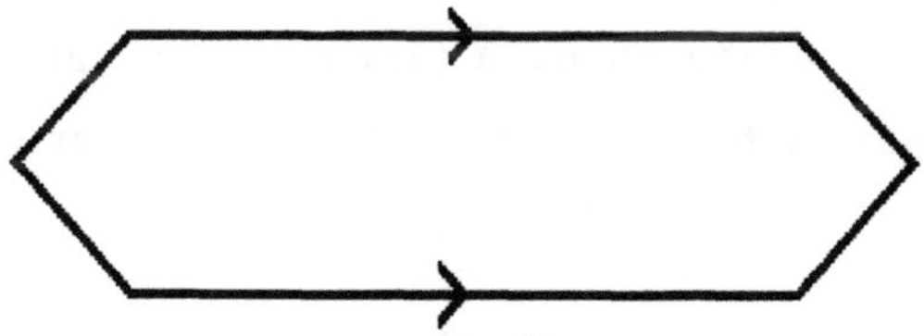

Figure 1. Network

For the purpose of these tests travel costs are assumed to be the sum of a fixed free flow travel time and a queuing delay.(table 1)

	Free Flow Travel Times	Saturation Flows (S_a)
Route 1	4 minutes	0.25 veh/sec
Route 2	5 minutes	0.35 veh/sec

Table 1. Network characteristics

Queues and delays are given by:

$$Q_a(\tau) = \max\{ [Q_a(\tau - d\tau) + (f_a(\tau) - S_a)d\tau)], 0 \}$$

$$D_a(\tau) = Q_a(\tau)/S_a$$

In order to illustrate the typical behaviour of the algorithms described bellow, figures 2 and 3 present two sets of results where two different demand patterns were tested on all alternative algorithms.

The demand patterns investigated here, consisted of three consecutive time periods.

The graphs show,for each iteration, the evolution of the convergence process, measured by δ (equation 2), and the resultant flow assigned to route 1 for each of the three time periods.

5.1 Algorithm 1

The first algorithm derived from the basic structure described above is based on the Method of Successive Averages, where λ is a predefined decreasing value dependent on the iteration number.

$$\lambda = 1/n$$

The objective of this formulation is to prove the viability of achieving equilibrium flows in this dynamic formulation through successive linear combinations of flows.

As expected, in analogy to the static case, convergence is achieved in the long term. As we can see from figures 2.1 and 3.1, even for simple situations, where we have three time intervals and two alternative routes, it took a large number of iterations to achieve a reasonable level of convergence.(16 iterations in figure 2.1 and 21 iterations in figure 3.1)

In large networks, extra iterations imply extra recurrence to time consuming dynamic loading and tree building procedures. It is therefore, highly desirable to improve the model by reducing the number of iterations needed to achieve convergence.

5.2 Algorithm 2

One natural way to improve the efficiency of the proposed basic algorithm is to search, at each iteration, for the λ value that would give a closer approximation to equilibrium. This leads to algorithms that resembles Frank-Wolfe's applications to static models.

However, in this formulation we cannot define an equivalent minimization problem based on the standard objective function Z. Due to the model's dynamic nature, flows in time interval (t) affect link costs in time interval (t+1), but the inverse relation is not true, consequently the Jacobian matrix of link travel times is highly asymmetric. This model's characteristic directed the search for a solution algorithm towards procedures currently suggested to solve static asymmetric assignment problems.

The algorithm presented in this section is based on Frank-Wolfe viewed as a variational inequality. Applications of variational inequalities for static problems have been studied by several authors like Dafermos(1980),Smith (1979) and Van Vliet (1987). In this formulation, the search for the optimum step length in each iteration consists in finding the λ value that results in cost equilibration between the current and all-or-nothing flows for the sum of all time periods. Some examples of this algorithm's behaviour is shown in figures 2.2 and 3.2.

$$\sum_{t}\sum_{pij}\left(C\left(h_{pij}^{t}\right)\times f_{pij}^{t}\right)=\sum_{t}\sum_{pij}\left(C\left(h_{pij}^{t}\right)\times g_{pij}^{t}\right)$$

Although the results from figure 2.2 may suggest that it might be possible to achieve equilibrium through this algorithm in few iterations, in many tests it presented an oscillating behaviour, giving no guarantees of leading to convergence (figure 3.2) This instability may be attributed to the fact that, in the process of equilibrating total costs, the algorithm takes no account of the chosen λ influence on individual time periods.

5.3 Algorithm 3

This algorithm applies a method that closely mirrors Frank-Wolfe to the dynamic assignment formulation. It is a descent algorithm that finds at each iteration the λ value that minimizes an objective function, here, the gap function (G) where:

$$G = \sum_{t}\sum_{pij}\left(f^{t}_{pij}\left(C^{t}_{pij} - C^{t*}_{pij}\right)\right)$$

In analogy to usual static applications, the descent direction choosen to minimize G is the vector of current minimum cost flows.The use of gap functions in static problems is discussed by authors like Hearn (1982).

Through the analysis of numerical tests(figures 2.3 and 3.3), we can conclude that low values of δ are achieved much faster than in algorithm 1. However in some tests, after a quick reduction the convergence process is stopped at a local minimum.(figure 3.3) Despite the intent of minimizing G the algorithm leads sometimes to situations where, although not in equilibrium, it is not possible to move flows towards the current all-or-nothing routes without increasing the objective function.

The mechanism that causes this problem is due to the fact that queues in time period (t) can have a very strong influence on delays in time period (t+1), hence an improved flow pattern in (t) may have a very negative effect in period (t+1).

5.4 Algorithm 4

The algorithm described in section 5.3 may be characterized by three elements:

.the objective functiόn to be minimized

.the choice of a descent direction

.the step length calculation

With the intent of persisting in the search for an efficient convergent algorithm, we propose in this section two basic modifications to the descent procedure described above. The ideas described in this section are derived from a paper presented by Smith (1983).

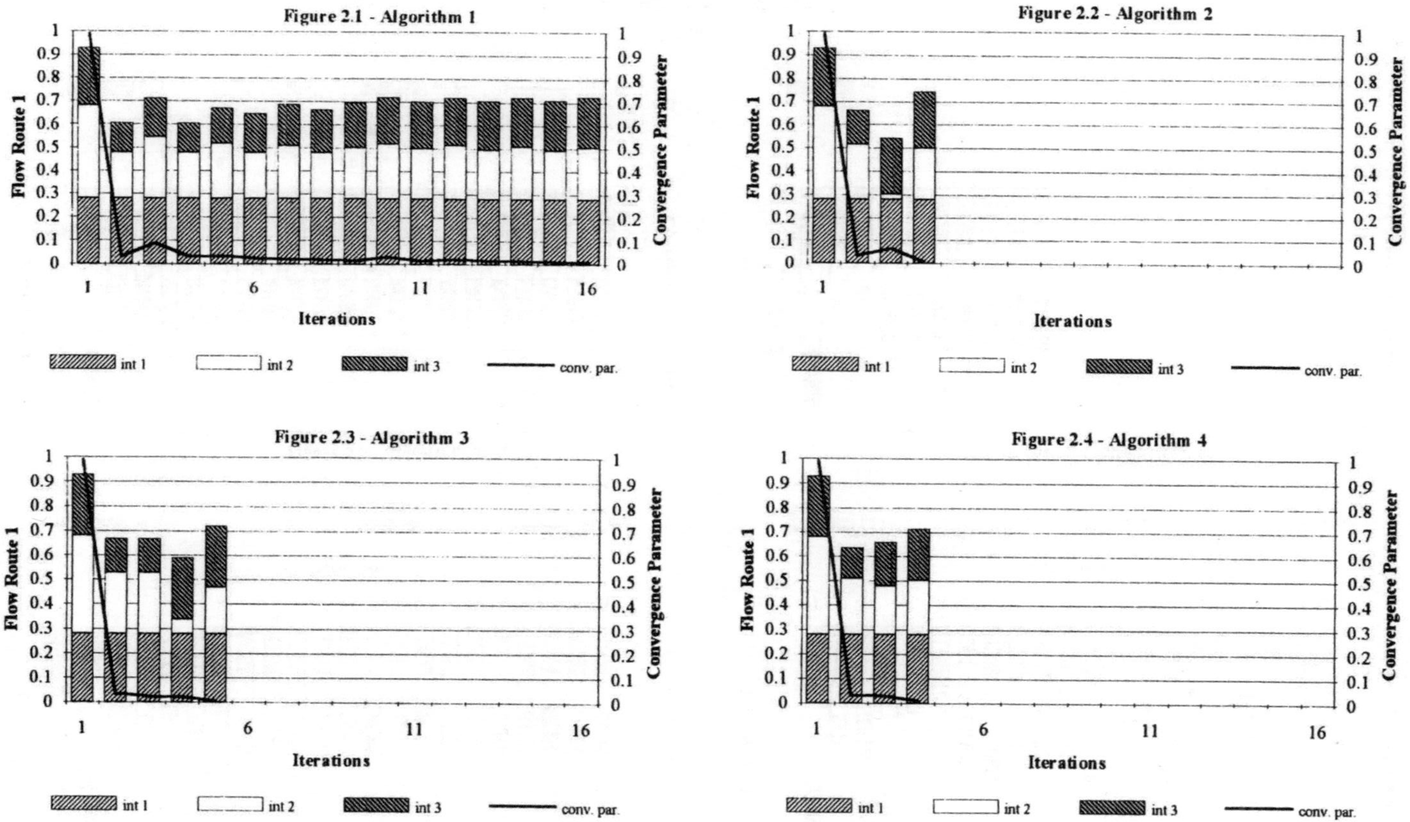

Figure 2
demand 0.28 0.40 0.25 (veh/sec)

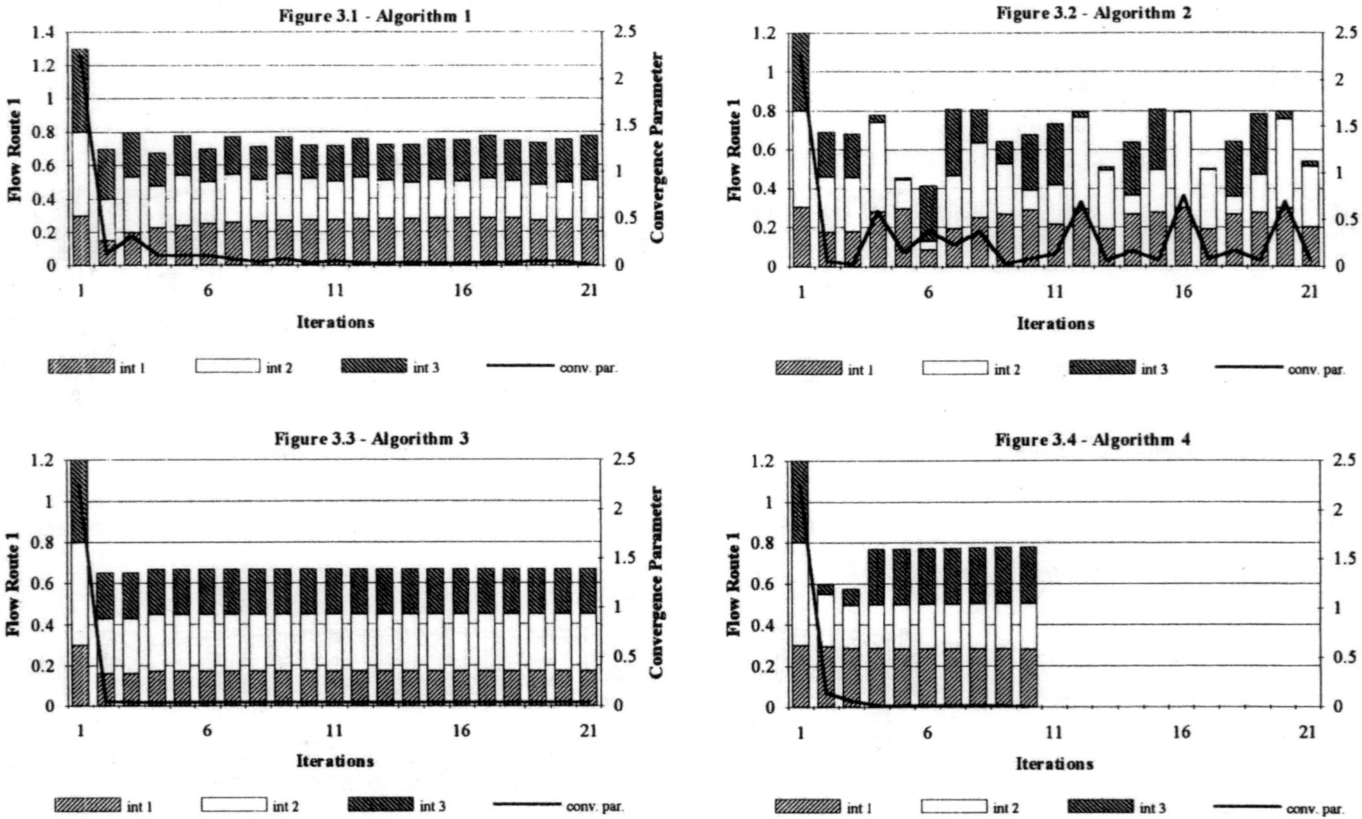

Figure 3
demand 0.30 0.50 0.40 (veh/sec)

(i) Modification of the descent direction - introduction of a "social pressure" term.

In this proposition, the rate at which trips are transferred to the cheapest route is not a constant factor of the route flow as in algorithms 1 to 3, but a variable value, proportional to a social pressure term.

The "social pressure" concept is intuitively based on the fact that drivers farther from equilibrium are more strongly inclined to shift route than those with travel times closer to the minimum cost. It can be viewed as a social measure of dissatisfaction for each route.

The pressure term for every path is assumed to be a function of route flows and the difference between current costs and the minimum, raised to some power m.

$$P^t_{pij} = (f^t_{pij} \cdot (C^t_{pij} - C^{t*}_{ij}))^m \qquad (4)$$

Due to the pressure term introduced,the modified flows equation (3), defined in step 9 of the basic algorithm, is substituted by:

$$h^t_{pij} = f^t_{pij} + \Sigma_{ij}(f^t_{pij}.\lambda.P^t_{pij}) \quad \text{for the minimum cost route} \quad (5)$$

$$h^t_{pij} = f^t_{pij}\,(1 - \lambda.P^t_{pij}) \quad \text{for all other routes} \quad (6)$$

It should be noticed, however, that λ must have units of inverse pressure, i.e. inverse (cost.flow). It is also no longer bounded in the range (0,1). In order to satisfy basic flow conditions for equations (5) and (6):

$$0 \le 1 - \lambda P^t_{pij} \le 1$$

therefore λ limits are set by:

$$0 \le \lambda \le 1/_{max}(P^t_{pij})$$

(ii) Modified objective function

The objective function G defined in section 5.3 has also been modified to:

$$G = \sum_{t}\sum_{pij}\left(f^{t}_{pij}\left(C^{t}_{pij} - C^{t*}_{pij}\right)\right)$$

for any integer $n > 1$.

There are also theoretical reasons that suggest a relationship between n and m (defined in (4)) where:

$$n = m + 1$$

The modification introduced to the function is meant to improve the convergence ability of the algorithm. It increases the contribution of elements with great cost differences and reduces the effect of flow shifts in routes closer to equilibrium, inducing a more homogeneous distribution of cost differences along different paths.

Figures 2.4 and 3.4 show some applications of this algorithm for $m=1$ and $n=2$. Different values of n (and m) have been tested but it was not found evidences that high exponent values actualy improve the algorithm's performance.

6. Conclusions and further work

The development of this research has concentrated on the investigation of a solution algorithm for the dynamic assignment model proposed in this paper.

Although it proved to be possible to achieve equilibrium through a heuristic approach derived from the Method of Successive Averages, the major research effort was directed towards the development of a more efficient method.

It seems that we have found,through the set of ideas described in section 5.4, an algorithm that allows, not only the solution of the model's present stage of development, but also to increase the modelling complexity.

The proposed algorithm has also been successfully tested for demand patterns defined by several time intervals and three links networks. Future work plans involve tests on multiple origin-destination networks.

In order to investigate this algorithm's behaviour in varying network conditions, it has also been applied to the static case, where it also seems to have interesting applications.

Acknowledgements

The author would like to thank the important suggestions provided by Dirck Van Vliet and Mike Smith to the development of this work.

References

[1] Carey,M. (1986) *"A constraint qualification for a dynamic traffic assignment model"* Transportation Research vol 20 no2

[2] Dafermos S. (1980) *"Traffic equilibrium and variational inequalities"* Transportation Science vol 14 no1

[3] Drissi-Kaitouni,O. and Hameda-Benchekroun,A.(1989) *"A dynamic traffic assignment model and a solution algorithm"* Centre de recherche sur les transports, Publication 649

[4] Hearn D. (1982) *"The gap function of a convex program"* Operation Research Letters, 1

[5] Jason,B. (1989) *"Assignment for urban road networks"* Transportation Research vol 25B no2

[6] Leonard,D.,Gower,P. and Taylor,N. (1989) *"CONTRAM: structure of the model"* TRRL RR 178

[7] Merchant,D. and Nemhauser,G. (1978) *"A model and algorithm for the dynamic traffic assignment problem"* Transportation Science vol 12 no3

[8] Smith,M. (1979) *"The existence, uniqueness and stability of traffic equilibria"* Transportation Research vol 13B

[9] Smith,M. (1983) *"The existence and calculation of traffic equilibria"* Transportation Research vol 17B no4

[10] Van Vliet,D. (1987) *"The Frank-Wolfe algorithm for equilibrium traffic assignment viewed as a variational inequality".* Transportation Research vol 21 no1

[11] Wie,B. (1989) *"Dynamic system optimal traffic assignment on congested multidestination networks"* Paper pres at the 5th Word Conference on Transport Research,

[12] Wie,B., Friesz,T. and Tobin,R.(1990) *"Dynamic user optimal traffic assignment on congested multidestination networks"* Transportation Research vol 24B no 6

APPROXIMATION TO OPTIMAL DYNAMIC TRAFFIC ASSIGNMENT OF PEAK PERIOD TRAFFIC TO A CONGESTED CITY NETWORK

M.O. Ghali and M.J. Smith

Department of Mathematics
University of York
York YO1 5DD
United Kingdom

We present a vehicle-by-vehicle deterministic queueing traffic assignment model for minimising total travel delay in a congested road network. The model routes drivers, one at a time, according to the marginal delay of each link. This is regarded as a combination of the delay caused by each driver to others travelling in the whole network, and the travel delay experienced by the driver himself. This yields an approximate system optimal routeing pattern and the corresponding road prices needed to cause the user equilibrium traffic pattern, which would arise from each driver minimising only his own travel delays, to be the approximate system optimal one. A key characteristic of the model presented is that it is applicable to (multi-commodity) networks having many origin-destination pairs and many bottlenecks. The model has evolved as a result of the initial study on this problem in Ghali and Smith [1], and basically extends the model in that paper. Computational results comparing network performance of applying the model of this paper against that of the previous paper as well as against network performance due to the natural user equilibrium networks are provided.

1. General remarks

There is currently a growing concern to curb the congestion problem arising from the ever increasing demand for transportation in urban cities around the world. This congestion problem is primarily manifested during the peak-hour period due to building up of queues.

This concern highlights the need for sound models which are aimed at the modelling and the replication of the rush-hour congestion period within a simulation environment, with due regards given to the inherent dynamical evolution of the traffic process over varying time frames.

This paper outlines a new algorithm intended to approximately solve the multi-commodity dynamic system optimum traffic assignment problem. It has grown out of the considerations given to this problem in Ghali and Smith [1], and essentially extends the algorithm described in that paper to model a network of a general topology.

The dynamic system optimum problem has been approached by many authors. From amongst those we mention Merchant and Nemhauser [2, 3], Carey [4, 5] and Friesz, Luque, Tobin and Wie [6]. All of these authors have mainly dealt with a particular case of the dynamic traffic assignment problem, where all drivers in the network travel to the same destination - the so-called single commodity case. Extension of these author's work to model a multi-commodity network where there several destinations and several origins in the network, has proved to be difficult. An account on the problems which arise in a multi-commodity network, is given Carey [5].

In Ghali and Smith [1] we approached the dynamic system optimum problem from a different view. We used simulation to avoid the problems highlighted by Carey [5] ("first-in, first-out" and non-zero "flow controls" (or holding back) perhaps). In [1] we particularly gave a definition for the link marginal delays of travel along a single link and used this definition to construct an algorithm which reroutes drivers according to these marginal delays to all users of using a particular link at a particular time. The definition in [1] is referred to as *the local marginal delay*. This, for each link and each time, was viewed 1) as the delay experienced by a vehicle on the link, and 2) the additional delay the vehicle use of the link causes to others **on the same link**.

Also in [1] it was pointed out that unlike the steady state case, where routeing of each vehicle along the route which has the least link (local) marginal delays determines the least total travel delay pattern of flow, this feature may not always carry over to the dynamic state. The paper concludes with optimality being ensured for the single-bottleneck-per-route case already highlighted in Smith and Ghali [8, 9] in their theoretical study on related dynamic equilibrium traffic assignment problems. Notwithstanding the restrictive conditions in [1], crucial insights were gained into the dynamic system optimal problem.

The model in this paper differs essentially from that in [1] in two respects: 1) a different traffic model is used, and 2) a different definition for the link marginal delay is given. The traffic model used in [1] is a time slice model, in which the planning horizon is segmented into a number of time slices and the queueing delays are approximated by a linear profile such as in CONTRAM (Taylor [10].) In this paper no time slices are involved.

As for the different definition of the link marginal delay, which we refer to later as *the global marginal delay*, it differs considerably from *the local marginal delay* in that, for each link and time, it takes account of delays experienced by vehicles on the link, and 2) the additional delays the vehicles' use of a link causes to others **throughout the whole network**, and not just the delays caused **on the same link**, as opposed to the definition of *the local marginal delay*.

The new algorithm outlined in this paper is based on these *global marginal delays*. It seeks to minimise total travel delay when holding back is not allowed.

2. The network model

In this section we give first the modelling assumptions made in developing the model of this paper, describe how we discretise the traffic demand and represent vehicles in our model, and indicate the way we translate from route flow to link flow. Then we define how the queues and delays, together with the marginal delays, are calculated on each link.

2.1. Modelling assumptions

The assumptions we make here should not be viewed as rigidly fixed, as the model we propose in this paper is at this stage in early development and can be easily extended.

Throughout this paper we shall suppose that links, representing real carriageways in the original network, have associated exit capacities, and that provided all the flows are less than or equal to this capacity, intersecting movements are readily able to filter through one another. This is often achieved on road networks by using fixed-time or flow-dependent signal-settings to reduce individual link capacities; insulating traffic movements which would otherwise conflict unacceptably. When two of the movements appear to cross, we suppose that there is no delay or constraint imposed on the two corresponding traffic streams. Clearly, in reality this may not be the case, but these assumptions are made here for ease of illustration.

We shall further be considering that travelling costs amount to travelling time that can be regarded as composed of running time (a constant which reflects the free-flow speed) and queueing delays. Inclusion of car-following and lane-changing mechanisms, and perhaps platoon modelling, is very possible in our model, as it is based on a vehicle-by-vehicle simulator and in which network performance is evaluated using simulation.

Also, we suppose that the model has a vertical queueing property, so that blocking back is left out of consideration. As an alternative to the vertical queueing property in our model, each link can be viewed as having a length that is sufficient to inhibit the blocking back of an upstream junction. (The vertical queueing or the sufficient link length assumption we have adopted in this paper is perhaps a shortcoming of our model. We intend to deal with this shortcoming in a different setting.)

Our final assumption is related to the demand for each origin-destination pair. Here we shall regard that demand is given, inelastic and time-varying; thus, we do not address in this paper elastic demand or variable time of departure.

2.2. Vehicle representation

Vehicles are represented in our model by discretising the traffic demand for all the origin-destination pairs in the network. Each vehicle in the network is given a *label* or an identity, so that vehicles can be distinguished and traced through the network.

In the route calculation, each vehicle is assigned to only a single route connecting the corresponding origin-destination pair. Hence, we do not allow for splitting of vehicles, and there is no such a value of 1/2 a vehicle in our model! We treat vehicles as indivisible entities. (In the steady state where the time variable is excluded, the demand flow rate for each origin-destination pair is split with respect to the flow variable to have multiple routeing. In this paper, we split the demand with respect to the time variable. Hence, multiple routeing is maintained.)

The vehicles' times of departure from each origin are determined from the corresponding origin-destination traffic demand. This is done such that each vehicle's time of departure is equal to the middle of the time interval matching the size and position of the vehicle in the corresponding demand profile.

2.3. From route flow to link flow

Given the route and label of each vehicle, the flow on a link is determined by storing the labels and times of entry of all the vehicles that happen to embark on that link, if the link is on their routes.

In the usual steady-state assignment framework, link flows can be easily determined in terms of route flows. In the dynamic case, this is not straightforward, and the time dimension affects the order of vehicle arrivals into each link. In section 3.2 below we describe a procedure for determining link flows from route flows while respecting the order of entry and exit of vehicles into and from each link.

2.4. *Queues and delays*

We follow a deterministic approach to queue modelling; a queue forms on a link due to excess input flow into the link as compared to its service rate, which is determined by the capacity of a bottleneck located along the link or, as is common, situated at the exit of the link. If the input flow is equal to or less than the capacity of the link, then we assume no queueing delays occur.

2.4.1. *Notations*

The following notations are needed.

p = The number of vehicles constituting the total demand from all origin nodes.

k = A vehicle number or an identification label.

i = Iteration counter.

$a^l{}_k$ = Time vehicle k enters link l.

r^l = Uncongested running time on link l.

s^l = Time taken to serve one vehicle at the bottleneck on link l. (For uncapacitated link, s^l can be taken zero.)

$d^l{}_k$ = Delay encountered by vehicle k on link l (the sum of both uncongested running time and queueing delay at the bottleneck on link l).

$q^l{}_k$ = Queue length on link l in front of vehicle k.

2.4.2. *Vehicle link delays*

The delays encountered by a vehicle k entering link l at time $a^l{}_k$, can be written as:

$$d^l{}_k = s^l + r^l + \max[0, d^l{}_{k-1} - r^l - (a^l{}_k - a^l{}_{k-1})], \qquad [3.1]$$

which is the sum of the uncongested travel time on the link and the queueing time at the link bottleneck, if any. The queueing time at the bottleneck includes the time needed to serve vehicle k, in addition to the time needed to discharge the queue ahead of vehicle k.

2.4.3. *Queue length*

The queue length on link l in front of a vehicle k entering the link at time a^l_k, can be calculated inductively from the expression:

$$q^l_k = \max[0, q^l_{k-1} + \min(1, k-1) - \min(1, k-1)(a^l_k - a^l_{k-1})/s^l] \qquad [3.2]$$

on the assumption that 1) $a^l_1 < a^l_2 < \quad < a^l_{k-1} < a^l_k < \quad < a^l_h$, where h is the last vehicle to enter link l, and 2) no overtaking takes place between the link entry and the link bottleneck (at the end of the link).

2.5. *Marginal delays*

Here we define (or describe how to calculate) the marginal delays which would arise, on *a single link*, from the entry of an additional single vehicle into the link at a given time. The calculation of *route* marginal delays is discussed later in section 3.3.

Prior to defining *the global marginal delay* we need first to give the definition of *the local marginal delay* of Ghali and Smith [1]. We use notations similar to those specified in section 2.4.1.

2.5.1. *The local marginal delay (LMD)*

The link local marginal delays m^l_k arising from a single vehicle k entering link l at time a^l_k can be written as:

$$m^l_k = d^l_k + \sum_{j=k+1}^{j=h} \max\{0, \min[s^l, d^l_{j-1} + a^l_{j-1} - (a^l_j + r^l)]\}, \qquad [3.3]$$

where $h = \{\min_v [k+1, k+2, .., k+v, .., n^l] \mid d^l_{k+v-1} - (a^l_{k+v} + r^l) \le 0\}$, or it is the time at which the queue dissipates. n^l is the last vehicle to enter link l.

The summation term in expression [3.3] is the additional delays experienced by the vehicles arriving between time a^l_k and a^l_h, due to the single vehicle k, entering link l at time a^l_k. Or, the summation term in expression [3.3] simply amounts to the remaining time for which the link has a positive queue at the bottleneck. In this sense, early arrivals at the link bottleneck thus cause more total delays **on the link** than later arrivals.

A property of *the local marginal delay* on each link is that it is always non-negative. This is so because the delays caused by one vehicle to another on the same link must be non-negative.

2.5.2. The global marginal delay (GMD)

For a certain link (say) l and time (say) t, the global marginal delay is simply the change in the total travel time **in the whole network** which would arise from an additional `single vehicle' entering link l at time t.

The global marginal delay includes travel-time changes throughout the whole network, and not just on link l. Thus, this accounts for:

i) the delays caused on link l,
ii) the delays encountered *later*, on *downstream links*, by each vehicle held back by (the consequences of) that `single vehicle' on link l, and
iii) the consequent delays encountered by other vehicles not traversing link l, but are delayed by the vehicles which were delayed on link l, due to the single vehicle.

As we are only concerned here with calculating the global marginal delay on a single link, it will be necessary to consider the additional vehicle as exiting the network at the exit of the link, having entered the link at some time, and participated in any queueing on the link. This ensures that the global marginal delay on a link only accounts for the delay due to an additional vehicle which entered the link. Further by insisting that the additional vehicle exits the link at its exit end, the additional vehicle thus will not contribute to the delays encountered by other vehicles on other links due to the vehicles which were held back by the consequences of the additional vehicle on link l; this is iii) above.

In contrast with *the local marginal delay*, *the global marginal delay* may be negative. This occurs if the additional vehicle entering a link causes a negative change in total travel time. Or, to put it differently, this occurs if the additional vehicle, by delaying some other drivers to exit the same link, **reduces** the total travel time on the network. In this case it would (at least in our theoretical model) be beneficial for additional vehicles to use the relevant link at the relevant time, and it would perhaps then be natural to consider "encouraging" drivers to traverse the link at or near this time (by "charging" an appropriate toll). Consequently, it would also be beneficial to encourage certain drivers to take routes which traverse the same links again (and again ...) at a later stage, so as to form loops, until the marginal delays have become non-negative. It would not **then** be beneficial to enter those links again and *the global marginal delay* must now be non-negative. An example-network in which negative marginal delays arise is given in Ghali and Smith [7].

As we can see, there is a considerable difference between *the local marginal delays* and *the global marginal delay*. Therefore, charging road users in accordance with *the global marginal delay* rather than *the local marginal delay* may produce completely different impacts on network users, and after re-routing has occurred (using the assignment algorithm described below), may produce completely different network flows and performances.

3. An assignment algorithm based on the global marginal delay

The algorithm presented here is a combination of a network loading procedure, for evaluating network performance under two natural "first-in, first-out" priorities (stated below), and a vehicle routeing procedure. The network loading procedure loads the network in a time-varying fashion which respects both link exit capacities and the two "first-in, first-out" priorities.

In our model, all vehicles' routes need to be stored, and the aim of the algorithm, for a given, initial route-entry-flow pattern, is to reroute each vehicle in order to reduce total travel-delay in the network.

3.1. Two natural first-in, first-out priorities

We will adopt the following two natural first-in, first-out priorities in loading the network.

On a certain link: Given two vehicles, the one joining the queue first, also exits first.

At a certain node: Given two vehicles embarking on the same link whose upstream end node is that certain node, the vehicle arriving at the node first, enters the link first. Here, we are implicitly assuming that give-way links do not exist.

3.2. Loading of the network

Given the routes of all vehicles, the network is loaded by moving the vehicles in the network according to the least time of departure a) from the origin node if the vehicle is, say, at time t is at the origin, or b) from its current location to the next downstream node if the vehicle is at t at some node. For each vehicle, the location and the time of arrival at that location have to be stored, and the vehicle whose time or clock is the least is moved first and its clock is then updated by advancing its clock by the amount of time taken to get to its current location. If a vehicle has arrived at its destination, then it is removed from the store or its store is declared empty. By the end of the planning horizon, all the vehicles will have arrived at their destinations, all the vehicle stores will have become empty, and the total costs will have been noted.

3.3. Routeing of a vehicle

The route calculation of each vehicle is performed with reference to *the global marginal delay* of each link, but the time of entry into and exit from each link represent the times when the vehicles actually enter and leave each link.

It should be noted that the global marginal delay of each link need not be calculated beforehand for all times and for all links.

With the routes of all other vehicles being fixed, a link's marginal delay of a rerouted vehicle is calculated at the time when the vehicle is about to enter that link. By employing a shortest path algorithm, such as Dijkstra's [11] algorithm, the algorithm determines the times as well as the links whose marginal delays are to be calculated, as it branches out from the origin node to the destination node for each vehicle.

To exemplify, consider a vehicle whose origin-destination is some o-d pair and whose route from o-d is to be determined, while the routes of all other vehicles are temporarily fixed. Suppose that the vehicle is at time t on some link l between the o-d pair. To find which link, downstream of link l, the vehicle will follow, so that it minimises the total network travel-delay, one has to calculate the GMD of each downstream link and place the vehicle on the link which brings about the least total travel-delay up. In calculating the total travel-delay, the vehicle is assumed, as we have already mentioned, to exit at the end of the link. In this vein, it requires as many separate implementation of "loading of the network" procedures as there are downstream links, in order to find out on which link the vehicle induces the least total travel-delay. Having found out which of the downstream links the vehicle must follow, the search then continues for another downstream link, from the last fanned in link. Finally, the vehicle arrives at its destination.

3.3.1. Other shortest path routeing algorithms

We explained above how a vehicle can be routed using Dijkstra's algorithm. Other shortest path algorithms could be possibly applied, such as the algorithm due to Bellman-Ford [12]. But, does it matter, other than in the sense of efficiency and speed? Dijkstra's algorithm has been suggested for constant link travel delay (or costs), or for time-invariant link costs. So its use in a dynamical setting such as in this paper, where the marginal delays are different at different times, should be given careful considerations.

Indeed, it turns out that Dijkstra's algorithm, being label-setting, may not always be able to find the optimal route when the link delays vary with time, except under some strict conditions on the link travel delay (see Kaufman and Smith [13]). These conditions are satisfied by neither *the local marginal delay* nor by *the global marginal delay* of this paper. An example which shows the failure of Dijkstra's algorithm to find the optimal route in a time-varying context such as this, is given in Ghali and Smith [7].

Another reason why Dijkstra's algorithm may not be able to find the optimal route is because, as we have already mentioned, *the global marginal delay* may give rise to negative marginal delays, which in turn may give rise to loops, forming for a period of time. So viewed from the perspective of GMD, Dikjstra's algorithm has the additional disadvantage in that the solution obtained contains no loops. This is because once a link is permanently labelled, it cannot be entered again, which is not exactly what the GMD suggests. (Loops which contain links that are entered again (and again) at a later stage, should not be confused with negative cycles. Though GMD gives rise to negative marginal delays, it does not however give rise to negative cycles.)

On the other hand, the Bellman-Ford algorithm is non-label setting; no link labels are considered final until all are - what we are probably looking for. But a question remains open: can the Bellman-Ford algorithm visit the same links again, without stopping? As far as we know, the Bellman-Ford algorithm stops when it detects a negative cycle.

Assuming that loops are ruled out for practical considerations, then the Bellman-Ford algorithm should be able to find the optimal route (subject, of course, to these routes containing no loops).

In the rest of the paper we use Dikjstra's algorithm, although, despite ruling out loops for practical considerations, it does not find the optimal route. The reason we have used Dijkstra's algorithm in this paper is due to its minimal implementation requirements and efficiency. We do accept however that this implementation cannot be relied on to find the optimal solution - but the problem is non-convex anyway!

3.4. *The GMD assignment algorithm*

For a given route-inflow pattern, the algorithm reroutes vehicles, one at a time, in search of least total travel-delay flow patterns.

3.4.1 Notations

The following notations are needed in addition to those introduced in Sect. 2.4.1.

i = Iteration counter.

ρ^i_k = Route of vehicle k in iteration i.

C^i = Total network delay in iteration i. This is equivalent to adding up, over all links, the total delays encountered on each link by each vehicle traversing the link. The delay encountered by each vehicle on a link is given by expression [3.1].

3.4.2. Steps of the algorithm

The steps of the algorithm are as follows.

1) $i = 0$
2) Assign all the vehicles to their least marginal journey delay routes while taking into account the "first-in, first-out" discipline by mixing the vehicles inflows and outflows as in Section 3.2. Let the route of each vehicle k be ρ^i_k.
3) Calculate total network delay, C^i.
4) Let $k = 1$, $i = i + 1$. For each vehicle k:
   a) Remove vehicle k from each link on the route ρ^{i-1}_k determined in iteration i - 1.
   b) Determine new least marginal journey cost route ρ^i_k, on the basis of the values of *the global marginal delay* on each link. Let TC be the total network delay when vehicle k has arrived at its destination.

c) Assign vehicle k to its new least marginal delay route ρ^i_k.

d) If k=p, then go to step 5). Otherwise, increment k by 1 and return to 4-a).

5) If TC < C^{i-1}, then let C^i = TC and return to step 4). Otherwise, the algorithm is terminated.

3.5. Discussion

The algorithm embodied in the above steps is an approximation to system optimal assignment, for two reasons. The first reason is because we used a label-setting algorithm (see Sect 3.3). It must be said however, even if we could use a minimum path search algorithm which is non-label-setting (to account for the different link marginal delays at different times) this may not be able to account for the loops which may be part of the system optimum flow pattern.

As for the second reason, this is because the algorithm considers rerouteing vehicles, one at a time, in any single iteration. It may be that if more than one vehicle was rerouted simultaneously at the same time, the total delays may be further reduced, or reduced if rerouteing one vehicle at a time did not result in any reduction. In fact this is likely to occur if the total travel delay objective function of the network being modelled is non-differentiable and non-convex, which we show this is the case for some networks in Ghali and Smith [7].

To get around the non-differentiability problem, one may have to employ a shortest path routine which is capable of rerouting more than one vehicle at a time, in a time varying context. To the best of our knowledge, today there is no routine which can reroute simultaneously more than one vehicle at a time on a single processor machine, while taking into account the changes brought by one vehicle on the network, and consequently on other vehicles. Even if we use a parallel machine, we cannot see how this can be done!

As for the non-convexity problem, again this is a difficulty which is hard to circumvent. This may result in local solutions which may be unattractive. It seems that one would have to consider different starting points in order to get around unattractive local solutions.

Comparing the algorithm given here with that in [1], the algorithm given here differs in two aspects. The first aspect is that here drivers are routed with reference to their GMD marginal travel costs rather than local marginal travel costs. The second aspect relates to the evaluation of the network performance. In our previous paper we evaluate network performance only after the route which has the least marginal delays has been calculated; the new route is accepted if network performance, compared before and after rerouteing the vehicle, has improved. Here, network performance is evaluated each time a downstream link is explored by the minimum path algorithm, and only those links which induce the least total travel delays are chosen.

3.6. Convergence of the algorithm

Convergence of the algorithm is guaranteed as the total network delays decline each time a vehicle is allocated a new route; it stops when no further delay reduction is possible by single-vehicle route swaps, and so no new routes are allocated in any single, complete iteration.

4. Computational results

To provide some numerical results, two networks were used. Each of these networks has more or less a size of about 30 links, 15 junctions, and 9 origin/destination nodes, but each has a different topology and different turning movements at the junctions. In both networks the demand between all origin-destination pairs is time-varying, with a peak during the morning rush hour.

The results are summarised in Graphs 1 and 2. Graph 1 gives the results of the first network, while graph 2 of that of the second network. These were obtained with a program called RONETS (ROad

NETwork Simulator), a vehicle-by-vehicle program developed by the authors.

In each graph, we plot network performance against the iteration number. Each graph is divided into three regions. These are: region UE, region LMD, and region GMD. Region UE corresponds to the network performance when vehicles are routed according to *the vehicle link delays* (expression [3.1]), region LMD to that of *the link local marginal delay* (expression [3.3], using the LMD assignment procedure given in Ghali and Smith [1], and region GMD due to *the global marginal delay* assignment procedure of this paper.

The way each graph is plotted within each region is as follows. In region UE we run RONETS with the link travel delays being the *vehicle link delay*, and used the resulting flow pattern as an initial pattern of flow for the LMD procedure in [1]. Similarly, on running the LMD assignment procedure the pattern of flow obtained is used as an initial route-inflow pattern for the GMD assignment procedure, where then the GMD procedure is started. The program RONETS automatically performs the three assignment procedure, and spells out network performance at the end of each iteration, until termination.

Graphs 1, 2 show clearly that substantial benefits may accrue from routeing vehicles according to the LMD or GMD rather than the UE. It seems from these graphs that the difference between the LMD and GMD strategies in terms of network performance may not be great, as it was thought before testing the GMD strategy. If this is always the case, then this is good in two ways. First, one would not perhaps need to determine the GMD pattern of flow if it turned out that the difference between the UE and LMD is substantial. Second, the GMD is extremely expensive in computer time, and this makes the LMD more attractive. However, the difference may not always be as subtle as is the case in graphs 1, 2.

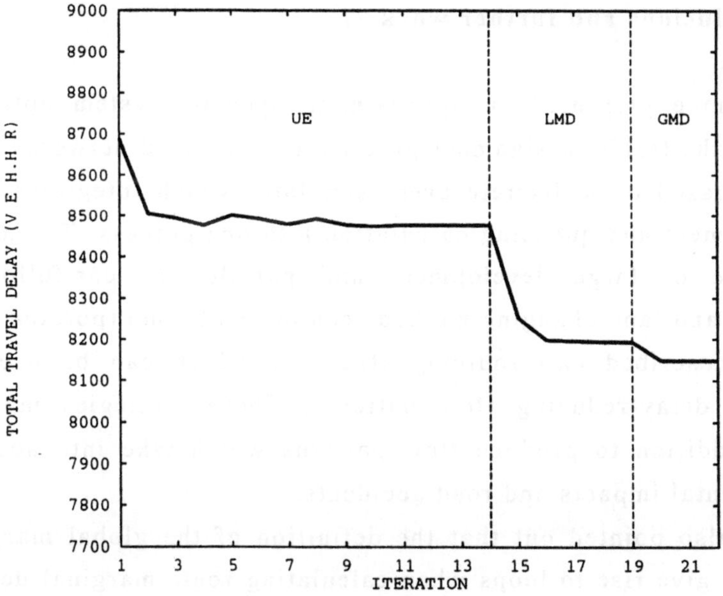

Figure 1.

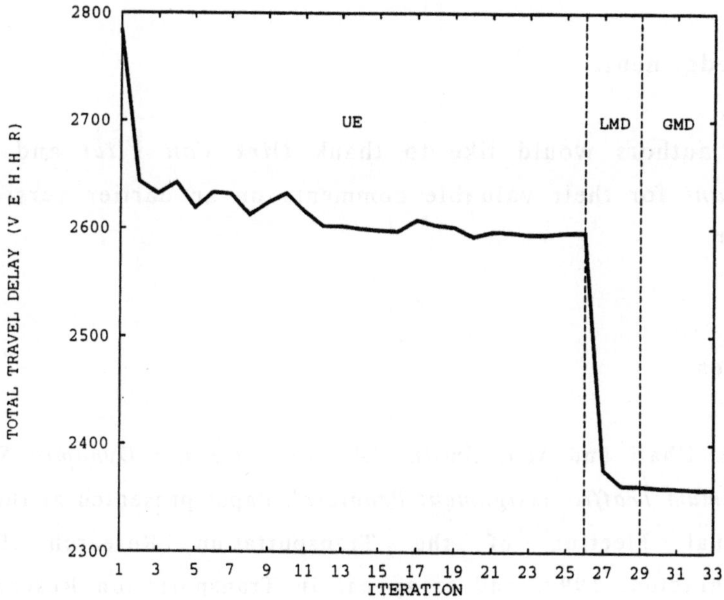

Figure 2.

Therefore, more results are needed to provide more information on this important point, in addition perhaps to using the Bellman-Ford routine rather than Dijkstra's, which we used to obtain the results here.

5. Conclusions and further work

We have presented an approximate dynamic system optimum model for the traffic assignment problem of congested networks. The model is based on a discrete event simulator which integrates both the assignment and queueing calculation into one process. The model is capable of large development and parallelism; car-following modelling and lane-changing methods can be easily incorporated.

We described two routeing strategies which can be used to determine delay-reducing flow patterns. These strategies may be used in addition to produce flow patterns which take into account environmental impacts and road accidents.

We also pointed out that the definition of the global marginal delay may give rise to loops when calculating route marginal delays. These seem to indicate some computational difficulties.

Acknowledgements

The authors would like to thank *Dirk Van Vliet* and *Hani Mahmassani* for their valuable comments on an earlier version of this paper.

References

[1] M.O. Ghali and M.J. Smith. *"A Model for the Dynamic System Optimum Traffic Assignment Problem"*. Paper presented at the 71st Annual Meeting of the Transportation Research Board, Washington, 1992, and to appear in Transportation Research-B, 1994.

[2] D.K. Merchant and G.L. Nemhauser. *"A Model and an Algorithm for the Dynamic Traffic Assignment Problem"*. Transportation Science, 12 (3), 183-199, 1978.

[3] D.K. Merchant and G.L. Nemhauser. *"Optimality conditions for a dynamic traffic assignment model"*. Transportation Science, 12(3), 200-207, 1987.

[4] M. Carey. *"A constraint qualification for a dynamic traffic assignment model"*. Transportation Science 20(1), 55-58, 1986.

[5] M. Carey. *"Optimal Time-Varying Flows on Congested Networks"*. Operations Research, 35(1), 58-69, 1987.

[6] T.L. Friesz, F.J. Luque, R.L. Tobin and B.W. Wie. *"Dynamic Network Traffic Assignment Considered as a Continuous Time Optimal Problem"*. Operations Research 35, 58-69, 1989.

[7] M.O. Ghali and M.J. Smith. *"Traffic Assignment, Traffic Control and Road pricing"*. Proceedings of the 12th International Symposium on Transportation and Traffic Theory. Editor: C. Daganzo. Elsevier, New York, 147-170, 1993.

[8] M.J. Smith and M.O. Ghali. *"The dynamics of traffic assignment and traffic control"*. Paper presented at the joint Italian/USA Seminar on Congested Urban Networks: Traffic Control and Dynamic Equilibrium, Capri, April 1989, and is published in Transportation Research, Vol. 24B, pp. 409-422, 1990.

[9] M.J. Smith and M.O. Ghali. *"Dynamic traffic assignment and dynamic traffic control"*. Proceedings of the 11th International Symposium on Transportation and Traffic Theory. Editor: M. Koshi. Elsevier, New York, London and Amsterdam, 223-263, 1990.

[10] N B Taylor. CONTRAM 5: *"An Enhanced Traffic Assignment Model"*. Transport Research Laboratory, Crowthorne, United Kingdom, 1990.

[11] E.W. Dijkstra. *"A Note on Two Problems in Connexion with Graphs"*. Numer. Math. 1, 269-271, 1959.

[12] R. Bellman. *"On a Routing Problem"*. Quart. Appl. Math., Vol 16, 87-90, 1958.

[13] D.E. Kaufman and R.L. Smith. *"Fastest Paths in Time-Dependent Networks for Intelligent Vehicle-Highway Systems Applications"*. IVHS Journal, 1(1), 1-11, 1993.

SENSITIVITY ANALYSIS OF A DYNAMIC EQUILIBRIUM MODEL FOR ROUTE AND ARRIVAL TIME CHOICE

Giuseppe Bellei

Dipartimento di Idraulica, Trasporti e Strade
Università degli Studi di Roma "La Sapienza"
Via Eudossiana 18 - 00184 Roma, Italia

Maurizio Bielli

Istituto di Analisi dei Sistemi ed Informatica - Consiglio Nazionale delle Ricerche
Viale Manzoni 30 - 00185 Roma, Italia

In order to assess the performances of information systems to road users and of demand management measures, it is necessary to simulate dynamics of traffic flows on the transport network and users' behaviour. This can be accomplished by taking into account several aspects such as users' information level, travel demand distribution over time and the expected smoothing effect of adopted measures and information systems. A computational procedure to perform this assessment for various traffic scenarios and hypotheses about users' behaviour is presented in this paper. Moreover, the indicators needed to quantify information systems and demand management effectiveness are identified together with parameters defining traffic scenarios. The results of an application of the procedure as a deterministic simulation tool are also presented and discussed.

1. Introduction

During the last years, the interest in road transport system analysis and mathematical modelling has been increasingly oriented to represent and evaluate system performance with respect to road users' information and demand management, rather than to infrastructure endowment and characteristics. Methodological transport research tries, in this way, to satisfy the growing demand for less capital intensive solutions to individual mobility needs.

Road information systems to users can be divided into three main categories:

- broadcasting of general information on transport and traffic;
- real-time route guidance to drivers, including in and out of vehicle support;

- trip planning support to users.

The aim of the last level of intervention is to provide the user with an effective support in planning trips taking place in the near future, and many different systems have been developed and tested in providing drivers with pre-trip information, such as Atis, Oracle in UK, Antiope-Route in France, Route-Tel by TRRL and Prestel by British Telecom, in general relying on existing media (e.g. Minitel) [1]. Moreover, within the Prometheus Project, a research is carried out on Pre-Trip-Planning with the aim of developing a demonstrator prototype system, able to meet user requirements both for short and long distance travels [2].

Many relevant research and testing activity is ongoing also with regard to systems belonging to the first two levels [3] [4] but, since it can be assumed that any kind of information may induce, through an improved learning mechanism, a trip and activity scheduling which is, on the average, better than the one based only on experience, an improved trip planning may be considered as a comprehensive goal and, at the same time, as a global index for the "average day" users' information performance, evaluating system benefits as better as performance is not randomly varying from day to day.

The goal of a better trip planning is also implicit at least in some of the "demand management" actions because:

- staggered work and school times enforce a change in trip planning;
- flexible work times, which are increasingly adopted as job type and organisation changes, widen trip planning opportunities;
- limited traffic zones operation rules often change during the day, thus affecting, in addition to mode and route choice, also trip planning;
- road pricing favours a trip planning behaviour, if tolls are variable in time, by compelling the individuals to compare pros and cons of different alternatives [5].

Thus, since the representation of trip planning behaviour may allow an aggregate evaluation of the whole range of non infrastructural remedies to road traffic congestion, it plays a central role. More generally, forecasting the effects of information systems and the motorists' reaction to demand management, requires to highlight their role in a dynamic framework and stimulated several studies on time-dependent aspects of traffic phenomena.

Some of these studies concentrate on modelling behavioural responses to various aspects of information production, transmission and utilization and represent these phenomena on schematic and idealized transportation systems, to avoid the burden of representing a general network when addressing conceptually complex issues [6]. A few reports, dealing with motorists' choices modelling and not only with technological soundness and viability of information systems, come from practical experience, but, also in this case, or reference is made to some corridors [7] [8] [9], either commuters behaviour is investigated without explicit reference to the network where their choices take place [10].

Other studies [11] [12] addressed dynamic route choice assuming it takes place in a full network and also previous work by the authors [13] followed such an approach. Here the relevant behavioural hypotheses of the discrete-time deterministic equilibrium model developed are tested, by simulating traffic conditions for a feasible range of parameters (related both with users' behaviour and with model application environment) to improve the analysis of trip planning, exploiting its value as a system performance index.

Because of the computational burden required by such an approach, and because deterministic equilibrium models keep their practical and theoretical value when applied to simulate average situations, the probabilistic and non-equilibrium features of models already appeared in scientific literature have not been considered [14]. Of course, criticism to deterministic equilibrium approach is well-founded when issues like day-to-day traffic fluctuations, real-time information and transient mobility patterns are at the centre of attention [15]. Nevertheless, if it can be assumed that average users' choices are influenced only by historical information, equilibrium models can be appropriate.

Also with respect to drivers' perception of time it is possible to define models where departure time is continuous [16] [17], but, leaving aside any consideration about computational effort, it is at least doubtful that a well-tuned discretization would lead to a worse behavioural model than a continuous time hypothesis.

Another feature of the discrete-time deterministic equilibrium model utilized is that arrival time choice instead of departure is considered, in contrast with usual assumptions. The idea behind this approach is that, when travel is planned, different choice mechanisms can be adopted in the real world by tripmakers, with planning of arrival time prevailing in the morning peak. It is quite likely that arrival time choice is followed by a travel time estimate and by a consequent departure time choice, but this process is implicit and departure time results to be an endogenous variable in this model.

In the following, what has been sketched here is given an analytical form, defining a discrete-time deterministic equilibrium problem and a heuristic solution algorithm, whose ability to approximately solve the problem has been tested. The simulation exercise has been carried out with respect to the percentage of users planning their arrival time choices, the level of traffic congestion and the shape of penalty function representing drivers' behaviour in relation to arrival time shifting; its results are described and final conclusions are drawn in relation to underlying modelling hypotheses and further research perspectives.

2. Problem formulation

The problem to be defined is an almost straightforward extension of the well-known deterministic user traffic equilibrium assignment problem for fixed demand. Within this problem the time of day is partitioned into periods, while the travel time, as a component of travel impedance, is considered as a positive real variable. The model for traffic delay in a link is the same as used for static traffic assignment, that is a volume-delay function; assuming that such an aggregate relationship between traffic flows and travel times holds separately for each period, it supplies variable-in-time travel times if variable-in-time flows are given as inputs. In such a way the problem is, at least heuristically, solvable without exceeding computational efforts and the need for input data can be satisfied almost completely within the limits of current traffic engineering practice.

2.1 The representation of transport network and time of the day

To establish the traffic model, the road network is represented, as usual, as a directed graph G(N,L) where N is a set of nodes and L a set of links connecting two nodes. Trips have their origin in a subset R of N and their destination in a subset S of N; each origin destination pair rs is an element of the set RxS (more briefly denoted as RS) and is connected by a set of paths P_{rs}, while the set of all acyclic paths in the graph is indicated with P; any path p being an ordered set of K(p) links $\{ l_1,...l_k,...l_{K(p)}\}$, where the initial node of l_1 is the origin r, the final node of $l_{K(p)}$ is the destination s, and the initial node of l_k coincides with the final node of l_{k-1}, for any k = 2 ,...,K(p).

The time horizon on which the evolution of traffic flows is represented is made up of Q periods i, with $i \in I$, of equal length T; since trips arriving at destination during this time horizon may need to leave their origin before it, another period (at least in principle of infinite length) must be considered before period 1, and the same should be done after period Q, if trips leaving from their origins during the time horizon, and arriving at destination after its end, are to be taken into account.

Departures and arrivals at intermediate nodes can then be located in any point of time and attributed to one of the Q + 2 periods $\{0, 1, \ldots, i, \ldots, Q, Q+1\}$, belonging to the extended period set I^+, while departures from origins, or arrivals at destinations, are constrained to take place at some a priori defined point (for instance, at the end of each period). Assuming that the last point of any period, but Q+1, belongs to it and fixing the origin of time axis at the end of interval 0, the Q+2 time periods correspond to the intervals $\{(-\infty, 0], (0, T], \ldots, ((i-1)T, iT], \ldots, ((Q-1)T, QT], (QT, +\infty)\}$ on real numbers set.

Several versions of dynamic assignment problems can be defined on this basis, but, since the one which is developed here simulates traffic conditions during the morning peak hour, trips are planned with regard to arrival times and, while total demand with arrival time during time horizon defined by interval set I is fixed, a certain degree of flexibility exists and is exploited by a fraction of users. To define this problem some further modelling assumptions, related to represented choice dimensions, have to be adopted.

2.2 The trip planning behaviour

Arrival time choice is modelled as a discrete choice because it is assumed that arrivals can take place only at the end jT of each interval j; arrival planning hence becomes the choice of the arrival interval, which can be described as the final outcome of the following three step process:

a) determination of a first approximation travel and activity schedule based on a priori limited knowledge about the implications of each alternative schedule and the travel times of trips between consecutive activities;
b) acquisition and/or consideration of more detailed information about travel times, costs and conditions at different times of the day, determination of departure times and full appraisal of consequences with regard to alternative schedules;
c) choice of the optimal schedule, defined as a sequence of activities (in their turn identified by locations, beginning and ending times) and trips from each activity location to the next,

together with estimated departure and arrival times of these trips.

Focusing on journeys to work during the morning peak, activities may be located at home or workplace and only a single trip is represented by its arrival time; step a) means only that official work start time is taken as first approximation arrival time, step b) can include the acquisition of additional information on travel times and the analysis of implications of early or late arrivals, in one with the evaluation of corresponding departure times (having breakfast close to workplace, or earlier waking up, may have pros and cons to be evaluated, like longer sleeping, the risk of a fine for being late, or the need to compensate for lost worktime), step c) reduces to the selection of the planned arrival time for journey to work and of the implicit departure time.

To represent travel during the morning peak to activities other than work, of course, the above sketched process should be modified giving different specification to first approximation arrival time in step a) and taking into account appraisal of other kinds of consequences for deviations from schedule in step b). To deal with different trip purposes then, several users' class should be defined, assuming as still valid the previous formalization of the decision process, but with different parameters for each class.

This framework is general enough to deal also with work start time stagger or flexibilities and implications on tripmakers of demand management measures like limited traffic zones and/or road-pricing, but here we assume that only two users' classes exist, one that goes through all the steps of the decision process and the other that, for lack of information, kind of trip purpose, type of job or simply out of habit doesn't take the opportunity of a schedule rearrangement as defined at step b).

Arrival time defined at step a) can be taken, as in the journey to work example, as approximately optimal, matching the beginning of the first activity of the day at the most suitable moment; the schedule rearrangement and arrival time change performed at step b) should then produce a net benefit to the individual by overcompensating a slightly less than optimal arrival time with some travel time saving. A behavioural model for arrival time should then measure the utility of an arrival time different from the approximately optimal one and evaluate its sum with travel related disutility.

Since optimality is only approximately defined at step a), it is reasonable to assume a lower threshold for arrival time and utility perception and measurement, so that an "indifference" time band around first approximation optimal arrival time exists, where also in step b) utility is kept constant and rescaled to zero, differences in utilities and arrival times within the band being not perceived by the tripmaker or measured by the analyst; it can also be assumed, with particular reference to the journey to work example, that the same deviation from approximately optimal schedule leads, outside indifference band, to a larger disutility in the case of late than in the case of early arrival. Disutilities, however, should increase, in both cases, together with deviation from approximately optimal schedule, up to a point where it is no more possible to realize some essential activity, like sleeping or working, above a minimum acceptance level. Larger deviation from schedule, and correspondingly earlier or later arrival times for the morning peak travel, should then be assumed as unfeasible; the whole range of arrival times which allow realizing essential activities will be named "feasibility band". This leads to model step b) of arrival time decision process as the evaluation of a penalty function, defined and taking values on discrete sets, as specified in the following.

2.3 Definition of the model for planned arrival time choice

Let j*T be a first approximation arrival time (FAT) and jT an alternative arrival time, candidate as planned arrival time (PAT); a penalty function pf (j, j*) being evaluated at step b) of decision process may be assumed equal to zero within the indifference band, proportional to arrival time differences absolute value, (with larger proportionality coefficients for late than for early arrivals) outside the indifference band but within the feasibility band and equal to plus infinity, or to a suitably large finite value, outside the feasibility band.

Since time of the day is represented by intervals of equal length T, penalty function is proportional to arrival interval differences (j - j*) and proportionality coefficients are the travel time equivalents of the disutility determined by a one interval arrival time shift

$$pf(j,j*)=\begin{cases}\infty & j<\max\{0,j*-e_f\}\\ \alpha(j,j*) & \max\{0,j*-e_f\}\le j<\max\{0,j*-e_i\}\\ 0 & \text{if } \max\{0,j*-e_i\}\le j\le \min\{Q,j*+l_i\}\\ \beta(j,j*) & \min\{Q,j*+l_f\}\ge j>\min\{Q,j*+l_i\}\\ \infty & j>\min\{Q,j*+l_f\}\end{cases} \tag{1}$$

where e_i, e_f are the early arrival indifference and feasibility band widths respectively and l_i, l_f their late arrival counterparts, while the maximum and minimum are taken to keep arrival times within time horizon. The set of arrival intervals $j \in I$ falling within the range ($j*-e_f$, $j*+l_f$) will be denoted as $I_{FA}(j*)$ and it is equivalent to the set of feasible PATs with j*T as FAT, and hence to feasibility band; function (1) is analogous, with some marginal difference in notations, with the definition of schedule delay disutility given by Ben-Akiva et al. [18]. In this case, however, randomness of utility is not taken into account and penalty function is defined on a discrete set; while randomness could be considered, discretization of arrival times is inherent to this approach and should not be taken as limitative, since arrival times are rounded up, when planning a trip, at least at the tenth minute.

To define a route and arrival time choice model we should then distinguish between several travel demand flows, to take into account both the division in two users' classes, (the ones who perform detailed evaluation of alternative schedules and the others who accept first approximation optimal schedule) and, for the first user class, the different PAT choices allowed to users with the same FAT.

Let then $g^1_{rs}(j_*)$ be the flow from origin r to destination s belonging to the first user class, with FAT j_*T; users out of this flow may contribute to flows $g^1_{rs}(j)$ of this class with PAT jT, in accordance with the result of alternative schedules evaluation. Denoting by $g^2_{rs}(j)$ the flow from origin r to destination s belonging to the second user class with FAT jT, which is by definition coincident with flow from r to s of this class with the same PAT; total flows from r to s with FAT j_*T and with PAT jT are respectively:

$$g_{*rs}(j_*) = g^1_{rs}(j_*) + g^2_{rs}(j_*) \qquad "rs \in RS,\ j_* \in I$$

$$g_{rs}(j) = g^1_{rs}(j) + g^2_{rs}(j) \qquad "rs \in RS,\ j \in I \tag{2}$$

By assuming that users belonging to the first user class are a fraction q of total users, only travel demand with reference to FAT $g_{*rs}(j_*)$ is needed as input (inducing, with respect to a standard OD survey, the need to know first approximation arrival time of each trip, easily satisfied, at least for home to work trips, by asking for official work start time) since $g^1_{rs}(j_*)$ and $g^2_{rs}(j_*)$ are obtained by multiplying $g_{*rs}(j_*)$ for q and (1-q) respectively, while also $g^2_{rs}(j)$ is equal to $(1-q)g_{*rs}(j)$ and $g^1_{rs}(j)$, as well as $g_{rs}(j)$, turn out to be endogenous variables, determined by the PAT choice model through evaluation of the global disutility $u^1_{rs}(j, j_*)$ of choosing a jT PAT when FAT is j_*T:

$$u^1_{rs}(j, j_*) = v_{rs}(j) + pf(j, j_*) \qquad "rs \in RS,\ j_* \in I,\ j \in I_{FA}(j_*) \tag{3}$$

where $v_{rs}(j)$ is travel time from r to s when arrival time at s is jT.

Denoting by $g^1_{rs}(j, j_*)$ the flow of users going from r to s, belonging to the first user class, with FAT j_*T and PAT jT, they are those out of the $g^1_{rs}(j_*)$ who plan to arrive at jT,and hence, summing over feasible arrival intervals:

$$\sum_{j \in I_{FA}(j_*)} g^1_{rs}(j, j_*) = g^1_{rs}(j_*) \quad \text{"}rs \in RS,\ j_* \in I \tag{4}$$

and the aggregate arrival time choice model to be defined reduces simply to (3-4), non negativity of endogenous flow variables:

$$g^1_{rs}(j, j_*) \geq 0 \quad \text{"}rs \in RS,\ j_* \in I,\ j \in I_{FA}(j_*) \tag{5}$$

and a set of equilibrium conditions, resulting from individual minimization of disutilities $u^1_{rs}(j, j_*)$:

$$\begin{aligned} &[u^1_{rs}(j, j_*) - u_{*rs}(j_*)]\, g^1_{rs}(j, j_*) = 0 \\ &\qquad \text{"}rs \in RS,\ j_* \in I,\ j \in I_{FA}(j_*) \\ &u^1_{rs}(j, j_*) - u_{*rs}(j_*) \geq 0 \end{aligned} \tag{6}$$

where $u^1_{rs}(j_*)$ is the equilibrium disutility for tripmakers from r to s with FAT j_*T; thus only minimum disutility arrival times are chosen as PATs

2.4 Basic assumptions and time transformation function

The second choice dimension to be considered is about the route, represented by a path on graph G, to be utilized to arrive at destination at planned arrival time; basic modelling assumptions about this choice are formally defined as follows:

i) vehicles travelling on a path and arriving at destination at the end of an interval are a platoon of negligible length;

ii) time spent by a platoon on a link depends by the arrival time at its final node;

iii) link travel times vary only from interval to interval;

iv) link travel time during any interval $i \in I$ depends on link flow during that interval, while it is equal to free flow travel time for intervals 0 and $Q + 1$;

v) *interval flow for any interval $i \in I$ is the sum of vehicles belonging to all platoons arriving at the final node of the link during the interval, divided by T;*

vi) *platoons don't overtake.*

To define dynamic traffic equilibrium conditions on the basis of these assumptions, link and path variables must be defined first (variables at origin-destination pair level have already been introduced for arrival time choice). Path and link flow variables are, ignoring the dependence from arrival times, identical to the ones used in static assignment problems and may be defined, respectively for flows and times, as follows:

$h_p(j)$	flow arriving at destination s at time jT travelling path $p \in P_{rs}$;
$f_l(i)$	flow arriving at final node of link l during interval i;
$z_p(j)$	time needed to arrive at destination s at time jT travelling path $p \in P_{rs}$;
$u_l(i)$	time needed to traverse link l during interval i.

In the dynamic case, however, the definition of relationships between link and path variables is less simple than in the static case and leads us to consider a function, named time transformation function. This function supplies the time when a platoon must be at the initial node of a link, if it is at its final node at time t and the vector of link travel times U_l, with interval link travel times $u_l(i)$ as elements for each $i \in I^+$, is given (in practice, only a few elements need be considered in computations), so that it can be specified as $tf(U_l,t)$.

Since each interval link travel time $u_l(i)$ depends from interval link flow $f_l(i)$, time transformation function can also be expressed as a function of link flows vector F_l, analogous to U_l, but we shall begin studying the case when link travel times are held constant; in this case $tf(U_l, t)$ depends only from the scalar t and it can be drawn on a Cartesian plane. This is an easy task, since for a zero travel time the graphic representation of $tf(U_l, t)$ is a straight line passing through the origin and with a 45° slope, while any positive travel time $u_l(i)$ would proportionally shift the line downwards along the ith interval.

Given a U_l vector as in Fig. 1, this graphic representation exhibits a discontinuous decrease at some interval end points in the area of increasing travel times, with the counter-intuitive implication that earlier arrivals could be achieved by starting later. In such cases, assumption vi) could be violated. To avoid that, time transformation function should be non-decreasing and one way to satisfy such a requirement is to modify it, taking, for any time t, the maximum between its value at t and values at times t' < t.

The resulting time transformation function $tf^n(U_l, t)$ is also represented in Fig. 1 and implies that times spent on the links are always reduced with respect to $tf(U_l, t)$. The rationale of assuming $tf^n(\cdot)$ instead of $tf(\cdot)$ lies in the minimum principle that rules the users' choices; if platoons which are at the final node of the link at any time t within sub-intervals S_0, S_1, S_2 of Figure 1 could arrive at time points 0, T, 2T respectively and stop until t, they would spend on the whole less time in the link than by arriving directly at t. Of course, stopping at a node is not the real-world behaviour of tripmakers and the advantage they would get by stopping is only a consequence of assumption iii).

Generalizing what has been observed about the example reported in Fig. 1, the time transformation functions can be defined, for $i' \in I^+$, such that $t \in i'$, as:

$$tf(u_l, t) = t - u_l(i')$$

$$\forall l \in L \qquad (7)$$

$$tf^n(U_l, t) = \max\{\, t - u_l(i');\ iT - u_l(i) \mid i \in I,\ i < i' \}$$

The two formulas for time transformation function correspond to two slightly different graphs representing a transport network in time, both derived with some change from a method initially presented by Ford and Fulkerson [19] to reduce a shortest time path problem with variable-in-time link times to an ordinary shortest path problem on an equivalent graph with fixed links weights. Ford and Fulkerson measured travel times as a discrete number of time units and, even if no theoretical limit may exist to use as small as needed time units, a practical one comes from the dimension of the time-expanded graph. This is not, however, the case for time transformation functions' approach, whose equivalent time-expanded graph has an infinite number of links, not explicitly taken into account. Time expansion methods go beyond the scope of the analysis; it is however worth of note that also the arrival time choice model described before can be represented by an equivalent graph.

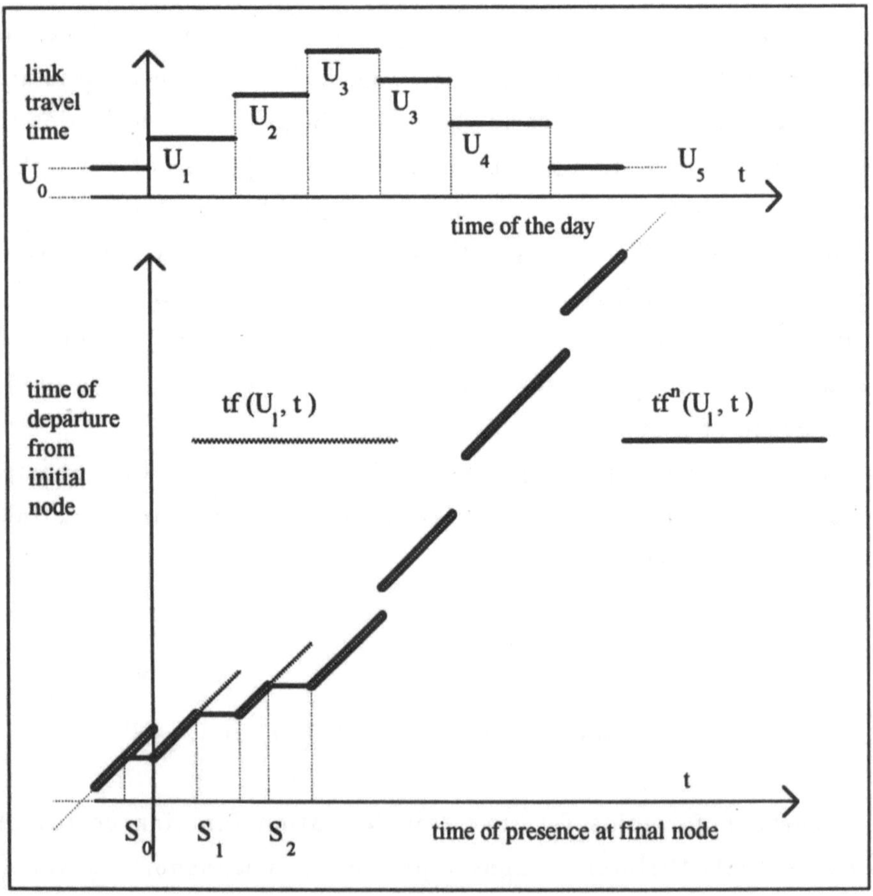

Figure 1. - Time transformation functions

Travel times represented by these expanded graphs don't depend by link flows; dependence of interval travel time by interval flow introduces in this framework a change in expanded graph structure and not only in link weights. An expanded graph structure invariant with link flows and times has been proposed in [20].

2.5 Route choice and integrated model definition

Both time transformation functions could be used in the following, with drawbacks and advantages in each case; route choice formulation model wouldn't change by that, while detailed formulation of the algorithm of course would. The implemented algorithm and the results obtained make reference to the non decreasing function $tf^n(U_1, t)$.

The relationship between link and path travel times is established deriving, through function (7), the departure times from initial nodes of any link l along a given path $p \in P_{rs}$, with arrival time at destination jT. If we define these departure times as $t_{lp}(j)$, the path travel time is given, for any $p \in P$, $j \in I$, by:

$$z_p(j) = jT - t_{1p}(j) \qquad (8)$$

where $t_{1p}(j)$ is obtained, assuming without any loss of generality, that path p includes, in the order, links (1,...,k,...,K) via the following recursive formula:

$$t_{kp}(j) = tf^n(U_k, jT) \text{ for } k = K$$
$$t_{kp}(j) = tf^n[U_k, t_{k+1,p}(j)] \quad \text{for } k = K-1,\dots,1 \qquad (9)$$

The inverse relationship is, consistently with assumption v):

$$\sum_{p \in P} \delta_{lp} \sum_{j \in I_{lp}(i)} h_p(j) = f_l(i) \qquad (10)$$

where δ_{lp} has the usual meaning of a binary variable equal to 1, if link l belongs to path p, and 0 otherwise, while $I_{lp}(i)$ is the set of arrival times at destination s of path $p \in P_{rs}$ such that $(i-1)T < t_{lp}(j) + u_l(i) \leq iT$. The set $I_{lp}(i)$ could be determined by repeatedly applying recursion (9), starting by j=1 and increasing it until $t_{lp}(j) + u_l(i) > iT$ or j = Q. Formalization of this procedure is not needed in the following; it is anyway worth noting that both the relationships between link and path variables are defined through function (7).

If the vectors of link travel times for all intervals $i \in I^+$ and of path flows for all intervals $j \in I$ are denoted by U and H respectively, the route choice model can be formulated as a deterministic equilibrium problem where it is required to determine path flows $h_p(j)$ for any $p \in P$ and $j \in I$ such that:

$$[z_p(j) - v_{rs}(j)]\, h_p(j) = 0$$
$$z_p(j) - v_{rs}(j) \geq 0 \qquad \forall p \in P_{rs},\ rs \in RS,\ j \in I \tag{11}$$

$$\sum_{p \in P_{rs}} h_p(j) = g_{rs}(j) \quad \forall rs \in RS,\ j \in I \tag{12}$$

$$h_p(j) \geq 0 \ \forall p \in P,\ j \in I \tag{13}$$

$$u_l(i) = u_l[f_l(i)] \ \forall l \in L,\ i \in I;\ u_l(0) = u_l^0 \text{ (free flow travel times)} \quad \forall l \in L \tag{14}$$

with $z_p(j)$ depending from U by (8) and (9), for any $p \in P$, $j \in I$ and $f_l(i)$ depending from H by (10), for any $l \in L$, $i \in I$.

Only interval 0 has to be considered in addition to intervals $i \in I$, since all trips are assumed to arrive before or at the end of interval Q. Conditions (11) on path flows and times extend Wardrop's first principle for static assignment to any arrival time and may be integrated with arrival time choice model by simply requiring all previously defined equations to hold together. This integrated route and arrival time choice model, hence, can be considered as a generalization of static user equilibrium, which would reduce to it by considering a single arrival period; this generalization, however, is not so easily amenable to a nonlinear programming problem as equilibrium conditions in the static case. Nevertheless, the close resemblance between the two cases has been exploited to develop an heuristic based on a static assignment algorithm, to be sketched in the next chapter.

3. The algorithm

The most widely known method to find flows and travel times satisfying Wardrop's principle is an adaptation of Frank-Wolfe's algorithm for nonlinear programming problems. This method exploits the network structure, as represented by its graph, to solve linear programming problems encountered at each iteration, through a minimum spanning tree search. The algorithm dealing with dynamic assignment problems exploits in a similar way the network structure of the problem, searching, at each iteration, for minimum spanning trees with link travel times variable with respect to time of the day.

3.1 The algorithm for travel time equilibrium

Since no equivalent nonlinear programming problem has been defined for equilibrium conditions (6) and (11), the generation of successive link flows, on which computation of link travel times and search for minimum spanning trees is based, cannot be obtained by a line search for the minimum value of an objective function. Instead, link flows at each iteration are obtained as the average of link flows determined by the assignment of travel demand to minimum spanning trees generated up to that iteration, with a procedure identical to the method of successive averages (MSA), also widely used in static equilibrium problems and proved to converge to the correct solution in that case [21]; A general description of the algorithm is given in the following:

STEP 0(Initialization) Find link flows $f_l^0(i)$, for each $l \in L$ and $i \in I$, obtained by path flows satisfying equations (12) and (13), through equation (10); set iteration index $D = 0$;

STEP 1(Travel times calculation) Compute tentative link travel times from current flows:

$$u^T_l(i) = u_l[f_l^D(i)] \ \forall l \in L \text{ and } i \in I;$$
$$u^T_l(0) = u_l^0 \ \forall l \in L;$$

STEP 2(Minimum spanning trees search) Find minimum spanning trees from all origins to any destination $s \in S$ for any arrival time $j \in I$, with link travel times for each interval given by $u^T_l(i)$;

STEP 3(Arrival time choice) Compare for each origin-destination pair rs and for each interval $j*$ the sum of OD travel times $v_{rs}(j)$, corresponding to arrival at the end of intervals $j \in I_{FA}(j*)$, and of the values of penalty function $pf(j, j*)$; determine, for each rs and $j*$ the arrival interval j^T minimizing total disutility;

STEP 4(Minimum spanning trees loading) Assign to minimum spanning trees found in Step 2, the travel demand $g*_{rs}(j*)$ for any $rs \in RS$ and $j* \in I$, taking as arrival times the end of interval j^T for travel demand $qg*_{rs}(j*)$ and the end of interval $j*$ for travel demand $(1-q)g*_{rs}(j*)$, to get tentative link flows $f^T_l(i)$;

STEP 5(Link flows updating) Calculate for any $l \in L$ and $i \in I$ the new current link flows as a linear combination of old current flows and tentative flows:

$$f_l^{D+1}(i) = \alpha f^T_l(i)+(1-\alpha)f_l^D(i), \text{ with } \alpha = 1/D+1;$$

STEP 6(Convergence check) Verify if a measure of the difference between current flows at successive iterations is lower than a predetermined value; if it is STOP, otherwise set D = D+1 and go back to Step 1.

It should be noted that the most effective way to get interval link flows $f_l^D(i)$ is to take one destination at a time and execute steps 2, 3 and 4 of the algorithm for that destination, minimizing memory requirements without affecting results, while initial flows $f_l^0(i)$ may be obtained, in analogy with static equilibrium algorithms, by setting link travel times equal to free flow travel times.

3.2 Static and dynamic algorithms

The main differences between this algorithm and a static MSA type assignment algorithm are the insertion of step 3 (which has no correspondence in static assignment), the fact that steps 1,4 and 5 have to be performed once for each time interval, and the minimum spanning tree search and path flows loading (steps 2 and 4 of the algorithm), where dependence of travel time from time of the day must be taken into account [22] [23]. The first two changes are quite obvious and don't need to be dealt with further on; with reference to the third one it can be observed, with regard to step 2, that only a slight modification of a static "label-correcting" algorithm is needed, making use of time transformation function $tf^n(U_l, t)$ to check, at each step if there is any out-of-tree link (denoted by its initial and final nodes m and n) such that its insertion into the tree improve the length of in-tree paths.

In our case the tree is rooted at a destination s for an arrival time jT and path length is given by travel time; at each step the algorithm should then check if:

$$tf^n(U_{mn}, t_n) > t_m \qquad (15)$$

where t_m and t_n are departure times from m and n to arrive at destination s at time jT.

When no link n,m for which (15) can be verified is found, the current tree is a minimum spanning tree. With regard to step 4, path flows are simply loaded to interval link flows in accordance with recursively determined arrival times to end nodes $t_{kp}(j)$. This procedure, combined with arrival time choice defined at step 3, can be shown to reproduce exactly a minimum disutility path tree search and loading on an equivalent graph.

3.3 An extension to generalized costs

The hypothesis that travel time is the only relevant variable for route choice may be easily removed by adapting, without substantial changes, the model and algorithm presented. We assume that a single disutility measure is relevant, defined as the sum of travel time $u_l(i)$ and of another impedance attribute $w_l(i)$, to be named "cost", expressed in travel time equivalents and depending by $f_l(i)$

$$w_l(i) = w_l[f_l(i)] \quad \forall l \in L,\ i \in I \qquad (16)$$

To take into account the cost component of travel disutility, travel cost $c_{lp}(j)$ from initial node of link l to destination s with arrival at time jT, along path $p \in P_{rs}$, and path travel cost $y_p(j)$ have to be defined, in analogy with travel times $t_{lp}(j)$ and $z_p(j)$ respectively and recursion (9) has to be modified, to compute these travel costs, since both time and cost are to be taken into account when determining minimum disutility arrival interval, while only time is relevant to supply the new starting point at each recursion step.

This is accomplished by first defining a function if (U_l, W_l, t), named arrival interval transformation function, supplying, for any $i' \in I^+$, the arrival interval i^* at initial node of link l if a platoon is at its final node at time $t \in i'$:

$$i^* = if(U_l, W_l, t) = \text{argmin}\,\{u_l(i') + w_l(i');\ u_l(i) + w_l(i) + (t - iT) \mid i \in I^+,\ i < i'\} \qquad (17)$$

The expression of (non-decreasing) time transformation function may then be simplified thanks to the arrival interval i^* obtained by (17); arrival times remain, however, implicitly dependent from both U_l and W_l:

$$tf^n(i^*, t) = \min\{t - u_l(i^*);\ i^*T - u_l(i^*)\} \qquad (18)$$

Path travel times are still given by (8), while path travel cost $y_p(j)$ is simply equal to $c_{1p}(j)$ and, assuming again path p coincident with the ordered set (1,...,k,...,K), the recursive formula supplying path travel times and costs is:

$$\begin{aligned} i^* &= if(U_k, W_k, jT) \\ t_{kp}(j) &= tf^n(i^*, jT) \quad \text{for } k = K \\ & \qquad\qquad\qquad\qquad\qquad\qquad (19) \\ i^* &= if[U_k, W_k, t_{k+1,p}(j)] \\ c_{kp}(j) &= c_{k+1,p}(j) + w_k(i^*) \\ t_{kp}(j) &= tf^n[i^*, t_{k+1,p}(j)] \text{ for } k = (K-1,\ldots,1) \end{aligned}$$

The complete definition of a dynamic assignment based on generalized cost would require equilibrium conditions be reformulated to include the cost component $y_p(j)$ of path disutility but, since these conditions are a straightforward extension of (11) and the generalized cost version of the model has not been utilized in simulations, they will be omitted here.

4. Algorithm validation and simulation framework

The algorithm has been utilized for preliminary tests to verify convergence to equilibrium conditions. The network and travel demand in this case, while not representative of any real-life situation and of a size smaller than expected in application, allow for OD pairs in the order of tens, each connected by several alternative paths. Travel demand sums up to 10,900 vehicles travelling in the morning peak, defined as 7,00 ÷ 9,00 a.m. and divided into eight fifteen-minute intervals. The demand profile over time presents such a peak that significant congestion phenomena occur and vehicles arriving at the end of a period are generally forced to begin their trip before that period.

The convergence to equilibrium has been tested, with reference to given arrival times, by inspecting origin-destination travel times on some paths chosen (or not) during the minimum path tree search, revealing a tendency to oscillate between alternative paths, likewise static assignment; a more favourable convergence has been obtained with reference to link flows, on which oscillations on different paths compensate, reaching a stable pattern after about ten iterations, and relative difference on total flows lower than 0,1% has been reached, after 21 iterations.

After these preliminary test on a small network, the software developed has been applied to a real, even if moderate, size network, to get some information on the sensivity of benefits deriving from the introduction of information systems and the application of demand management measures briefly discussed in the introduction.

This kind of application was not intended to give numerical values for relevant variables describing traffic environment, information systems diffusion and effects of demand management. It is instead a sort of deterministic simulation by which the relationship between travel time savings achievable and some strategic variables are explored, verifying if any point exists where noteworthy behaviour is exhibited and contributing to individuate some "likelihood" requirements, together with the order of magnitude of parameters needed to specify the model.

Simulation data were obtained, with some minor integration, from a case study developed as a part of the National Research Council Special Project on Transportation Research, where road network is represented by a 244 nodes (38 of which are centroids), 705 links graph, corresponding to the urban road network of a medium-sized Italian town (about 150.000 habitants). Even if some of information systems and demand management measures producing an increased amount of trip planning (arrival time choice) are usually applied to larger towns, this choice can be justified because the graph's size may be representative also of larger areas (at least in a schematic and aggregate way). Mobility demand was determined by an OD survey which took place in 1984, making available an auto driver mode OD matrix, together with the distribution of trips between macro zones over time, utilized only to derive an overall distribution, taken as valid for all OD pairs.

This distribution was available from 6,30 am to 8,30 pm, but only the peak period distribution (6,30-9,30 am) was considered, in accordance with the arrival-constrained model adopted, computing from the original 30-minutes interval another 10-minutes interval time profile. On the whole, despite the crudeness of some integration of lacking data, the data base built up may be considered as realistic with two possible exceptions:

- the 10 minute period chosen may be questionable for the purpose of describing arrival time flexibility, but, since intervals shorter than 5 minute can be considered as unrealistic with regard to trip planning oriented time perception, 10 minutes appears to be a satisfactory compromise between realism and data availability;
- the level of congestion was so low in the case study, that the need for information systems and demand management seemed highly improbable, thus the actual demand was multiplied by increasing factors, to choose the one best suited to simulation needs.

Denoting link physical capacity by s_l, congestion can be

quantified through an index c(i), defined, for every interval $i \in I$, as:

$$c(i) = 100 \frac{\sum_{l \in L} f_l(i) \frac{f_l(i)}{s_l(i)}}{\sum_{l \in L} f_l(i)} \tag{20}$$

Congestion index is thus the weighted average of flow/capacity percentage over all links, taking as weights the links flows or, in other words, the average congestion experienced by tripmakers. Assigning the mobility demand from 6,30 to 9,30 am without arrival time flexibility, a 9% value in the interval of maximum congestion was computed, reduced to 8.2% with arrival time flexibility, assumed as induced by trip planning and having the shape of penalty function and the percentage of trip planners determined, in absence of experimental data, by rule-of-thumb hypotheses. This result was not unexpected, since these simulations said only the obvious: information systems and demand management measures are not suited to medium sized lightly congested towns.

Some data input had then to be changed in such a way to make congestion high enough to be not marginally affected by trip planning, while not so high to be unrealistic. It was found that the actual demand multiplied by a factor of 4 resulted in a 28.6% saturation index during the maximum congestion interval without trip planning (fixed arrival times), which was reduced, by simulating the same situation with trip planning, to 27.3%. If we consider that computed indexes are average values and that even at a simple four-arms two phases traffic light regulated node only slightly more than half effective green split is available even for main traffic flows, we conclude that saturation index values higher than 35-40% are not likely to be found in reality. Thus, this mobility demand multiplication factor is such that a congestion situation is simulated, at least in the worst ten-minute interval and around, while not being too high to be realistic.

The congestion indexes for these four simulations are shown in Figures 2 and 3.

Figure 2 Saturation index fox actual demand simulation

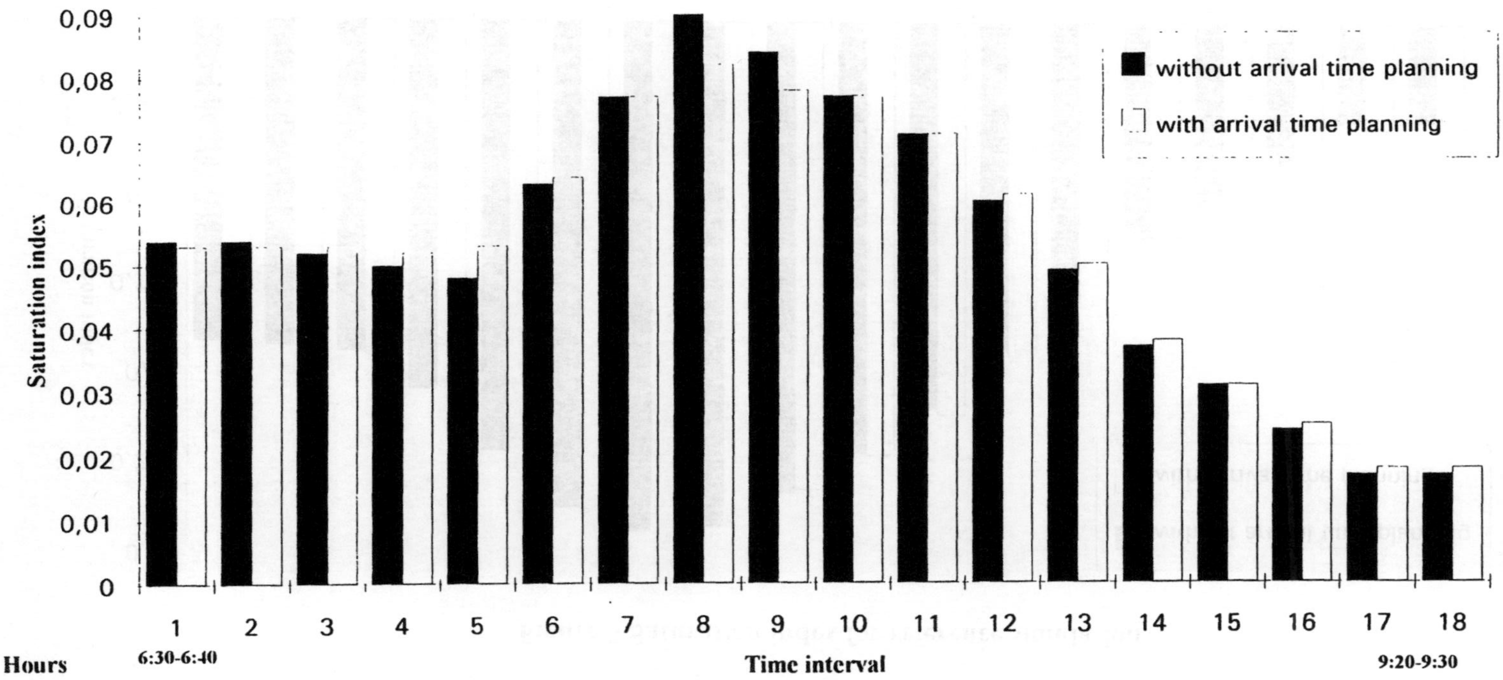

Figure 3 Saturation index for reference simulation

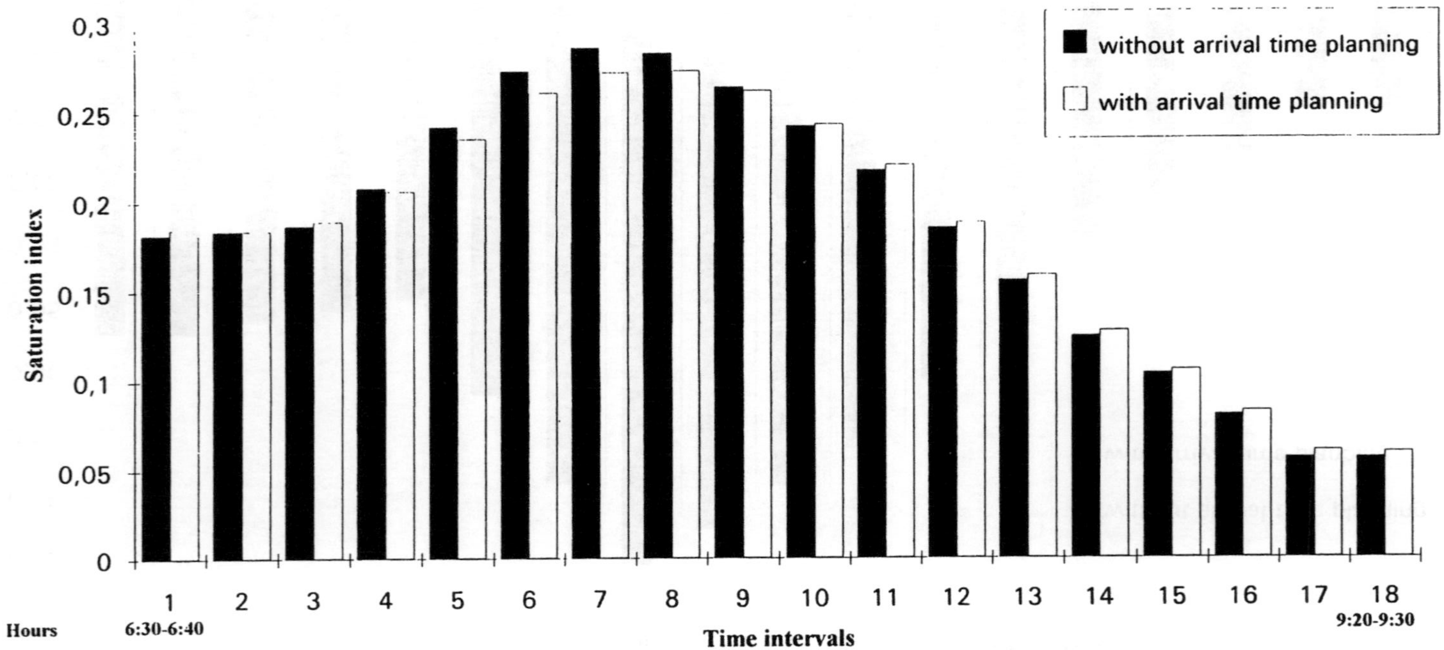

5. Simulation plan and results

Congestion indexes give a picture of traffic which is aggregated with regard to space, but still too detailed for sensitivity analyses, involving the evaluation of traffic patterns in correspondence to many different possible inputs, so that performance indexes have to be defined as scalars, aggregated also with regard to space. The performance indexes are thus identified as travel time savings deriving, both to trip planners and to non planners, from the choice of arrival time. For a generic assignment total travel time T is easily calculated as the summation over OD pairs rs of OD total travel times $v_{rs}(j)g_{rs}(j)$; denoting by T^B the total travel time in base case (dynamic assignment without arrival time choice / no trip planning), and by T^{TP} the total travel time for trip planners obtained by a dynamic assignment with arrival time choice (trip planning), travel time percentage savings for trip planners S^{TP}, non planners S^{NP} and total users S^T are respectively:

$$S^{TP} = 100\frac{qT^B - T^{TP}}{qT^B}; \quad S^{NP} = 100\frac{(1-q)T^B - T^{NP}}{(1-q)T^B}; \quad S^T = 100\frac{T^B - T^{NP} - T^{TP}}{T^B} \tag{21}$$

In the following, Figures 5 and 6 show the graphics of S^{TP} and S^{NP} and T^{TP} when mobility demand amount and composition are varied, while in Figure 7 only S^{TP} and S^{NP} are reported in correspondence with different penalty function shapes.

All results must be read keeping in mind that travel times for trip planners and non planners for the base case are fictitious, since no trip planning is simulated in the base case; they represent total travel times for future trip planners and non planners before any degree of freedom with regard to arrival time is introduced and are computed, in accordance with (21), as total travel time fractions, with the same trip planners percentage assumed in trip planning simulations.

5.1 Reference simulation and simulation plan

While mobility demand in the reference simulation was set at the level determined by above mentioned multiplication factor 4, the shape of penalty function (1) and the value of trip planners percentage q were set like in the simulation of trip planning for actual situation. Briefly, we should say that penalty function for reference simulation has no penalty for arrivals occurring one interval earlier or later than wished, while arrivals up to three intervals earlier or later than wished are admitted with suitable penalties to represent, for example, a flexible work time situation where marginal shifts may be not perceived at all, while late arrivals cause only a correspondingly later departure from workplace and early ones are subject to a lesser penalty to give a value to time made available for daily activity. A graphical representation of penalty function is given in Fig. 4 where, since penalty function is pointwise, it is not represented by dashed line, which is only a visual aid.

The percentage of trip planners has been set equal to 10% for the reference simulation with the aim of representing the crucial phase when information systems begin to spread in the market, or, with reference to demand management, the equally crucial phase when a not marginal share of drivers is allowed (or forced) to rethink its activity schedule. Starting from reference simulation, the simulation plan implies taking one simulation input at a time and varying it with a suitable step, measuring in each point the travel time savings to make a comparison with the base case with no trip planning. To clarify how simulation inputs and results vary, their exposition is divided into the two previosly defined groups.

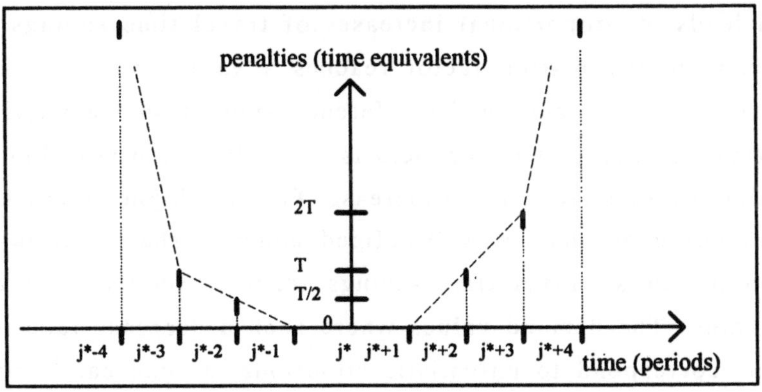

Figure 4. - Penalty function

5.2 Mobility demand simulations and results

Mobility demand inputs are, in addition to mobility demand multiplication factor, sharpness and skew of demand profile over time. Multiplication factor was varied from 0,5 to 8 with a 0,5 step, while several demand profiles, with different sharpness and symmetry were obtained from actual profile by linear transformations preserving total mobility and tested. The testing of demand profile shape influence on network performance has shown that profile sharpness has an influence on network performance similar, but less pronounced, with respect to demand absolute value, while profile skew influence is almost negligible; neither the complete definition of the way these simulations were performed or the detailed analysis of their results is thus deemed necessary.

Every simulation involving variations of mobility demand, as the ones whose results are reported in Figure 5, required a double run of the program, one to get T^B, differing from reference simulation one because the demand is varied, the other to get T^{TP} and T^{NP}. The main result of these simulations is that sensitivities of travel time savings to demand are quite strong, both for trip planners and non planners, even if the growth takes place at a slightly decreasing rate.

In quantitative terms one can say that increasing mobility

demand leads to proportional increases of travel time savings until the demand multiplication factor reaches a values of about 2.5-3; around factor value chosen for reference simulation, sensitivity of travel time savings of trip planners is only slightly diminishing and travel time savings continue to increase. For still higher factors, trip planners curve assumes a well-defined concave shape, while total and non planners' travel time savings reach a maximum at a 6.5 factor value. The demand values where such a decrease take place, however, correspond to unrealistic situations, as one can verify by maximal saturation indexes reported in Table 1.

Sensitivity of travel time savings to demand composition has been explored by varying trip planners percentages in the range from 5 to 95% with a 5% step; the output of trip planners percentage simulations reported in Figure 6, were quite interesting both in terms of general trends and quantitative values; trip planners travel time savings approach a non-zero value (about 11%) when user percentage goes to zero, then grow steadily up to about 15% when user percentage approaches 100%, while non planners travel time savings start from zero, when trip planners percentage is zero, then grow faster then trip planners ones, but at a decreasing rate, up to about 11% (total savings lay, as obvious, between the two, approaching non planners at 0 and trip planners at 100%).

From the point of view of information systems and demand management assessment these results are very promising because they show that travel time savings are quite substantial for trip planners from the very beginning, while also non planners begin to experience not marginal travel time savings at a relatively early stage (5% when trip planners are 20%). The effects of this favourable start don't vanish, since travel time savings continue to grow up, indicating that growth momentum is presumably not going to cease before most part of social and economic benefits have been attained.

Figure 5 Travel time savings with respect to mobility demand

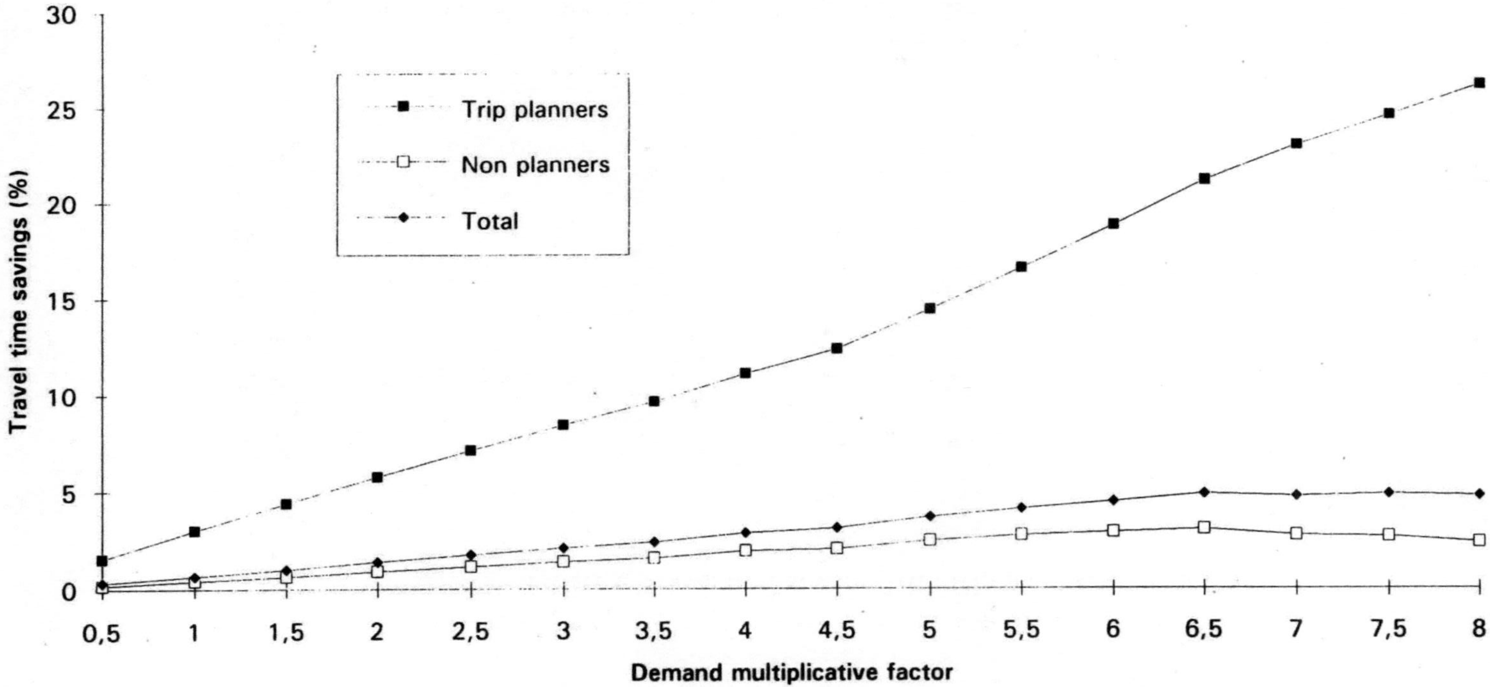

Figure 6 Travel time savings with trip planners percentage

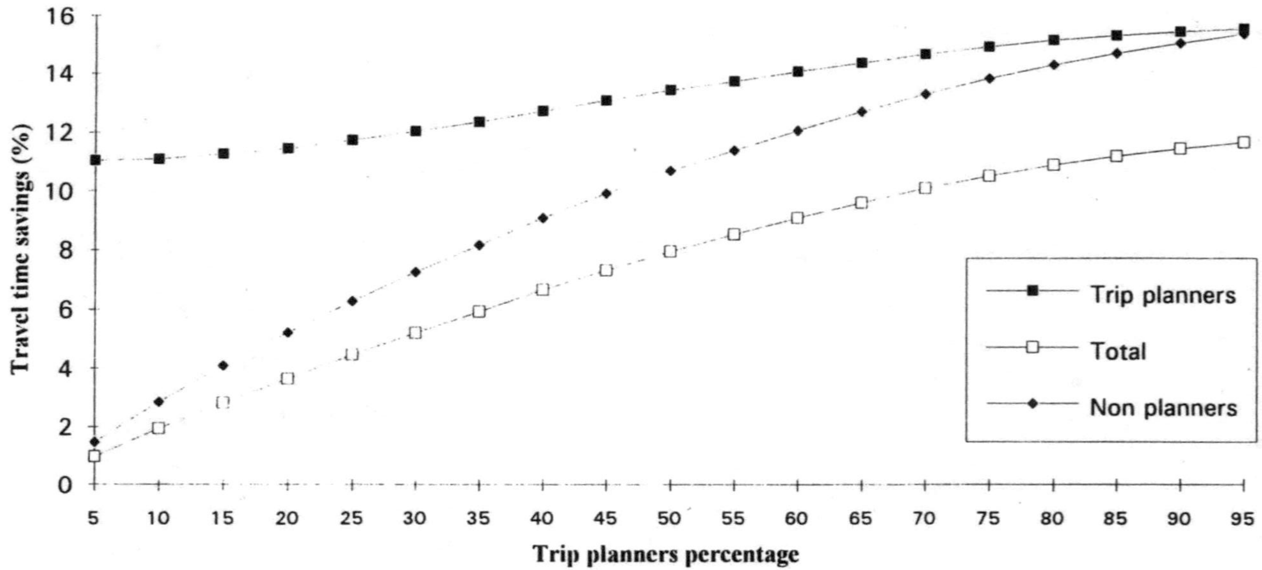

5.3 Penalty function shape simulations and results

As for the demand profile, the penalty function shape can vary in infinitely many ways and a somewhat arbitrary choice has to be made about what kind of variations from the reference simulation function to test. Here indifference band width (IBW), the slope of the function outside this band and the ratio between earliness and tardiness penalty were taken as the most distinctive features of function shape.

More precisely, no arrival time choice has been penalised for a number of intervals up to 16, symmetrically located before and after the first approximation arrival interval, while the arrival outside this band has not been allowed at all, to test consequences of various IBWs. As regards the slope of penalty functions, values in the range from 0 to 16 have been tested for tardiness penalty, with a step of 2, while at the same time keeping the 1 to 2 ratio between earliness and tardiness penalty as fixed; the variation of this ratio has been tested by keeping the tardiness penalty at 10 minutes per interval as in reference simulation, while earliness penalty varied as before.

The results of these simulations are summarised in Figure 7, which shows, as obvious, an increase of time savings with IBW, but whit evident saturation effects, more marked for trip planners than for non planners, as IBW approaches total freedom of arrival time choice for tripmakers (a situation not represented in the figure , being unrealistic an arrival time choice model with IBW equal to or larger than time horizon). A reduction in trip planners' savings sensitivity to IBW becomes however evident when it is about 6 intervals (1h.), which can be thought as a border between commuters' and occasional tripmakers' behaviour. Even if on a smaller scale, also the simulations carried out varying the measure and the balance of penalty show a saturation effect, opposite in sign to the one related with IBW (time savings decrease with increasing penalties), with simulations results quite rapidly convergent to the 2 interval indifference band case. From a quantitative point of view, one can say that, after accepting that a 2 interval indifference band

Figure 7 Travel time savings with penalty function shape

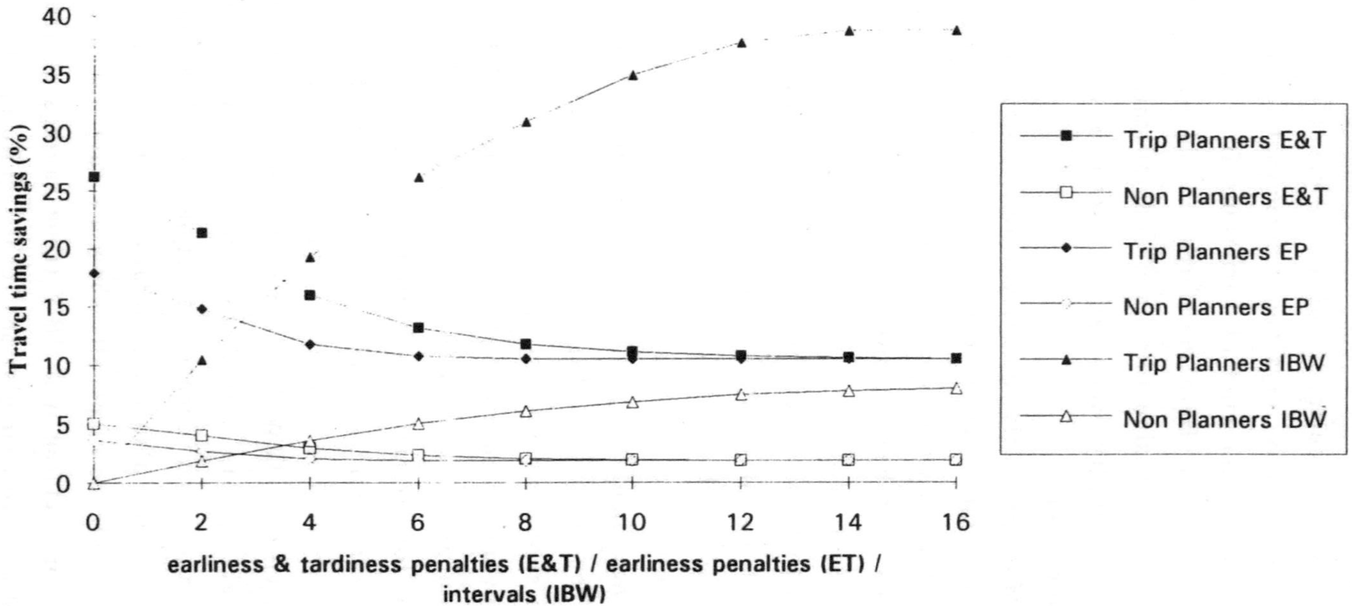

exists, a minimum of 12% travel time savings for trip planners and 2.5% for non planners is guaranteed, independently from additional flexibility. If flexibility is added, travel time savings increase only up to about 21% for trip planners and 4% for non planners if both earliness and tardiness penalty are varied down to a minimum of 2 and 1 minutes per interval and to about 15% for trip planners and 2.5% for non planners if there is a fixed 10 minute per interval tardiness penalty and earliness penalty is varied as above.

By comparing penalty function shape simulations anyway, IBW results clearly as the single most important feature of tripmaker behaviour as described by the penalty function, determining variations in travel time savings (for reference simulation) an order of magnitude higher than other simulation inputs. In addition, it should be noted that saturation effects for penalties appear when their values decrease under the ones assumed for reference simulation; thus time savings deriving from additional flexibility are small, and penalty function assumed as a reference point is representative. Two more simulations were performed, aimed at testing relative weight of IBW and schedule flexibility outside it, by taking the same slopes of penalty function as in the reference simulation and assuming no indifference band in the first and a one-interval indifference band only for early arrivals in the second.

As it was expected, the resulting values, reported in Table 2 confirm that IBW or, in behavioural terms, utility perception precision, is the main factor influencing travel time savings. This happens up to the point that assuming no IBW, substantial results for reference simulation become almost marginal (at least for the key group of trip planners, who may not appreciate a 2.3% saving as a result of trip planning, while in social terms, a reduction of around 1% of time lost in traffic has a certain weight). Assumption of asymmetric indifference band, even if accompanied by its halving, leaves on the contrary a good margin of opportunity for trip planners in the risk adverse behaviour (less heavy earliness than tardiness penalty) hypothesis made.

Multiplicative factor	Base case		Reference simulation	
	Max index	Interval	Max Index	Interval
1	9.0	8	8.2	8
2	16.1	8	14.1	8
3	22.6	8	21.4	8
4	28.3	7	27.3	7-8
5	33.3	7	32.5	7
6	37.3	6	36.7	6
7	41.2	6	40.3	6
8	44.2	5	43.6	5

Table 1. Maximal saturation index and mobility demand

Simulation	Travel time (mins.)			trip planners time equivalent penalty (mins.)
	Total	Trip planners	Non planners	
Reference	1427047	124608	1308120	5681
int.early IBW	1453567	134683	1318884	6429
Zero IBW	1477777	145707	1332070	9062
Base	1491338	149134	1342204	

Simulation	Travel time savings			trip planners time equivalent penalty (%)
	Total %	Trip planners %	Non planners %	
Reference	43.1	16.4	25.4	4.6
int.early IBW	25.3	9.7	17.4	4.8
Zero IBW	9.1	2.3	7.6	6.2

Table 2. Additional simulations results

6. Conclusions

The simulation exercise carried on allows some indications to be drawn out on information systems and demand management measures feasibility and effectiveness in different situations which can be summarized as follows:

- the level of congestion at which an intervention aimed at increasing trip planning induces travel time savings interesting for tripmakers, information and transport systems managers, public authorities, can be found in metropolitan areas;
- information systems development and demand management look quite smooth, without intrinsic critical points, even if generally loosing comparative advantage when spreading among tripmakers;
- main concern about behavioural aspects is IBW, because savings induced by trip planning are quite substantial only assuming a travel time and/or utility perception threshold in the order of ten minutes.

Determining in a certain context the true extension of IBW, such as other behavioural parameters, needs field analysis and model calibration, which would be the most effective research to pursue further. Some useful elements could be anyway obtained also by introducing randomness, reducing interval length and applying the method, still as a simulation device, to a large, congested, metropolitan area.

From the point of view of model calibration, the percentage of trip planners to assume as existent in a given situation has to be determined, because it looks unlikely that trip planning (as defined above) is not performed unless some information system is built up or some demand management action is taken.

References

[1] Bonsall P., Bell M., 1987, "*Information technology applications in transport*", VNU Science Press.

[2] Bielli M., 1990, "*Characteristics and requirements of trip planning systems*", Preprints of Prometheus Conference, London.

[3] Blum Y., Van Aerde M., 1988, "*Comprehensive approach to in-vehicle route guidance using Q-route*", Transportation Research Record 1232, TRB Washington.

[4] Robb C.M., 1987, "*Route Information Systems for Motorists*", *Transport Review*, vol.7, n.3, pp. 259-275.

[5] Hills P., Thorpe N., 1991, "*Traffic demand management in urban areas - The role of road pricing*", Proceedings of DRIVE Conference, Bruxelles, pp.658-664, Elsevier.

[6] Arnott R., A. de Palma, R. Lindsey, 1991, *"Does providing information to drivers reduce traffic congestion?"*, *Transpn. Res. A*, vol. 25A n.5 pp.309-318.

[7] Mahmassani H.S., Chang G.L., 1986, *"Experiments with departure time choice dynamics of urban commuters"*, *Transpn. Res. B*, Vol. 20B, n.4, pp.297-320.

[8] Mahmassani H.S., G.L. Chang, 1988, *"Travel Time Prediction and Departure Time Adjustment Behaviour Dynamics in a Congested Traffic System"*, *Transp. Res. B*, Vol.22B pp.217-232.

[9] Shirazi E., Anderson S., Stesney J., 1988, "*Commuter's attitude toward traffic information systems and route diversion*", Transp. Res. Rec. 1168, TRB, Washington.

[10] Mannering F.L., Hamed M.M., 1990, *"Occurrence, frequency and duration of commuters' work-to-home departure delay"*, *Transpn. Res. B*, Vol 24B, n.2, 99-109.

[11] Hamerslag R., 1989, "*Dynamic assignment in three-dimensional time space*", Transportation Research Record 1220, TRB- USA.

[12] Janson B.N., 1991, *"Dynamic Traffic Assignment for Urban Road Networks" Transp. Res.* 25B pp.143-161.

[13] Bellei G, Bielli M., 1990, "*Dynamic assignment for trip planning*

systems assessment", Proceedings of the ANIPLA Congress on Automation of Transport Systems, Trieste, 520-542.

[14] Cascetta E., Cantarella G.E., 1991, "*A day-to-day and within-day dynamic stochastic assignment model*", *Transpn. Res. A*, vol.25A, n.5, pp.277-291.

[15] Dehoux, Toint L., "*Some comments on dynamic traffic modelling in the presence of advanced driver information systems*", Proceedings of DRIVE Conference, Bruxelles, pp.964-980 Elsevier.

[16] Vythoulkas P.K., 1990, "*A dynamic stochastic assignment model for the analysis of general networks*", *Transpn. Rés. B*, vol.24B, n.6, pp.453-469.

[17] Wie B.W., 1991, "*Traffic assignment, Dynamic*", in Concise Encyclopedia of Traffic and Transportation Systems, pp.521-524, Papageorgiou M. Editor, Pergamon Press.

[18] Ben-Akiva M.,Cyna M., de Palma A., 1984, "*Dynamic model of peak period congestion*", *Transpn Res B*, Vol. 18B, n.4/5, pp 339-355.

[19] Ford L.R., D.R. Fulkerson, 1962, "*Flows in Networks*", Princeton University Press.

[20] Drissi-Kaïtouni O. A. Hameda-Benchekroun, 1992, "*A Dynamic Traffic Assignment Model and a Solution Algorithm*", *Transportation Science*, 26, pp. 119-128.

[21] Sheffi Y., 1985, "*Urban Transportation Networks*", Prentice Hall.

[22] Gallo G., Pallottino S., 1988, "Shortest path algorithms", *Annals of Operations Research*, vol.13, pp 3-79, Baltzer AG.

[23] Aronson J.E., 1989, "*A survey of dynamic network flows*", *Annals of Operations Research*, vol.20, pp 1-66, Baltzer AG.

AN ALGORITHM FOR EXTREMALS CALCULATION IN OPTIMAL CONTROL PROBLEMS WITH APPLICATIONS TO THE DYNAMIC TRAFFIC ASSIGNMENT PROBLEM

Esteve Codina, Jaume Barceló

Departament d'Estadística i Investigació Operativa
Facultat d'Informàtica
Universitat Politècnica de Catalunya, Spain

In this paper an algorithm to calculate extremals for the dynamic traffic assignment problem (DTAP) using a deterministic optimal control problem formulation is shown. This algorithm is based on a general method to integrate extremals for optimal control problems with inequality constraints in control variables. The DTAP is formulated and discretized and the consideration of using the method for extremals integration is made, showing how its steps would result for the DTAP, specially its inner loop, that becomes a network flow problem very similar to the stationary traffic assignment with separable costs objective function. Also, for the algorithm, it is remarked the need of techniques to update solutions in traffic assignment problems when, on the same network and the same cost functions, the demand varies slightly from one problem to the next. Finally aspects of the algorithm still under study are outlined. These aspects must be specialized for the DTAP in order to make possible the writing of competitive code specially suited to the treatment of medium-large urban networks

0. Introduction

In the last years dynamic behaviour of urban traffic flows under route guidance systems has been modeled by some authors (Ran B., Boyce D., Wie B.W. ..) using an optimal control approach formulation . This formulation has emerged as a dynamic extension of wardropian equilibrium principles (Wardrop 1952) widely used in urban transportation planning. Deterministic optimal control based formulation was stated initially by J.F. Luque and T.L. Friesz [6] , [7] , [9] (1980) coming from a Merchandt and Nemhauser model (1978) for networks with a single destination and have been extended to the case of multiple destinations by W. Wie Y.L. Friesz and R.L. Tobin [8]. Although theoretical properties of the stated problem are well known, research of efficiently specialized solution algorithms can not be yet considered closed, mainly for the case of large urban networks and recently Ran B., Shimazaki T. and Boyce D. [1] (1989) have developped solution algorithms based on time discretization of the continuous formulation and dualization and J.K. Ho [2] has developed an algorithm based on partial linearization and a multilevel application of the Dantzig - Wolfe decomposition principle. D.E. Boyce , B. Ran and L.J. Leblanc [15] (1991) presented the application of the Frank-Wolfe algorithm for the "directly discretized" multidestination DTAP showing how the continuous model is converted into a network flows problem.

In this paper an algorithm is described in order to calculate extremals for the dynamic traffic assignment problem (DTAP) when it is formulated as a deterministic optimal control problem (OCP). The algorithm is based on a general method to integrate extremals for OCP's with inequality constraints on the controls, and the inner loop of this algorith m is justified by two theorems. In the present paper, these theorems are enunciated without proof, being the second one restricted to quadratic functionals. Next and as a reference, the DTAP is formulated and discretized reproducing solution schemes already used [15] . After that, the integration method described in the first two sections is applied for the extremals of the DTAP and the problem solved in the inner loop is examined in order to show the characteristics of the resulting networks. As this problem is very similar to the well known stationary traffic assignment with separable cost objective function, it is remarked the need of a "link consecutive assignments" technique [13] as the one presented in [15] where the RSD algorithm is used. When compared with the direct discretization scheme presented in section 4 of this paper and used in [15], the extremals integration algorithm appears clearly as a network decomposition method allowing the calculation of the solutions for each time subinterval sequentially. This paper is organized in the following way : in section 1 the general algorithm for extremals calculation is shown. Section 2 the extends to the case of inequality constraints in the controls when the problem has quadratic functional. In section 3 the optimal control formulation of the DTAP is shown and in section 4) the application of the previous algorithms to the DTAP is discussed . In section 5 the need for "assignments linkage" techniques is discussed and in section 6 some extensions of the algorithm shown are considered and conclusions are given .

1. An algorithm to calculate extremals for Optimal Control Problems.

Let us consider the problem OCP1:

$$Min_{x,u}\ J[x,u] = \int_{t_0}^{T} f_0(x(t),u(t))\,dt \qquad u(t) : [t_0, T] \to R^m$$

$$\dot{x} = f(x(t), u(t)) \qquad x(t) : [t_0, T] \to R^n$$

with time interval $[t_0, T]$ fixed and initial conditions $x(t_0) = x_0$, free for terminal conditions and let´s assume that the hamiltonian $H = -f_0(x(t),u) + \lambda^T(t) f(x(t),u)$ is locally concave over an extremal $x(t), u(t), \lambda(t)$ (being $\lambda(t)$ the adjoint variables for (OCP1)), so the problem $\mathrm{Max}_u \; -f_0(x(t),u) + \lambda^T(t) f(x(t),u)$ defines locally a function $u = \hat{u}(x,\lambda)$ (the previous problem is the one used in the well known Pontryagin´s Maximum Principle). In these conditions extremals are given as solutions of the following ODE system (ODE1) (ODE= Ordinary Differential Equation (s))

$$\dot{x} = f(x, \hat{u}(x,\lambda))$$

$$\dot{\lambda} = -H_x \,|_{\hat{u}(x,\lambda)}$$

$$x(t_0) = x_0, \qquad \lambda(T) = \lambda_T (= 0)$$

We shall show a method to integrate system ODE1 when the terminal condition $\lambda(T)=\lambda_T$ is substituted by the initial one $\lambda(t_0)=\lambda_0$, being possible the use of the backward sweeping method (Ref [16]) or the single shooting method (Ref [17]) in order to reach the terminal condition $\lambda(T)=0$ (associated to $x(T)$ free). The method has as kernel the following optimitzation problem (OCP1):

$$\begin{array}{ll} Min_{u,x}\delta & f_0(x,u)-\lambda'^T x' \\ \delta & f(x,u)+x'-x=0 \quad |\lambda_0(L.M.) \end{array}$$

($\lambda', x', \delta=(T-t_0)/N$ are parameters in this problem)

If we write Kuhn Tucker conditions for problem (ΔOCP1) we obtain:

$$\begin{array}{l} \delta(f_{0u}-f_u^T) \quad |(u^*,x^*))=0 \\ \delta f_{0x} \quad -\lambda'=\delta \quad f_x^T\lambda^*-\lambda^* \end{array} \qquad (1)$$

It is clear that (1) along with the constraint $\delta \quad f(x,u)+x'-x=0$ in problem (ΔOCP1) are an approximation to the ODE system (ODE1) . Therefore it seems appealing, due to the similitudes shown between K-T conditions for problem (ΔOCP1) and the system (ODE1) , to look for an algorithm that, given the values $\tilde{x}(t-\delta)$, $\tilde{\lambda}(t))$ for an extremal, loops on λ' in order to:

a) Approximate the value $\tilde{x}(t)$ by x^* and the value $\tilde{\lambda}(t+\delta)$ by λ'_∞

b) Reproduce the value $\tilde{\lambda}(t)$; i.e. $\tilde{\lambda}(t)\equiv\lambda^*$

If this two properties are satisfied, then it is possible to think of reexecuting the algorithm from the approximation obtainded x^*, as parameter x' for (ΔOCP1) and λ'_∞, the new value to be matched by λ^*. In this way an approximation to an extremal would be obtained. The whole process is summarized as follows:

Algorithm AL0 (To integrate the system **ODE1**, with initial conditions $x(t_0)=x_0$, $\lambda(t_0+2\delta)=\lambda_2$)

0) Solve the problem:

$$\begin{array}{ll} Min_{u,x}\delta & f_0(x,u)-\lambda^2 x' \\ \delta & f(x,u)+x_0-x=0 \end{array} \quad \Big|\lambda^{(x_1,u_1,\lambda_1)} \; ; \qquad \text{Set } k=2$$

1) 1.0) Calculate initial prediction for λ^{k+}, for instance, by using :

$$\lambda^{k+1,0}=\lambda^k+\delta(f_{0x}-f_x^T \quad \lambda^k)|_{(x^{k-1},u^{k-1})}; \qquad \ell=0$$

1.1) Iteration over λ^{k+} and calculate (x^k, u^k) ; Solve the problem (P1):

$$\begin{array}{ll} Min_{u,x}\delta & f_0(x,u)-\lambda^{k+1,\ell}x' \\ \delta & f(x,u)+x_{k-1}-x=0 \end{array} \rightarrow (x^{k,\ell},u^{k,\ell},\lambda^{k,\ell}) \qquad \text{(P1)}$$

(As starting point for (P1) use, for instance:

a) if $l=0$ an updating of (x^{k-1},u^{k-1})

b) if $l>0$ use $(x^{k,\ell-1},u^{k,\ell-1})$

-if $\left\|\lambda^k-\lambda^{k,\ell}\right\|\le\varepsilon$ then goto 2) otherwise goto1.2)

1.2) $\lambda^{k+1,\ell+1}=\lambda^k+\delta(f_{0x}-f_x^T \quad \lambda^{k,\ell})|_{(x^{k,\ell},u^{k,\ell})}$

$\ell=\ell+1$; GOTO 1.1)

$\lambda^{k+1}\leftarrow\lambda^{k+1,\ell}$

2) $(x^k,u^k)\leftarrow(x^{k-1},u^{k-1})$

$k\leftarrow k+1; GOTO \quad 1$

Until the final time subinterval is reached.

In algorithm AL0, if $\hat{\ell}$ is the last iteration of the inner loop for a given time subinterval i, then $\lambda^{i,\hat{\ell}}\approx\lambda^i$ (the approximation of $\tilde{\lambda}(t)$), $i\geq 2$, linking the set of Kuhn Tucker equations for problem (ΔOCP1) in time subinterval i-th with the set of Kuhn Tucker equations for problem (ΔOCP1) in time subinterval i+1-th.

Inner loop iteration can be proved to be convergent. The proof relies on the well known theorem of "fixed point iteration" for contractive maps. We do not include here the proof and just enunciate the theorem.

Theorem 1 (convergence of inner loop iteration for problem (ΔOCP1))

Let us define $\phi_\delta(\nu) = \lambda^k + \delta(f_{ox} - f_x^T\lambda)\ ](x^*(\nu), u^*(\nu), \lambda^*(\nu))$,for a given time subinterval k, being $(x^*(\nu), u^*(\nu), \lambda^*(\nu))$ the solution of (ΔOCP1) with $x' = x^{k-1}$ and $\lambda' = \nu$. Also, let's supose that $-H(x_{k-1}, u, \lambda_{k-1})$ is locally strictly convex on $\hat{u}(x^{k-1}, \lambda^{k-1})$, thus defining $\hat{u}(.,.)$ in a neighbourhood of (x^{k-1}, λ^{k-1}). Otherwise stated, the problem $Max_u \quad H(x_{k-1}, u, \lambda_{k-1})$ has second order sufficiency conditions in its solution and defines $\hat{u}(.,.)$). In these conditions there exists an open interval $]\,0, \hat{\delta}\,[\subset R$ such that :

a) $\forall \ \delta \in]0, \hat{\delta}[$, $\phi_\delta(\nu)$ is a contractive map over a neighbourhood E of λ^k and $\phi_\delta(E) \subset E$. Therefore ϕ_δ has a fixed point ν^* i.e. $\nu^* = \phi_\delta(\nu^*)$

b) $\lambda^*(\nu^*) \equiv \lambda^k$

It can also be seen that the algorithm described can be adapted to handle an ODE system $\dot{y} = \varphi(y, t)$ that can be partitioned as: $\dot{x} = f(\lambda, x, t)$, $\dot{\lambda} = g(\lambda, x, t)$ and then the integration scheme becomes:

$$x_{k+1} = x_k + \delta f_k(x_{k+1}, \lambda_{k+1})$$
$$\lambda_{k+1} = \lambda_k + \delta g_k(x_k, \lambda_k) \qquad k = 0, 1, 2, 3, \ldots$$

It is easy to check that the above scheme is consistent with the ODE system $\dot{y} = \varphi(y, t)$, strongly stable and convergent and it matches algorithm AL0 when x_0, λ_2 are given as initial conditions instead of x_0, λ_0.

2. An extension when there are inequality constraints on the control variables, $u \geq 0$.

We shall handle the case of a quadratic functional, $f_0(x,u) = 1/2Q(x,u)$ being $Q(x,u) = xXx + uUu + 2au + 2bu$, and $f(x,u)$ linear, so that the problem **(OCPQU2)** with initial known state we are to deal with in this section is :

$$\begin{aligned} &Min_{x,u} \quad 1/2\int_{t0}^{T} Q(x,u)dt \\ &\dot{x} = Au + B(t) \quad |\lambda(t) \\ &u(t) \geq 0 \quad |\zeta(t) \end{aligned}$$

We assume that X and U are positive definite matrices, so that problem (OCPQU2) adjusts to the sufficiency theorems of A. Seierstad and K. Sydsaeter [19]. By discretizing it directly, an approximation of an optimal extremal would be obtained with terminal conditions for the adjoint variables. The following discretization, Δ OCPQU(λ_T), has as solution an approximation of an extremal ending with $\lambda(T) = \lambda_T$.

$$\begin{aligned} &Min_{x,u} \quad 1/2\delta\sum_{i=1}^{N} Q(x_i,u_i) - \lambda' x_{N+1} \\ &x_i - x_{i-1} = \delta A u_i + \delta B_i \\ &u_i \geq 0 \end{aligned}$$

Let $(x_i^*, u_i^*, \lambda_i^*, \zeta_i^*) = \omega_i$, i = 1, ... , N , a Kuhn Tucker point for problem Δ OCPQU(λ_T) and let's assume that second order conditions and strict complementarity hold . If we want to extend algorithm AL0 for problem Δ OCPQU(λ_T) the kernel problem is now :

$$\begin{aligned} Min_{x,u} \quad & 1/2\delta Q(x,u)-\lambda' x \\ & x-x0=\delta Au+\delta B \quad |\lambda \\ & u\geq 0 \quad |\varsigma \end{aligned}$$

where on each time subinterval i we would substitute $x_o \equiv x^*_{i-1}$ and $B \equiv B_i$, and inner loop iteration on algorithm AL0 would be sintetized as : $\lambda'_{l+1} = F_i(\lambda'_l) = \lambda^*_i + \delta Xx(\lambda'_l)$ (where $x(\lambda'_l)$ comes from the solution $(x(\lambda'), u(\lambda'), \lambda(\lambda'), \zeta(\lambda'))=\omega(\lambda')$ for problem (Δ OCPQU2) with $\lambda' \equiv \lambda'_l$ .The following theorem states that the above iterative process is convergent and that $\lambda^*_{i+1} = \lambda'_\infty$.

Theorem 2 (convergence of $\lambda'_{i+1} = F_i(\lambda'_i)$ for problems (Λ OPCQU2))

If the above conditions for problem (Λ **OPCQU**) hold, then there exists a neighbourhood E for the fix point of $F_i(\lambda') = \lambda^*_i + \delta Xx(\lambda')$, λ^*_{i+1} , (where λ^*_{i+1} comes from the solution $(x^*_i, u^*_i, \lambda^*_i, \zeta^*_i) = \omega_i$ of problem Δ OCPQU(λ_T)) , verifying:

a) $\forall \lambda'_1, \lambda'_2 \in E$ there must hold $A^*_u(\lambda'_1) = A^*_u(\lambda'_2)$
($A^*_u(\lambda')$ is the active set for constraints $u \geq 0$ in problem ΔOCPQU2 with parameter λ'.)

b) $\left\|\frac{\partial F_i}{\partial \lambda'}\right\| \leq 1, \quad \forall \lambda' \in E$ for $\delta \in I_\delta = \left\{\delta \in \left]0, \hat{\delta}\right[\quad |P_6(\delta) < 1\right\}$
(with $\hat{\delta}$ and $P_6(\delta) = a_6\delta^6 + a_4\delta^4 + a_2\delta^2$ being a polynomial whose coeficients depend on the euclidean norms of matrices A, U and X)

c) Iteration $\lambda'_{i+1} = F_i(\lambda'_i)$ converges to λ^*_{i+1} and $\lambda'_\infty \equiv \lambda^*_{i+1}$ if $\lambda'_0 \in E$.

(This results can be also extended without problem to the case where additional constraints of the type Cu(t) = D (t) are present in problem (OPCQU). Of course, the algorithm can be used when inequality contraints in the state variables are present but remain inactive for the extremal being calculated.) . Let's see the use of algorithm AL0 with a little example :

$$Min_{x,u} \quad \int_0^1 (x^2(t)+u^2(t))dt$$
$$\dot{x} = d(t)-u(t)$$
$$u(t) \geq 0$$

with d(t) defined as d(t)=-10(t-0.15) if 0<= t <= 0.15 and d(t)=0 if 0.15<= t <= 1.0 :

$$H(x,\lambda,u) = -x^2 - u^2 + \lambda(d(t)-u) \quad \Rightarrow \hat{u}(x,\lambda) = max\{0,-\lambda/2\}$$

The ODE system for extremals is $\dot{\lambda} = 2x, \quad \dot{x} = d(t)+\lambda/2$: . Let us consider the extremal with x(0)=1, $\lambda(0) = -1.4$, that on time subinterval takes the form:

$$\lambda(t) = 2(Ae^t - Be^{-t})$$
$$x(t) = Ae^t + Be^{-t}$$

With A=0.2035398 and B=0.9091712. $\lambda(t)$ vanishes at t=0.7483357. The inner loop of AL0 consists of solving successive problems of th form:

$$Min_{x,u} \quad \delta(x^2+u^2) - \lambda' x$$
$$\delta(d-u) - x + x' = 0$$
$$u \geq 0$$

The following tables show the use of algorithm AL0 for two values of δ : δ =0.05 and δ =0.1

TABLE 0 (values for the extremal)

t	x(t)		
0.65	0.8645179	----	----
0.7	0.8613600	-0.0832042	0.0416021
0.75	0.8603560	0.0028637	0.0
0.8	----	0.0889382	0.0
0.85	----	0.1752361	0.0

TABLE 1 $\delta = 0.05$

	t = 0.65,0.7,0.75			t = 0.7,0.75,0.8
	l = 1	l = 2	l = 3	l = 1
x'	0.8645179	0.8645179	0.8645179	0.8624377
λ_i'	0.0031	0.0030397	0.0030395	0.0892832
x^*	0.8624393	0.8624377	0.8624377	0.8624377
u^*	0.0415719	0.0416020	0.0416020	0.0
λ^*	-0.0831439	-0.0832040	-0.0832042	0.0030395
λ^i	-0.0832042	-0.0832042	-0.0832042	0.0030395
λ_{i+1}'	0.0030397	0.0030395	0.0030395	----

TABLE 2 $\delta = 0.05$ t = 0.65,0.7,0.75

x'	0.8645179	0.8645179	0.8645179
λ_i'	0.005	0.0030444	0.0030395
x^*	0.8624866	0.8624379	0.8624377
u^*	0.0406245	0.0415996	0.0416020
λ^*	-0.0812486	-0.0831993	-0.0832042
λ^i	-0.0832042	-0.0832042	-0.0832042
λ_{i+1}'	0.0030444	0.0030395	0.0030395

TABLE 3 $\delta = 0.05$ t = 0.7,0.75,0.8

x'	0.8624377	0.8624377	0.8624377
λ_i'	0.0	0.0890681	0.0892832
x^*	0.8602869	0.8624377	0.8624377
u^*	0.0430114	0.0	0.0
λ^*	-0.0860289	0.0028244	0.0030394
λ^i	0.0030395	0.0030395	0.0030395
λ_{i+1}'	0.0890681	0.0892832	----

TABLE 4 $\delta = 0.1$ t = 0.65, 0.75, 0.85

	l = 1	l = 2	l = 3
x'	0.8645179	0.8645179	0.8645179
λ_i'	0.0	0.1740553	0.1757672
x^*	0.8559583	0.8645179	0.8645179
u^*	0.0855958	0.0	0.0
λ^*	-0.1711916	0.0011517	0.0028637
λ^i	0.0028637	0.0028637	0.0028637
λ_{i+1}'	0.1740553	0.1757672	----

In table 1 a good initial approximation of $\lambda(0.75)$ is used and the step 1.0) of algorithm AL0 is followed in order to enter the inner loop in time intervals [0.7, 0.75] , [0.75, 0.8] . In table 2 a bad initial approximation of $\lambda(0.75)$ for time intervals [0.65, 0.7] , [0.7 , 0.75] . is used. In spite of that the convergence is also achieved. Notice that the control variable vanishes. In table 3 a bad initial approximation for $\lambda(0.8)$ in time intervals [0.7, 0.75] , [0.75, 0.8] is used . However the convergence is achieved.In table 4 a step length of δ =0.1 is used and the only pair of intervals are [0.65, 0.75] and [0.75, 0.85] a bad initial approximation of $\lambda(0.85)$ is used and also the convergence is achieved.

3. Formulation of the Dynamic Traffic Assignment Problem (DTAP) as an Optimal Control Problem.

The algorithms described can be applied to calculate extremals of the dynamic traffic assignment problem (DTAP) when it is formulated as a deterministic optimal control problem in the way done by authors like Ran B., Boyce D., Wie B.W. and others ([1], [6], [7], [8], [9], [15]) . We reproduce here this formulation:

Let's assume a graph G = (N, A) given by an incidence matrix B with n nodes and m links. In [1] a multicommodity formulation is used in which each commodity are trips ending in a common destination of the network. Throughout this section the graph G represented by the incidence matrix B shall be called the original graph or just the B-graph. There follows the most used notations:

[t0,T]	-	The time interval.
xa(t)	-	vehicles in link a Î A of B-graph at time t [t0,T].x(t) is an array of number of links (m) components.
ua(t)	-	incoming flow in link a Î A of B-graph at time t [t0,T].u(t) is an array of m components.

va(t) - outgoing flow from link a Î A of B-graph at time t [t0,T].v(t) is an array of m components.

O - subset of nodes in G that are origins.

D - subset of nodes in G that are destinations.

xq(t), uq(t), vq(t) analogous to x(t), u(t), v(t) but specific for destination q Î D.

K' = O x D be set of possible O-D pairs.

Let p(t)³0 be a function of |K'| components defined by: pk(t) is the number of trips per unit of time at time t that enter the network at node ℓ Î O and have destination m Î D (K = (ℓ,m) ÎK').

It is then possible to define the set of active O-D pairs, K Í K' as :

K ° { k Î K' | pk(t)>0 and xk(t0)>0, t Î [t0,T] }

In this model input flows to network, pk(t), are known but outputs grom network, sk(t) are unknown and must be determined by solving the dynamic problem. (sk(t) are trips that exit the network at time t at destination m Î D coming from node ℓ Î O, k = (ℓ,m) ÎK')

In the following formulations dk(t), k= (ℓ,m) ÎK is a vector of n componets defined by:

$$(d^k(t))_i = \begin{cases} 0, & i \neq \ell \quad and \quad i \neq m \\ p^k(t), & i = \ell \\ -s^k(t), & i = m \end{cases}$$

Let Fa (xa,ua,va) = Ua (ua) + Va (va) + Xa (xa)

$$(\mathrm{U}_a(u_a) = \int_0^{u_a} U_a(\omega)d\omega, \quad \mathrm{V}_a(v_a) = \int_0^{v_a} V_a(\omega)d\omega, \quad \mathrm{X}_a(v_a) = \int_0^{x_a} X_a(\omega)d\omega)$$

if F(x, u, v) = å a ÎA Fa(xa, ua, va) then the dynamic trafic assignment problem can be formulated as:

$$Min_{u^k, v^k, x^k, s^k} \quad \int_{t0}^{T} F(x,u,v)dt$$

xk(t) = uk(t)-vk(t) (2)

B+uk(t)-B-vk(t) = dk(t) (3)

xk(t)³0,uk(t)³0,vk(t)³0,sk(t)³0,"kÎK

xk(t0)=x_o^k (4)

xk(T) free (boundary cond.)

Through (2) and (3) the following redundant equation is obtained:

$$\sum_{a \in A} x_a^k(t) = p^k(t) - s^k(t)$$

4. Application of algorithm AL0 to (DTAP).

It has to be pointed out that the order of contact p of the constraints $x(t) \geq 0$ is p=1 and as the time interval $[t_0, \ T]$ is fixed, this implies that $u^q(t)$, the control variables, $\lambda^q(t)$, the adjoint variables for the dynamic system equations $\dot{x}^q = u^q - v^q$ and, $\dot{\eta}^q(t)$ the Lagrange multipliers for the constraints $x^q(t) \geq 0$, are all them continuous along $[t_0, \ T]$ (See [14]) . Therefore as there shall not be jumps in the adjoint variables, $\lambda^q(t)$, and in the Lagrange multipliers $\eta^q(t)$, we can discretize problem DTAP1 in N time subintervals $[t_o, t_1], [t_1, t_2], \ldots, [t_{N-1}, T]$ of equal length δ The mathematical programming problem (ΔDTAP1) formulated below has as solution an approximation of an extremal of DTAP1 with the property $\lambda^q(T) \approx \lambda_T^q, \quad \forall q \in D$.

$$\mathrm{Min}_{x^q, v^q, u^q, s^q} \quad \delta \sum_{i=1}^{N} \left\{ \sum_{a \in A} F_a(x_a, u_a, v_a) \right\} - \sum_{q \in D} \lambda_T^q x_{N+1}^q$$

$$x_i^q - x_{i-1}^q = \delta (u_i^q - v_i^q), \ i = 1, \ldots, N \ , \in Q$$

$$B_+ u_i^q - B_- v_i^q + e_q s_i^q = p_i^q$$

$$x_i^q \geq 0, u_i^q \geq 0, v_i^q \geq 0, s_i^q \geq 0$$

In [15] the Frank - Wolfe algorithm is used to solve directly this formulation. Also, in [15] the network structure for (ΔDTAP1) is clearly shown, consisting in as many stages as time subintervals are considered. Another particularity shown in [15] is that the network "seen" varies slightly from destination to destination. If we apply algorithm AL0 to calculate an extremal for problem DTAP1 we should have to carry out, at each time subinterval i, the inner iteration 1.1→ 1.2 1, solving the following problem (ΔDTAP2):

$$\mathrm{Min}_{x^q, v^q, u^q, s^q} \quad \delta \sum_{a \in A} F_a(x_a, u_a/\delta, v_a/\delta) - \sum_{q \in D} \lambda_T'^q x^q$$

$$x^q - x_{i-1}^q = \mathbf{u}^q - \mathbf{v}^q, \qquad q \in D$$

$$B_+ \mathbf{u}^q - B_- \mathbf{v}^q + e_q \mathbf{s}^q = \delta \dot{p}_i^q$$

$$x^q \geq 0, \mathbf{u}^q \geq 0, \mathbf{v}^q \geq 0, \mathbf{s}^q \geq 0$$

with:

$$\mathbf{v}^q = \delta v^q, \ \mathbf{u}^q = \delta u^q, \ \mathbf{s}^q = \delta s^q$$

$$(v_a = \sum_{q \in D} v_a^q, \quad u_a = \sum_{q \in D} u_a^q, \quad x_a = \sum_{q \in D} x_a^q)$$

(at each iteration 1.1→ 1.2 1. λ′ would be "refined")

It is worthwhile to examine the graph structure for the constraints of (ΔDTAP1) and (ΔDTAP2) with a little but illustrative example: let G = (N, A) the graph depicted below.

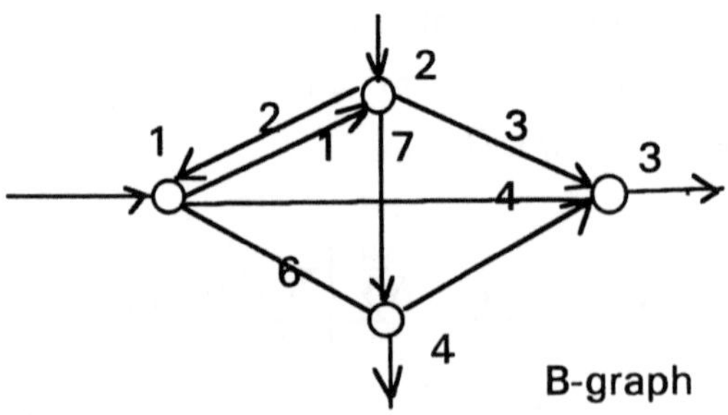

Figure 1. The original graph G of the example.

This is a problem with two destinations: D={3,4}. Problems (ΔDTAP1) and (ΔDTAP2) have separability for the destinations; let us focus on destination q = 3. The figure 2 shows the "time expanded network" structure seen for destination q = 3 with N = 3 time subintervals, for the example graph G.

Let's examine the network structure in (ΔDTAP2) for the example graph of figure 1. and let $x_{i-1}^{3,T} = (x_1^3, \dots, x_7^3)$ be the state variables for the previous time subinterval and δp_1^3, δp_2^3 flows entering the network at time subinterval i on nodes 1, 2 respectively. Clearly, problem (ΔDTAP2) is an equilibrium problem that for destination q = 3 uses the network depicted in figure 3.

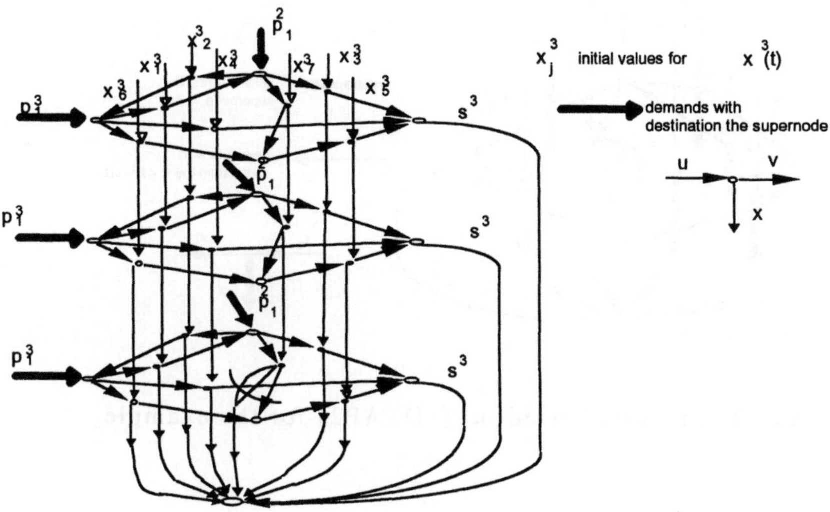

Figure 2 The "time expanded" network for the graph G.

There are four kinds of links in the depicted network, the u-links , the v-links, the x-links and the s-link going from destination 3 to the "supernode"; costs for u-links and v-links would be

$\frac{\partial F_a}{\partial u_a}$, $\frac{\partial F_a}{\partial v_a}$, respectively, but for x-links costs would be

$\delta\frac{\partial F_a}{\partial x_a} - \lambda'^q_a$ and are destination dependent. For destination q = 4 the network would be the same excluding link s^3 . Instead of this link, the network would have link s^4 going from node 4 and entering the supernode. Also, demands would change being x^4_{l-1} and δp^4_1, δp^4_2, respectively.

In the loop *1.1→ 1.2→ 1.* of algorithm AL0 only the objecive function changes from iteration to iteration, therefore the optimum solution for iteration l -1 , say $(\mathbf{x}^*_{l-1}, \mathbf{u}^*_{l-1}, \mathbf{v}^*_{l-1}, \mathbf{s}^*_{l-1})$, can be used as starting point for iteration l and thus, for each iteration, few steps of the solution algorithm for (ΔDTAP2) are needed to obtain $(\mathbf{x}^*_l, \mathbf{u}^*_l, \mathbf{v}^*_l, \mathbf{s}^*_l)$.

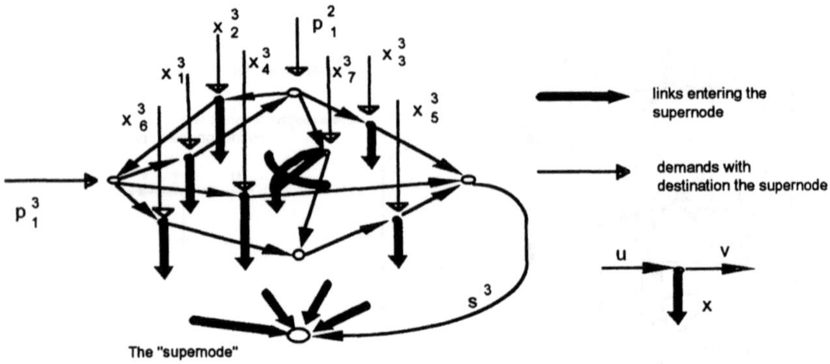

Figure 3. The graph used in (ΔDTAP2) for the example.

5. The need for "assignments linkage" tecniques and the Simplicial Decomposition Algorithm for DTAP.

In the previous section it has been shown that in the inner loop *1.1→ 1.2→ 1.* only the objective function of (ΔDTAP2) changes from iteration to iteration. Also it has been shown that at each time subinterval the state variables x^q and the demands change (because of the derivability properties for the state variables [14], these changes are expected to be "small") and that the state variables for time subinterval i-1 can be interpreted as demands or flow injections on L-nodes (see Fig. 1) . With the help of some smoothness property for the actual demands $p^q(t)$ it seems necessary in order to reduce computer costs the use of a technique to find a good starting point $(\mathbf{x}_o^*, \mathbf{u}_o^*, \mathbf{v}_o^*, \mathbf{s}_o^*)$ calculated with low computational, cost from the point obtained at time interval i-1 to start the inner loop *1.1→ 1.2→ 1.* at time interval i . In [13] two methods are described to link consecutive stationary traffic assignments problems for large size networks when, from one problem to the next the only difference are "slight changes" in the O-D matrix. The methods can be used advantageously when the RSD algorithm [3], [4], [5] is used for the separable cost traffic assignment problem TAP because they are based on the storage of a reduced subset of the O-D paths in the network for TAP k -th and the updating of flows on these paths for the next problem k + 1 th . These methods can be easily readapted for the case of solving consecutive problems i -th , i +1 th of the type ($\wedge$DTAP2)

6. Aspects to study and conclusions.

In the previous sections an algorithm to integrate extremals for OCP's with inequality constraints on the controls has been shown and some of the implications of its use on DTAP have been highlighted . The algorithm, when applied on DTAP, allows the partition of the "time expanded network" implying a significant reduction of its size. The computational cost of the process can be reduced if a tecnique for "assignment linkage" is used to update solutions from one time subinterval to the next. The RSD algorithm appears specially suited to be used for this purpose . However there can be other factors that can influence the performance of the global process of extremal calculation and that must be studied previously to the elaboration of a code in order to obtain computational results:

- As the extremal calculation process is , in fact , an integration method for ODE's , then it is necessary to balance the approximation errors with errors involved in solving (ΔDTAP2) Therefore it appears necessary the elaboration of a stopping criterion for (ΔDTAP2) specially adapted for this purpose and the connection between this stopping criterion and the time interval length δ must be studied.

- If algorithm RSD is to be used to solve (ΔDTAP2) then it is known that, for the stationary traffic assignment problem, the computer cost involved splits into ~80 % for the subproblem stage (shortest path calculations) and ~20% for the master problem using algorithms with high order of convergence as [12] or few evaluations of the objective function and its derivatives [5] . For the shortest path calculations, the network structure strongly influences the computational performance of the algorithm as shown in [10] , and in [10] and [11] and [19] efficient shortest path methods are described for networks with characteristics close to the urban networks. In the examples of section 4 of this paper it is shown also that when algorithm AL0 is applied to the DTAP the decomposition of the "time expanded" network results not only in size reduction but also makes the networks for (ΔDTAP2) closer in structure to the original graph. It is currently under development a shortest path algorithm for the type of networks used in (ΔDTAP2) with computational performance relying on the performance of any shortest path algorithm for the original graphs with the only restriction that all the nodes in the original network must be reached before the algorithm stops.

- The problem DTAP has inequlity constraints in the state variables and , in practice, this constraints can become active. Even more, if $A^*_{x,i}$ is the set of indexes for the active components of the vector of state variables for time subinterval i then in general

$\overset{*}{A}_{x,i} \neq \overset{*}{A}_{x,i+1}$. An extension of algorithm AL0 is currently being developped in order to handle the inequality constraints on the state variables by using penalties.

- Problem $^{\Delta}$DTAP2 is a convex network flow problem with uniqueness in x_a, u_a, v_a, s_a but not in $x^q{}_a, u^q{}_a, v^q{}_a, s^q{}_a$; the addition of a quadratic term of the type $\varepsilon \Sigma_{a\in A} \Sigma_{q\in D}\left\{ (x_a^q)^2 + (\mathbf{u}_a^q)^2 + (\mathbf{v}_a^q)^2 + (\mathbf{s}_a^q)^2 \right\}$ is proposed thus being εobtained an "ε-equilibrium ". The addition of a quadratic term is only needed for uniqueness in the case of more than one destination in the network. It must be noted that the sequence of solutions obtained by using algorithm AL0 with the "ε-term" for the N time subintervals can be used as a good initial and feasible solution of the direct discretized problem ΔDTAP1.

- If terminal conditions are to be satisfied , then it must be also studied the inclusion of algorithm AL0 as part of a single shooting algorithm [17] (or in the backward sweeping method [16]). Terminal conditions of the form $\lambda_T = 0$ can be achieved by integrating the variational equations as the extremal is being calculated and using a zero finding algorithm to solve $\lambda_T(\lambda_0) = 0$ having a good approximation of $\left(\frac{\partial \lambda_T}{\partial \lambda_0}\right)$.

References

[1] B. Ran, T. Shimazaki, D.E. Boyce. " *A Dynamic User Equilibrium Traffic Assignment Model and its Extensions: The Optimal Control Approach*". Working paper.

[2] J.K. Ho " *Solving the Dynamic Traffic Assignment Problem on a Hypercube Multicomputer*" Transportation Research Vol 24 B 6 1990.

[3] D.W. Hearn, S. Langphowpanich, J.A. Ventura. *"Restricted Simplicial Decomposition: Computation and Extensions"* Mathematical Programming Study 31 p. 99-118 1987.

[4] D.W. Hearn, S. Langphowpanich, J.A. Ventura. *"Optimization ALgorithms for Congested Network Models"* Dept. of Ind. and Systems Eng. Research Report 87-3

[5] S. Langphowpanich, D.W. Hearn. *"Restricted Simplicial Decomposition with Application to the Traffic Assignment Problem"* Ricerca Operativa 38 1986 p. 97 - 120

[6] B.W. Wie, T.L. Friesz, R.L. Tobin. *"Dynamic User Optimal Traffic Assignment: A Control Theoretic Formulation"* US - Italy Joint Seminar on Urban Transportation Networks, Capri Italy June 20-23 1989.

[7] B.W. Wie, T.L. Friesz, R.L. Tobin. *"Dynamic User Optimal Traffic Assignment on Congested Multidestination Networks"* Transportation Research 24 B 6 December 1990.

[8] T.L. Friesz, F. J. Luque, R.L. Tobin, B.W. Wie. *"Dynamic Network Traffic Assignment Considered as a Continuous Time Optimal Problem"* Op. Research Vol 37-6 pags 58-69 USA

[9] T.L. Friesz, F. J. Luque, R.L. Tobin, B.W. Wie. *"Dynamic Network Traffic Assignment Considered as a Continuous Time Optimal Problem"* TIMS/ORSA Joint National Meeting, Washington USA 1980

[10] G.Gallo, S. Pallotino *"Shortest Path Methods in Transportation Models "* Proceedings in the Course given at the ICTS (Amalfi, Italy) October 11-16 1982

[11] U. Pape *"Implementation and Efficiency of Moore Algorithms for Shortest Route Problem"* Math. Progr. 7 (1974) 212-222

[12] D. Bertsekas *"Projected Newton Methods for Optimization Problems with Simple Constraints"* SIAM J. Control and Optimization. Vol 20, no. 2, March 1982.

[13] E.Codina, J.Barceló *"An Algorithm for Dynamic Traffic Assignement Model based on Optimal Control Approach"* Preliminary version First Triennial Symposium on Transportation Analysis (TRISTAN) Montreal, Canada June 1991.

[14| D.H. Jacobson, M.M. Lele and Speyer *"New Necessary Conditions of Optimality for Control Problems with State-Variable Inequality*

Constraints." Journal of Mathematical Analysis and Applications 35, p 255 284 (1971)

[15] D.E. Boyce, B.Ran, L.J. Leblanc. *"Dynamic User Optimal Traffic Assignment: A new Model and Solution Technique"* First Triennial Symposium on Transportation Analysis (TRISTAN) Montreal, Canada June 1991.

[16| Bryson A.E. , Ho Y-C *"Applied Optimal Control"* John Wiley & Sons, New York (1975)

[17| Dahlquist G. Björk A, Anderson N. *"Numerical Methods"* Prectice Hall Inc. 1974

[18] G.Gallo and S. Pallotino *"Shortest path Algorithms"* Annals of Operations Research, Vol. 13 (1988), pp. 3-79

[19] A. Seierstad and K. Sydsaeter *"Sufficient Conditions in Optimal Control Theory"* International Economic Review, Vol. 18, No.2 , June 1977

DYNAMIC CONGESTION PRICING AND THE PRICE OF FIFO

Malachy Carey

Faculty of Business and Management, University of Ulster, Northern Ireland, BT37 0QB

Congestion costs and optimal tolls have long been computed from static (single period) traffic assignment models. Only recently have efforts been made to derive congestion prices from dynamic traffic assignment models (i.e., models having time–varying traffic flows). We consider this problem here, and find that in the dynamic context, congestion costs imposed by users consist of a capacity cost plus a component associated with the first–in–first–out (FIFO) property of road traffic.

To show this, we set out a dynamic system optimizing model with FIFO constraints, derive expressions for system marginal costs, externalities and hence optimal congestion prices or tolls, identify the components of these prices which are due to FIFO, and derive various properties of these components.

An interesting feature of the FIFO components of dynamic congestion prices, for the models presented here, is that for some users they are negative (i.e., subsidies) and for others they are positive, and the sum of the resulting negative and positive FIFO components of toll revenues is zero. Subsidies are not actually paid, as the FIFO components simply reduce or increase the tolls which would otherwise be charged. In effect, some users compensate others to induce them to travel on other links or at other times.

1. Introduction

To derive optimal congestion prices for road traffic, it is necessary to explicitly or implicitly set out a system optimizing (cost minimizing or benefit maximizing) model of traffic flows. If travel demands and traffic flows vary over time (e.g., over the peak period), the resulting model is said to be a "dynamic assignment" model. It has been shown elsewhere that in such traffic network optimization models additional constraints are needed to ensure that the solution adheres to a first-in-first-out (FIFO) property, which is an inherent physical characteristic of road traffic (see [3]).

The congestion pricing implications of dynamic traffic models with FIFO constraints do not appear to have been explored previously. Congestion pricing has normally been based on static assignment models, and the FIFO problem does not arise explicitly in static assignment (SA) models for the following reason. SA models are single period models and implicitly assume that, in this single period, the traffic of each type on each link enters and exits at a uniform rate and is uniformly spread along the link. SA models, do not keep track of the time at which each component of traffic enters or leaves links, — hence the question of 'first' in or out does not arise. Even when congestion pricing has been raised in a dynamic context (e.g., [4]) a *single* traffic type or single destination has been assumed. so that maintaining the FIFO order between different traffic types does not arise.

It is well known that under congested traffic conditions each road user imposes a congestion cost (externality) on other users, and that by charging each user a congestion price (toll) equal to this externality the users may be induced to take account of this costs and hence adopt a 'system optimizing' pattern of flows. These congestion prices are computed from system optimizing models of traffic flows, and have been extensively investigated for static (single period) assignment models. However, we find that in the case of dynamic assignment the form of the externality (and hence optimal toll) is different, with different interpretations and properties. A component of this new externality and toll is due to the FIFO constraints. This component ensures that users take account of the costs which their route choices impose on others as a result of the FIFO characteristic of road traffic.

In Section 2 we consider dynamic congestion costs and tolls for a simple network example. We extend the discussion in Section 3 to a more general multi-period congested network. We find that, in the dynamic context, the congestion cost imposed by each user on other users (hence the optimal toll) consists of a capacity cost plus a cost component associated with the first–in–first–out (FIFO) property of road traffic. From the optimality conditions, we derive expressions for both of these components. Setting tolls equal to these congestion costs ensures that a system optimal solution will also be a user equilibrium: that is, ensures that costs experienced on all utilized paths from any point to the destination will be equal.

To indicate intuitively why the "FIFO cost" component arises, consider two (or more) types of users entering a link (or queue) at the same time. Suppose that for one type of user each additional minute of delay on the link is more costly than for the second type of user. If the users are otherwise the same, then they delay each other to the same extent. But the second (lower cost user) imposes a higher 'externality', since he delays a higher cost user: the externality is the cost one user imposes on others. A system optimizer could charge a toll to the second type of user to reflect this difference in externality. Or the system optimizer could impose a toll on one user and a subsidy to the other user. Even if these tolls or subsidies are not imposed in practice, it would still be useful to compute the externalities, since such externalities should be considered when computing the costs and benefits of providing alternative routes or facilities for each type of traffic, estimating the benefits of alternative traffic systems, flexitime, etc.

The above illustration is based on two types of users having different costs per minute delayed on a link. There are several reasons for such cost difference. It may be that the traffic types have different money values of time, or different cost or benefit functions associated with departure times at their origin or arrival times at their destination, or they may simply have have different target times for their departure from their origin or arrival at their destination. Or it may be that on exiting from the present link, the traffic types plan to follow different paths which have quite different congestion patterns over time. For example, one type of traffic may use a path on which congestion is fairly constant over time, whereas the other type may use a path (to a different destination) on which congestion is rapidly increasing over time. For the latter users, each additional minute spent on the current link is more costly, as the delay means that the rest of their journey will be undertaken at a more congested time.

Of the above reasons for different traffic types having different costs per minute delay on a link, the only one that also arises in static (single period) traffic network assignment models is the first one, namely traffic types having different money values of time. It can be shown that even in static assignment models users having different values of time gives rise to each type of user causing different externalities, and hence having different optimal tolls.

Another way of thinking about the FIFO component of marginal costs, externalities and optimal tolls is as follows. Such marginal costs (and hence tolls) are obtained by writing out the first order optimality conditions for a system optimizing traffic assignment model. If any additional constraints (e.g., FIFO constraints) are introduced into a continuous optimization model, then we get additional terms (Lagrange multipliers) in the resulting optimality conditions.

Since there is a FIFO constraint introduced for every link and every user type, we therefore get an additional marginal cost term associated with each user type traversing each link. And, as usual in optimization models, these Lagrange multipliers can be interpreted as the marginal costs of the corresponding constraints. Also, since these terms are not among the cost terms perceived by users, they would have to be imposed as tolls if user perceived costs are to equal system marginal costs.

2. A simple network example

To illustrate the effects which introducing FIFO constraints has on marginal costs, externalities, and hence optimal tolls, we here set out a simple multiperiod network example, having only two types of traffic, two destinations, and three links. (At least two periods and two traffic types are needed to illustrate FIFO.) We derive the congestion toll, including the FIFO component, which will be needed to shift users to a system optimal pattern of flows. This example is a special case of the more general model in Section 3 below, hence a reader may wish to go directly to the latter.

Let the two traffic types, A and B, represent users bound for destinations A and B respectively. The network is set out in Figure 1 and consists of four directed links, $j = 1, ..., 4$, with one origin and two destinations A, B. Time is divided into periods $t = 0, 1, 2$, each of length ℓ. For simplicity we assume that travel demands, capacities, etc. are such that all traffic can get to its destination by the end of period 2.

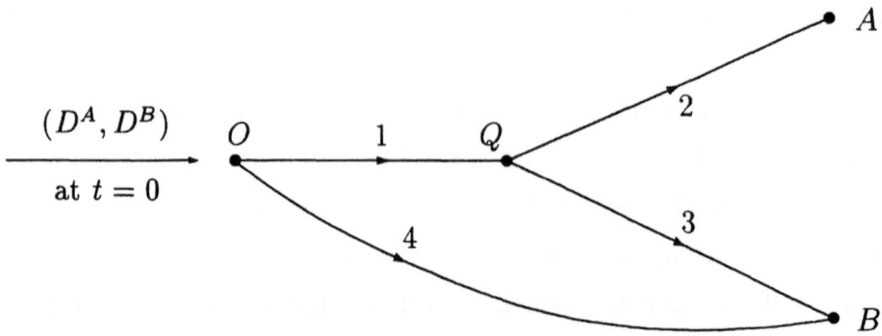

Figure 1. Simple network example.

Let the traffic inflows (travel demands) be D^A and D^B at the origin O in period 0. All traffic on link 1 arrives at queuing mode Q in period 1. From there, traffic exits in periods 1 and 2 for destinations A and B. Traffic for destination B can also use link 4, setting out in period 0. Let x_{tj} denote the inflow to link j in period t, hence the complete set of link inflows are, $(x_{01}^A, x_{12}^A, x_{22}^A)$, $(x_{01}^B, x_{13}^B, x_{23}^B)$, and x_{04}^B. Let the volumes in queue Q in periods 1 and 2 be (q_1^A, q_1^B) and (q_2^A, q_2^B) respectively. Let c_{tj} be the link trip time (or cost) incurred by each user embarking on link j in period t, and let ℓ be the cost to a user of spending one period in the queue.

We wish to find the "system optimal" tolls to impose on users on each of the possible paths from the origin to the destination. These are tolls just equal to the costs (externalities) which each user's choices imposes on other users. (Imposing such tolls makes users pay the true system cost of their decisions, so that their resulting "user optimizing" choices will also be "system optimizing" choices.) To find these externalities/tolls we proceed as follows.

2.1. System marginal costs

A system optimizing (SO) formulation for the above example is set out in an Appendix. This minimizes total travel time/cost subject to meeting travel demands, link capacities and FIFO constraints. To characterize a solution of this SO problem we write out the Kuhn–Tucker optimality conditions. By combining and rearranging various of these conditions, and recalling $(D^A - x_{12}^A) = x_{22}^A$, we obtain the following system marginal costs for traffic using paths 1, 2 or 3 (defined below) from the origin to destination B.

$$[c_{01}^B + c_{13}^B] + \alpha_1 + (\theta_2 - \theta_1)x_{22}^A \geq \lambda_0^B, \quad \text{for path 1} \tag{1}$$

$$[c_{01}^B + c_{23}^B + \ell] + \alpha_2 - (\theta_2 - \theta_1)x_{12}^A \geq \lambda_0^B, \quad \text{for path 2} \tag{2}$$

$$[c_{04}^B] \geq \lambda_0^B, \quad \text{for path 3.} \tag{3}$$

If a path (1, 2 or 3) is actually used by traffic, the "$\geq$" becomes a strict "=" in the above three equations. For traffic for destination A, using paths 5 and 6 respectively, the system marginal costs are,

$$[c_{12}^A] + \alpha_1 + \theta_1 x_{01}^B \geq \lambda_0^A, \quad \text{for path 5} \tag{4}$$

$$[c_{22}^A + \ell] + \alpha_2 + \theta_2 x_{01}^B \geq \lambda_0^A, \quad \text{for path 6.} \tag{5}$$

The variables θ_1 and θ_2 are the Lagrange multipliers associated with the FIFO constraints, and α_1 and α_2 are the Lagrange multipliers associated with the outflow capacity of queue Q in periods 1 and 2.

For traffic for destination B there are three possible time–space paths.

Path 1: exit Q into link 3 in period 1, hence avoid queuing.
Path 2: wait at Q in period 1, and exit into link 3 in period 2.
Path 3: use link 4.

For traffic for destination A there are two possible time-space paths:

Path 5: exit Q into link 2 in period 1, hence avoiding queuing.
Path 6: wait at Q in period 1, and exit into link 2 in period 2.

2.2. FIFO marginal costs and tolls

We can now interpret the system marginal cost expressions (1)–(5), to obtain expressions for user cost externalities and optimal tolls. In each of the path marginal cost equations (1)–(5), only the expressions in square brackets are perceived as costs by users. Hence, if users are to perceive and take account of the full system marginal cost of using a particular time-space path, then the additional marginal cost terms in (1)–(5) have to be imposed on users as congestion tolls or prices.

The α_1 and α_2 marginal cost terms are due to the capacity constraints, hence should be imposed on all users who exit from node (queue) Q in periods 1 and 2 respectively. (These α_1 and α_2 terms would be present even if there were no FIFO constraints.)

The remaining marginal cost terms in (1)–(2) and (4)–(5) are due to the FIFO constraints. These "FIFO marginal costs" are due to users exiting from node Q for destinations B and A respectively in periods 1 and 2 respectively. Note that the "FIFO marginal costs" for users using the queue Q to get destination B (i.e., $(\theta_2 - \theta_1)x_{22}^A$ and $-(\theta_2 - \theta_1)x_{12}^A$) are larger the larger the number of users using node Q to get to the *other* destination (i.e., the larger are x_{22}^A and x_{12}^A). Similarly, the "FIFO marginal costs" for users using node Q to get to destination A (i.e., $\theta_1 x_{01}^B$ and $\theta_2 x_{01}^B$) are larger the larger are the numbers (x_{01}^B) using Q to get to the other destination (i.e., destination B). This is intuitively reasonable, since FIFO constraints imply that each type of user using a facility (a queue or link) obstructs or delays other types of users using the facility. The FIFO cost imposed by a user therefore depends on the number of other users who are affected.

3. Congestion costs and tolls for a multiperiod network model

In this section we extend the discussion to a more general network. We take the single traffic type, single destination model in [5], extend this (Section 3.1) to multiple traffic types or destinations, and introduce FIFO constraints. In Section 3.2 we set out optimality conditions needed in Section 3.3 to obtain system marginal costs and optimal tolls. Properties of these tolls are derived in Section 3.4.

3.1. The model

This is a multi period network model in which travel times are assumed constant on each link, and congestion occurs in the form of deterministic queues at traffic signals and controlled intersections.

Consider a network consisting of a set of nodes $A = \{\ldots, j, \ldots\}$ connected by directed links (j,k), $j \in A$, $k \in A$. We will sometimes refer to this as the "spatial" network to distinguish it from the "time-space," or "time-expanded", network below. Let each node in A represent a queue at a road junction, turning point, etc. For convenience below we replace each queuing node j with two nodes j and j' connected by a new "artificial" link (j,j').

Let the time span to be modelled be subdivided into $t = 1, \ldots, T$, time periods. Let C denote a set of destinations or traffic types: the traffic bound for each destination is treated as a separate type of traffic. We also introduce the following notation. In defining and using this notation we do not always explicitly distinguish between the spatial network and the time-expanded network, since the intention should be clear from the context.

Variables.

$x^c_{tj'k}$ = the inflow in period t to spatial link (j',k) (or equivalently, to the time–space link $((t,j'),(t+\tau,k))$).

x^c_{tjj} = the volume waiting in queue j in period t (or equivalently, in the time–space network, the "flow" from node (t,j) to node $(t+1,j)$).

$x^c_{tjj'}$ = the outflow from queue j in period t (or equivalently, the inflow/outflow of time–space link $((t,j),(t,j'))$ in period t).

Constants.

D^c_{tj} = exogenous inflow or outflow at node j in period t. This is ≥ 0 if j is an origin node, ≤ 0 if it is a destination node and $= 0$ if it is neither.

$\tau_{tj'k}$ = number of time periods taken to travel from node j' to neighbouring node k, arriving at node k in period t. Hence this flow sets out from node j' at time $\eta(tj'k) = (t - \tau_{tj'k})$.

c_{tj}, $c_{tj'k}$ = cost incurred per unit of x^c_{tjj} and $x^c_{tj'k}$ respectively. (If costs are proportional to elapsed time then $c_{tj} = k\ell$ and $c_{tj'k} = k\ell\tau_{tj'k}$ where ℓ is the length of each time period and k is the value of time.)

$\bar{x}_{tj}$ = outflow capacity for node j in period t. That is $x^c_{tjj'} \leq \bar{x}_{tj}$.

Let $A(j')$ be the set of nodes which are immediate successors of node j' (connected to j' by a link pointing out of j'), and let $B(j)$ be the set of nodes which are immediate predecessors of node j. Let T_j be the set of periods for which we consider inflows/outflows at node j. The simplest definition is $T_j = \{1, \ldots, T\}$ for all nodes $j \in A$.

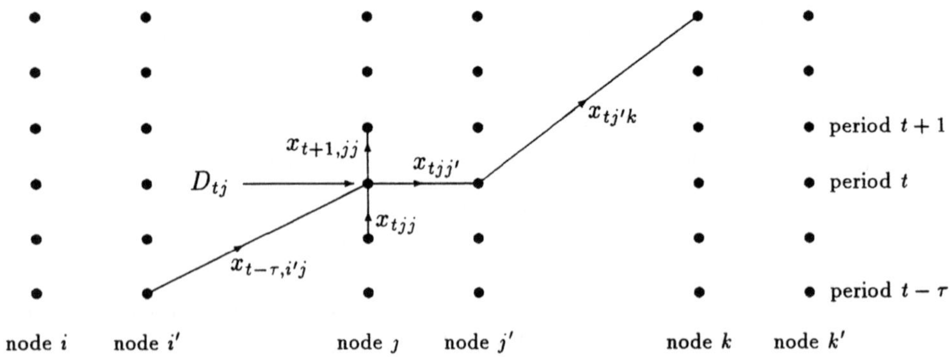

Figure 2. Time-expanded network, showing "travel links" $((t-\tau, i'), (t, j))$, "queuing links" $((t, j), (t+1, j))$, and "queue outflow links" $((t, j), (t, j'))$.

Flow conservation constraints.

For each traffic type c at time t, the *inflow* to queue j plus the net change in volume in the queue must equal the *outflow* $x^c_{tjj'}$. The net change in the queue volume is $(x^c_{t-1,jj} - x^c_{tjj})$. The inflow consists (see Figure 2) of the exogenous demand D^c_{tj} plus the sum of the inflows which set out from predecessor nodes

$B(j)$ in earlier periods. The latter is $\sum_{i\in B(j)} x^c_{\eta(tij)ij}$, where $\eta(tij) = (t - \tau_{tij})$, since the flow which arrives at queue j at time t set out from node i at time $(t - \tau_{tij})$.

Thus,

$$x^c_{tjj'} = (x^c_{t-1,jj} - x^c_{tjj}) + \sum_{i\in B(j)} x^c_{\eta(tij)ij} + D^c_{tj} \qquad \text{for all } c \in C,\ t \in T_j,\ j \in A. \quad (P.1)$$

On the other hand, the *inflow* $x^c_{tjj'}$ to node j' at time t must equal *outflow* from node j' to successor nodes $A(j')$ in later periods. Thus,

$$x^c_{tjj'} = \sum_{k\in A(j')} x^c_{tj'k} \qquad \text{for all } c \in C,\ t \in T_j,\ j \in A. \qquad (P.2)$$

(Note that the left-hand-sides of (P.1) and (P.2) are the same, hence we could eliminate $x^c_{tjj'}$ and combine (P.1) and (P.2) into a single conservation equation, for each c, t and j. However, we retain $x^c_{tjj'}$ as an explicit variable, for convenience and as it helps retain the multicommodity network form of the model and simplifies equation (P.4) below.)

FIFO constraints.

The FIFO requirement can be enforced, at least approximately, as follows. Require that, for each time period, the proportion of each traffic type in the outflow from each queue is the same as the proportion of that traffic type present in the queue (see [3]). Though this is an approximation we will use it here.

[FIFO could modelled more accurately by labelling all traffic types in each queue according to when they entered that queue, and then letting them exit only in that sequence. This can be done by introducing zero-one integer 0-1 variables to model logical either–or constraints (e.g., see [6] Ch. 1, or [8]). However, this is a sequencing problem which leads to a large scale 0–1 integer programming problem. Such a programme is hard to solve and, as it is discontinuous, there are difficulties interpreting its optimality conditions.]

$$\frac{x^c_{tjj}}{\sum_{c\in C} x^c_{tjj}} = \frac{x^c_{tjj'}}{\sum_{c\in C} x^c_{tjj'}} \qquad \text{for all } c \in C,\ t \in T_j,\ j \in A. \qquad (P.3)$$

Flow control constraints.

Let the maximum permissible outflow from node j in period t be $\bar{x}_{tj}$, hence

$$\sum_{c\in C} x^c_{tjj'} \leq \bar{x}_{tj} \qquad \text{for all } j \in A, t \in T_j. \qquad (P.4)$$

Also, there may be upper limits on the flows on some travel links or waiting links, thus $\sum_{c\in C} x^c_{tjk} \leq \bar{x}_{tjk}$ and/or $\sum_{c\in C} x^c_{tjj} \leq \bar{x}_{tjj}$.

Objective function and model formulation.

The costs incurred by the flows x^c_{tjj} and x^c_{tjk} are $c_{tj}x^c_{tjj}$ and $c_{tjk}x^c_{tjk}$. Thus the problem of finding the set of link flows which minimize the sum of the travel costs plus queuing costs over all links and time periods can now be stated as,

$$P: \text{minimize } \; Z = \sum_{t\in T_j} \left[\sum_{j\in A} c_{tj} \sum_{c\in C} x^c_{tjj} + \sum_{j'\in A} \sum_{k\in A(j')} c_{tj'k} \sum_{c\in C} x^c_{tj'k} \right] \qquad (P.0)$$

subject to, for all $c \in C$, $j \in A$ and $t \in T_j$,

$$\{\lambda^c_{tj}\}, \qquad x^c_{tjj'} = (x^c_{t-1,jj} - x^c_{tjj}) + D^c_{tj} + \textstyle\sum_{i'\in B(j)} x^c_{\eta(ti'j)i'j} \qquad (P.1)$$

$$\{\mu^c_{tj}\}, \qquad x^c_{tjj'} = \textstyle\sum_{k\in A(j')} x^c_{tj'k} \qquad (P.2)$$

$$\{\theta^c_{tj}\}, \qquad x^c_{tjj} / \textstyle\sum_{c\in C} x^c_{tjj} = x^c_{tjj'} / \textstyle\sum_{c\in C} x^c_{tjj'} \qquad (P.3)$$

$$\{\alpha_{tj} \geq 0\}, \qquad \textstyle\sum_{c\in C} x^c_{tjj'} \leq \bar{x}_{tj} \qquad (P.4)$$

$$(x^c_{tj'k}, x^c_{tjj'}, x^c_{tjj}) \geq 0 \qquad \text{for all } k \in A(j') \qquad (P.5)$$

The λ^c_{tj}'s, μ^c_{tj}'s and α_{tj}'s are Lagrange multipliers or dual variables associated with the corresponding constraints: they will be used in the next section below.

Excluding the FIFO constraints, the above programme is linear, and is a multicommodity least-cost capacitated network programme. The latter (excluding FIFO) can be solved as an LP or by using one of the various available efficient network computer codes (e.g., [1], [7]).

3.2. Optimality conditions

The Kuhn–Tucker conditions for Programme P are needed to obtain marginal costs and optimal tolls in the next section. If we ignore the FIFO constraints then the Kuhn–Tucker conditions for Programme P can be stated as:

constraints (P.1)–(P.5) together with, for all $t \in T_j$ and $j \in A$,

$$(x^c_{tjj'} \geq 0), \qquad \lambda^c_{tj} - \mu^c_{tj} \leq \alpha_{tj} \tag{K.1}$$

$$(x^c_{tjj} \geq 0), \qquad \lambda^c_{tj} - \lambda^c_{t+1,j} \leq c_{tj} \tag{K.2}$$

$$(x^c_{tj'k} \geq 0), \qquad \mu^c_{tj} - \lambda^c_{(t+\tau_{tj'k})k} \leq c_{tj'k} \qquad \text{for all } k \in A(j') \tag{K.3}$$

$$(\alpha_{tj} \geq 0), \qquad \textstyle\sum_{c \in C} x^c_{tjj'} \leq \bar{x}_{tj} \tag{K.4}$$

and, "complementary slackness" in (K.1)–(K.4) and (P.4).

If we include the FIFO constraints in Programme P, then the K–T conditions are the same as above, except that:

(a) the α_{tj} in (K.1) becomes $(\alpha_{tj} + \tilde{\theta}^c_{tjj'})$, where

$$\tilde{\theta}^c_{tjj'} = (\theta^c_{tj}/x_{tjj'}) - \sum_{c \in C} (\theta^c_{tj} x^c_{tjj'})/(x_{tjj'})^2 , \tag{K.5}$$

(b) the c_{tj} in (K.2) becomes $(c_{tj} + \tilde{\theta}^c_{tjj})$, where

$$\tilde{\theta}^c_{tjj} = -(\theta^c_{tj}/x_{tjj}) + \sum_{c \in C} (\theta^c_{tj} x^c_{tjj})/(x_{tjj})^2 . \tag{K.6}$$

where $x_{tjj} = \sum_{c \in C} x^c_{tjj}$ and $x_{tjj'} = \sum_{c \in C} x^c_{tjj'}$.

We are assuming that the K–T conditions necessarily hold at an optimum. To guarantee this the constraints in problem P should satisfy a 'constraint qualification' (see any mathematical programming text, e.g., [2]). However, we do not discuss this issue further here.

3.3. System marginal costs and tolls

The Lagrange multipliers λ^c_{tj} associated with (P.1) can be interpreted, in the usual way, as the rate of change of the system costs (P.0) with respect to change in the travel demand D^c_{tj} (the constant term in (P.1)). Hence λ^c_{tj} can be interpreted as the system marginal cost for any "utilized" path from queue j at time t to the destination. (Utilized paths are paths on which there is nonzero flow on all links on the path).

An expressions for the sum of system marginal costs along any path can be obtained by performing recursive substitutions in (K.1)–(K.6) over all links along the path. For comparison, we first obtain this expression for system marginal costs ignoring (dropping) the FIFO constraints (P.3): this reduces the K–T conditions to (K.1)–(K.4). Let $P_{\tau i}$ denote the set of paths $p_{\tau i}$ from queue (node) i

at time τ to the destination. Then from recursive substitution along any path $p_{\tau i} \in P_{\tau i}$, the system marginal cost of using the path is:

$$\left[\sum_{tjj \in p_{\tau i}} c_{tj} + \sum_{tj'k \in p_{\tau i}} c_{tj'k}\right] + \sum_{tjj' \in p_{\tau i}} \alpha_{tj} \geq \lambda^c_{\tau i} \tag{6}$$

where, in simplified notation, $(tjj \in p_{\tau i})$, $(tj'k \in p_{\tau i})$ and $(tjj' \in p_{\tau i})$ denote the three types of links in the traffic network: that is, queuing links, queue outflow links, and travel links respectively. Equation (6) simply states that the path marginal cost is the sum of the cost on all links on the path. The first two cost terms (in square brackets) are those experienced or perceived by users, and the third is a charge for using the limited queue outflow capacities.

If $p_{\tau i}$ is a utilized path, the "$\leq$" in (6) becomes "$=$".

If we now <u>include</u> the FIFO constraints (P.3) in the SO model then the system marginal costs are obtained from (K.1)–(K.6). These are the same as in (6) above, except that the α_{tj} in (6) is replaced by $(\alpha_{tj} + \tilde{\theta}^c_{tjj'})$ and c_{tj} is replaced by $(c_{tj} + \tilde{\theta}^c_{tjj})$. That is, the marginal costs for paths $p_{\tau i} \in P_{\tau i}$ are:

$$\sum_{tjj \in p_{\tau i}} (c_{tj} + \tilde{\theta}^c_{tjj}) + \sum_{tj'k \in p_{\tau i}} c_{tj'k} + \sum_{tjj' \in p_{\tau i}} (\alpha_{tj} + \tilde{\theta}^c_{tjj'}) \geq \lambda^c_{\tau i} \tag{7}$$

Of the five terms in the above marginal cost equation (7), only the c_{tj} and $c_{tj'k}$ terms are perceived as costs by the individual road user. Hence, if users are to perceive and take account of the full system marginal cost of using any time-space path, then the three additional marginal cost terms in (7) have to be imposed on users as congestion tolls or prices.

The α_{tj} marginal cost terms are due to the capacity constraints (P.4), which restrict the exit rate from queue j in period t. Hence these costs should be imposed as tolls on all users who exit from node (queue) j in period t. (This is similar to the well known optimal toll for capacitated links which applies in the case of static (or single period) traffic assignment.) The α_{tj} terms (and tolls) would be present even if there were no FIFO constraints.

The final marginal cost terms in (7) are due to the FIFO constraints (P.3). These "FIFO marginal costs" are $\tilde{\theta}^c_{tjj}$ and $\tilde{\theta}^c_{tjj'}$. The first of these (i.e., $\tilde{\theta}^c_{tjj}$) should be charged to users in queue j in period t. The second (i.e., $\tilde{\theta}^c_{tjj'}$) should be charged to traffic *exiting* from queue j in period t.

3.4. Properties of the FIFO tolls

If the FIFO component of the toll rate for queue j in period t is $\tilde{\theta}^c_{tjj}$ then the FIFO component of total revenue collected from the x^c_{tjj} users in queue j in period t is $\sum_c \tilde{\theta}^c_{tjj} x^c_{tjj}$.

Proposition 1: For each queue j in each time period t, the sum of the FIFO toll revenues collected from all traffic types is equal to zero. That is,

$\sum_c \tilde{\theta}^c_{tjj} x^c_{tjj} = 0.$

This implies that, for each queue j in each period t, some of the optimal tolls $(\tilde{\theta}^c_{tjj})$ are positive and some are negative, or they are all zero.

Proof: Multiplying (K.6) by x^c_{tjj} and summing over all c gives $\sum_c \tilde{\theta}^c_{tjj} x^c_{tjj} = [-\sum_c \theta^c_{tjj} x^c_{tjj}/x_{tjj} + \sum_c \theta^c_{tjj} x^c_{tjj}] = 0$. Since the factors x^c_{tjj} in this zero sum are all nonnegative, it follows that some of the $\tilde{\theta}^c_{tjj}$ are negative and some positive, or all are zero. □

Corollary: If there are only two traffic types c in a queue then *per capita* the FIFO components of their externalities (and tolls) are inversely proportional to the numbers in each traffic type in the queue.

Proof: If there are only two traffic types 'a' and 'b' then from the proposition, $\tilde{\theta}^a_{tjj} x^a_{tjj} = -\tilde{\theta}^b_{tjj} x^b_{tjj}$, hence $\tilde{\theta}^a_{tjj}/\tilde{\theta}^b_{tjj} = -x^b_{tjj}/x^a_{tjj}$. □

To put the corollary another way, for each traffic type the *per capita* externality (and toll) is larger the larger the numbers of the *other* type of traffic. This is intuitively reasonable, since FIFO constraints imply that each type of user using a facility (a queue or link) obstructs or delays other types of users using the facility. Hence the FIFO cost imposed by a user depends on the number of other users who are affected.

Proposition 2: The sum of the FIFO toll revenues collected from all traffic types *exiting* from queue j in time period t is zero. That is,

$\sum_c \tilde{\theta}^c_{tjj'} x^c_{tjj'} = 0.$

This implies that some of the optimal tolls $(\tilde{\theta}^c_{tjj'})$ are positive and some are negative, or they are all zero.

Proof: From (K.5), as in proof of Proposition 1. □

Corollary: Same as corollary to Proposition 1, but for the FIFO components of tolls on traffic *exiting* from the queue.

Proof: As for corollary to Proposition 1. □

The remarks following the corollary to Proposition 1 also apply here.

These two propositions show that the FIFO components of the optimal tolls consist of pure transfer payments between traffic types. The FIFO components are self-financing and generate no net revenue. The above propositions can be explained as follows. FIFO tolls are imposed on some traffic types, to discourage these traffic types from joining the queue. Negative FIFO tolls (i.e., subsidies) are paid to some other traffic types, to encourage these traffic types to join the queue. These tolls or subsidies are paid only by (or to) those who do actually join the queue. It may be wondered, why charge a toll to discourage traffic from joining a queue if the traffic has already joined the queue? However, the knowledge that they would have to pay the toll discourages others from joining the queue.

As usual in SO models, it is assumed that, before they set out, travellers have full information regarding tolls, and all other costs, and the levels of congestion which they would experience on all paths available to them. It is important that the tolls be known in advance to road users, otherwise they can not take account of them in making their decisions about which time–space route to choose. Only if the tolls are known in advance will they have the desired effect of discouraging or encouraging traffic to join a queue, or exit from a queue, at the particular times indicated by the SO model.

Proposition 3: For *each* traffic type c, queue j, and period t, the FIFO toll revenue collected from users *in* the queue equals the subsidy paid to users *exiting from* the queue, and *vice versa*, so that the net revenue collected is zero. That is,

$$\tilde{\theta}^c_{tjj'} x^c_{tjj'} = -\tilde{\theta}^c_{tjj} x^c_{tjj}.$$

Proof: Multiply the right-hand-side (r.h.s.) of (K.5) by $-(x^c_{tjj'}/x^c_{tjj})$, use equation (P.3) to substitute $(x^c_{tjj}/\sum_{c\in C} x^c_{tjj})$ for $(x^c_{tjj'}/\sum_{c\in C} x^c_{tjj'})$ throughout, and simplify. This reduces the r.h.s. of (K.5) to the r.h.s. of (K.6), hence $\tilde{\theta}^c_{tjj'}(x^c_{tjj'}/x^c_{tjj}) = \tilde{\theta}^c_{tjj}$, and the proposition follows. □

Corollary 1 of Proposition 3: If the FIFO toll $(\tilde{\theta}^c_{tjj})$ imposed on traffic type c *in* queue j in period t is negative (positive) then the toll $(\tilde{\theta}^c_{tjj'})$ imposed on the same traffic type *exiting* from the same queue in the same period t is positive (negative).

Proof: Follows from the proposition since x^c_{tjj} and $x^c_{tjj'}$ are nonnegative.

Corollary 2 of Proposition 3: For each traffic type c, queue j and period t, the ratio ("queue toll" / "queue exit toll") is *inversely* proportional to the ratio (queue volume / queue exit flow). That is,

$$(\tilde{\theta}^c_{tjj}/\tilde{\theta}^c_{tjj'}) = (x_{tjj'}/x_{tjj}).$$

Note that this ratio is the same for all traffic types.

Proof: Use the FIFO constraints (P.3) to substitute for $(x^c_{tjj'}/x^c_{tjj})$ in the proposition. □

This corollary indicates that if the exit flow rate from a queue is small relative to the volume in the queue then the queue "exit toll" will be large, and *vice versa.*

An explanation of Corollary 1 is simple. If traffic type c exiting from a queue imposes a cost on other users, then to reflect this we could either impose a toll on each unit of traffic type c exiting from the queue, or give a negative toll (a subsidy) to each unit of traffic type c remaining in the queue, or both. The optimal solution involves both.

4. Concluding remarks

In this paper we are concerned with deriving marginal costs, externalities and optimal tolls for a model of dynamic traffic assignment which explicitly includes FIFO constraints. The latter constraints are needed to ensure that the traffic flows respect the first-in-first-out property of road traffic. We find that the resulting costs, externalities and tolls include 'FIFO components' which have interesting properties.

For example, the FIFO components of costs and optimal tolls are sometimes positive and sometimes negative. The FIFO toll revenue collected from some users is exactly equal to the reduction in toll revenues from other users. As a result, the *sum* of the FIFO components of toll revenues is zero, for each link in each period. However, this does *not* mean that there is no point in including the FIFO component in tolls. If the FIFO component of the toll is not charged to users then each user's perceived travel costs would not equal the system marginal costs, hence users would not adopt a system optimal pattern of flows.

There is scope for much additional work on the topic considered in this paper. Such work would include exploring how the particular results may be affected by using different forms of dynamic traffic assignment models than used here.

Acknowledgements. This is a revised version of the paper presented at the conference: it corrects and replaces that version. The author would like to thank participants in the conference for their comments on the version presented there.

References

[1] A.I. Ali and J.L. Kennington (1989). *MODFLO User's Guide.* Technical Report 89-OR-03, Department of Operations Research and Engineering, Southern Methodist University, Dallas, TX.

[2] M.S. Bazaraa, H.D. Sherali and C.M. Shetty (1993). *Nonlinear Programming: Theory and Algorithms.* Second edition. John Wiley and Sons, New York.

[3] M. Carey (1992). "Nonconvexity of the dynamic traffic assignment problem". *Transportation Research 26B(2)*, 127–133.

[4] M. Carey and A. Srinivasan (1993). "Externalities, average and marginal costs, and tolls on congested networks with time-varying flows". *Operations Research 41(1)*, 217-231.

[5] M. Carey and A. Srinivasan (1992). "Solving a class of network models for dynamic flow control". Forthcoming in *European Journal of Operational Research* in 1994.

[6] R.S. Garfinkel and G.L. Nemhauser (1972). *Integer Programming.* Wiley, New York.

[7] J.L. Kennington and R.V. Helgason (1980). *Algorithms for Network Programming.* John Wiley and Sons, New York.

[8] H.P. Williams (1990). *Model Building in Mathematical Programming.* John Wiley, Chichester.

Appendix: SO model and optimality conditions for example in Section 2.

We set out here a system optimizing (SO) formulation, and optimality conditions, for the simple traffic network example presented in Section 2. These optimality conditions are needed in Section 2. The network, notation and data are set out there, hence we do not restate them here. (For simplicity we can assume that travel demands, capacities, etc. are such that all traffic can get to its destination by the end of period 2, so that queues are empty at the end of period 2.)

The total cost of link trip time for traffic for destination A is $(c_{12}x_{12}^A + c_{22}x_{22}^A)$, and for traffic for destination B is $(c_{01}^B x_{01}^B + c_{13}x_{13}^B + c_{23}x_{23}^B + c_{04}x_{04}^B)$. The cost of time spent in the queue Q, from period 1 to period 2, is $(\ell q_1^A + \ell q_1^B)$. The cost minimizing (system optimizing) problem for this example can now be stated as:

$$\textbf{Minimize } C = (c_{12}x_{12}^A + c_{22}x_{22}^A) + \ell q_1^A + (c_{13}x_{13}^B + c_{23}x_{23}^B + c_{01}^B x_{01}^B + c_{04}x_{04}^B) + \ell q_1^B. \tag{A1}$$

Subject to:

Conservation equations for traffic for destination A,

$$(\lambda_0^A) \qquad x_{01}^A = D^A \tag{A2}$$

$$(\lambda_1^A) \qquad x_{01}^A - x_{12}^A = q_1^A \tag{A3}$$

$$(\lambda_2^A) \qquad q_1^A - x_{22}^A = q_2^A \tag{A4}$$

Conservation equations for traffic for destination B,

$$(\lambda_0^B) \qquad x_{01}^B + x_{04}^B = D^B \tag{A5}$$

$$(\lambda_1^B) \qquad x_{01}^B - x_{13}^B = q_1^B \tag{A6}$$

$$(\lambda_2^B) \qquad q_1^B - x_{23}^B = q_2^B \tag{A7}$$

Outflow capacity limit k for queue Q in each period,

$$(\alpha_1 \geq 0) \qquad x_{12}^A + x_{13}^B \leq k \tag{A8}$$

$$(\alpha_2 \geq 0) \qquad x_{22}^A + x_{23}^B \leq k. \tag{A9}$$

FIFO constraints. In this example the FIFO requirements can be stated as follows (see Carey (1989)). There are two types of traffic in the queue Q (i.e., traffic for destination A and traffic for destination B), and these join queue Q in only one period (period 0). Hence, for FIFO, the proportions of the two types of traffic in the inflows to Q should be the same as the proportions of these two types of traffic in the outflows from Q in each period. This can be stated more formally as follows.

$$\frac{x_{01}^A}{x_{01}^B} = \frac{x_{12}^A}{x_{13}^B} \quad \text{and} \quad \frac{x_{01}^A}{x_{01}^B} = \frac{x_{22}^A}{x_{23}^B}.$$

By assumption, $x_{01}^A = D^A$ is fixed, hence the FIFO constraints reduce to,

$$(\theta_1) \qquad x_{01}^B x_{12}^A = D^A x_{13}^B \qquad (A10)$$

$$(\theta_2) \qquad x_{01}^B x_{22}^A = D^A x_{23}^B. \qquad (A11)$$

Nonnegativity constraints.

$$(x_{12}^A, x_{22}^A, x_{01}^B, x_{13}^B, x_{23}^B, x_{04}^B) \geq 0 \qquad (A12)$$

The Lagrange multiplier associated with each of the constraints in the above optimization model is shown in brackets beside the constraint. These have the usual interpretation of Lagrange multipliers. In particular, λ_0^A and λ_0^B can be interpreted as the rate of change of the objective function (total system costs) with respect to a (sufficiently small) change in the origin-destination travel demands D^A and D^B respectively. Hence λ_0^A and λ_0^B can be interpreted as the system marginal cost of having an additional user on any "utilized" path from the origin to the destination. A "utilized" path is any path which is actually used in the optimal solution.

To characterize the optimal solution of the above problem (minimize (A1) subject to (A2)–(A12)) we use the Kuhn-Tucker conditions. In doing this we assume for simplicity that the data (the travel demands D^A and D^B, and the queue exit capacity k) are such that all traffic exits from node Q in periods 1 and 2, with no queue held over to period 3.

AN EQUILIBRIUM ASSIGNMENT MODEL BASED ON SIMULATED DELAYS

Andres Regueros

Faculty of Civil Engineering, Technion, Israel Institute of Technology, Haifa, Israel

Joseph N. Prashker

Professor of Civil Engineering, Technion, Israel Institute of Technology, Haifa, Israel

David Mahalel

Professor of Civil Engineering, Technion, Israel Institute of Technology, Haifa, Israel

Ron Aharoni

Professor of Mathematics, Technion, Israel Institute of Technology, Haifa, Israel

This paper presents an equilibrium assignment methodology that can obtain delay values from a simulation program. The proposed assignment procedure is compared with the Frank and Wolfe algorithm with fully specified volume-delay functions and with the method of successive averages (MSA) suggested by Sheffi to be used in stochastic assignment problems. Some results of the application of the proposed algorithm on a real network are given, and a proof of convergence for the simple case of a two-link network is provided.

Introduction

Efficient assignment algorithms exist to solve problems in which flow-delay relationships are defined as continuous and twice-differentiable functions. These relationships represent only a crude abstraction of the real dependence of travel time on actual traffic volumes, the physical conditions of the transportation network and the behavior of drivers. The use of these coarse, aggregate, link-performance functions can be justified for medium and long-range transportation studies in which the details of network representation, actually travel times and the final traffic flows need not be very accurate.

Greater accuracy is required for assignment results to be used as a support tool for traffic-management decisions. Recently, several investigations in the field of Intelligent Vehicle Highway Systems (IVHS) research have identified the impact of various in-vehicle navigation and route- guidance systems on drivers' behavior and network assignment (Halati 1992). Several works combine the assignment with network simulation in order to achieve more realistic results from the assignment model (Hall 1980, Hu 1992, Stephanedes 1990).

This paper presents a new methodology to solve for equilibrium assignment by obtaining travel times through a simulation model. The basic problem is that the flow-delay relationships are not continuous, differentiable functions. The output of the simulation model can be seen as a pointwise flow-delay relationship (a function that is defined only at discrete points). The algorithm presented here was developed by following the Frank and Wolfe method, and is based on a linear approximation of the real flow-delay function. At each iteration of the algorithm, a new flow-delay point was generated for each link in the network and a straight line calculated that passes through the flow-delay point of the previous iteration and the current one. The current straight line is assumed to represent the current flow-delay relationship. A succession of these linear approximations, combined with the standard steps of the Frank and Wolfe algorithm, constitutes the proposed assignment method.

The method described in this paper was tested on different networks, including that of the city of Jerusalem. The rate of convergence and the accuracy of the method were compared to other methods, with favorable results.

Assignment and Flow-Delay Functions

Underlying all considerations of traffic assignment, there is the assumption that the basis of all route choices may be represented by the travel cost. Such a relationship constitutes a primary input for capacity-restraint methods of assigning traffic to networks. The equilibrium assignment model requires that the flow-delay relationship satisfy some properties in order to converge to a solution (i.e., the functions should be monotone, non-decreasing, continuous and differentiable).

Empirical flow-delay models include only a limited number of variables, and therefore are not realistic enough for congested urban areas. In order to obtain more realistic assignments, the delay models involved must be improved. This is, however, only one side of the problem. The assignment of traffic by using improved and sophisticated delay models may be very difficult or almost impossible with the standard user equilibrium assignment methodology. In order to overcome the disadvantage of using coarse and approximate delay functions, some assignment models employ fine-scale simulation of the delays.

Assignment - Simulation Methods

A number of heuristic approaches exist to combine traffic assignment and detailed network simulation, the best known of such models being SATURN (Hall et al. 1980). This model was developed mainly to handle transportation planning problems. A second model was developed to assist in traffic control problems. The common characteristic of all these models is the existence of an iterative loop between an entire converged equilibrium assignment phase and a detailed network simulation phase. This iterative loop is repeated until a convergence criterion is satisfied.

SATURN Model

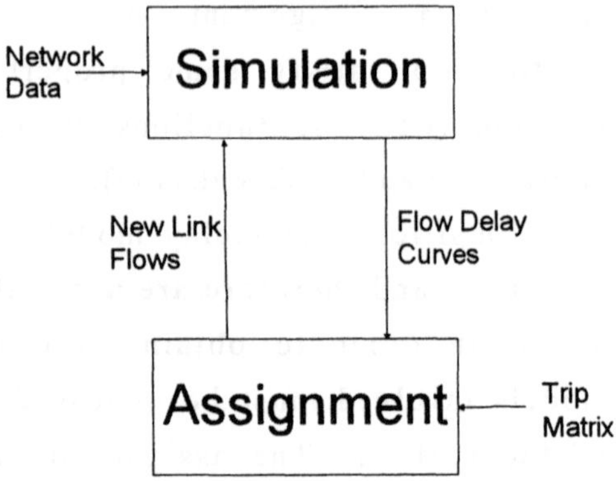

Figure 1: Simulation and Assignment Phases in SATURN

The SATURN model (Simulation and Assignment of Traffic to Urban Road Networks) consists of two sub-models (see Figure 1), combined to achieve realistic assignment results. The first sub-model is a TRANSYT-type simulation model that uses cyclic flow profiles to represent the platoons of vehicles on the network elements. Inputs to the simulation model are the physical and topological characteristics of the network, on the one hand, and vehicular volumes on each element of the network, on the other hand. The outputs are estimates of queues, delays and capacities of the network elements. Traffic-flow data comprise the necessary input for the simulation stage. The traffic flow volumes are obtained by a separate assignment sub-model. The two sub-models are used in an iterative manner as shown in Figure 1, a process that is repeated until "satisfactory" results are obtained. It must be noted that this process does not necessarily converge to a stable solution; therefore, it is usually repeated a predetermined number of iterations.

The objective of the simulation phase is to provide flow-delay relationships from a given pattern of traffic flows in the network. The method used in the SATURN program to generate these functions is to evaluate traffic delays at three different flow levels: zero flow; current flow (the results of the last assignment procedure); and capacity flows, with all other flows (i.e., opposing traffic) fixed. A polynomial flow-delay relationship is fitted on the basis of these three points.

This flow-delay equation is used during the current assignment phase, which is a fully converged equilibrium assignment. The reason for the iterations is that the delay curves used by the assignment phase are based on fixed values of opposing traffic; however, as each assignment gives different opposing flows, the simulation also gives different delay curves.

Minneapolis Traffic Control Model

Stephanedes et al. (1990) developed a Simulation-Assignment model based on an iterative process between an assignment and a simulation. The assignment phase distributes trips to a corridor network, and the simulation phase provides detailed information about the performance of the network, given its geometric and operational characteristics. As in the SATURN model, the loop terminates when the travel times of the links between two successive iterations reach reasonably stable values.

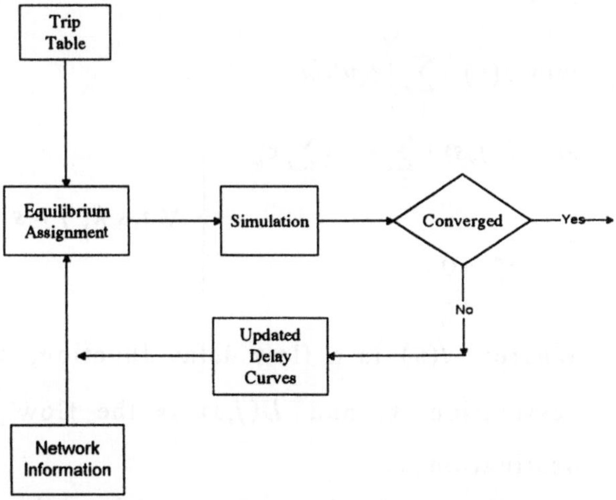

Figure 2: Simulation and Assignment Loop (Minneapolis Model)

The objective of the simulation phase is to provide detailed information about link travel times resulting from a given traffic flow pattern. This information includes various flow-delay points, which are used by a statistical estimation procedure to fit the link performance functions. These fitted link performance functions are then used by the assignment phase to distribute the traffic over the network. As in the SATURN model, the assignment phase of this model consists of a completely converged assignment procedure.

Current Equilibrium Assignment Practice

Beckman *et al.* (1956) formulated the user equilibrium (UE) problem as a convex, non-linear objective function and a set of linear constraints:

$$\min\ z(x)=\sum_{ij}\int_0^{x_{ij}} t(u)du$$

$$\left.\begin{aligned} st\text{: } & D(j,s)+\sum_i x_{ij}^s=\sum_k x_{jk}^s \\ & x_{ij}^s>0 \end{aligned}\right\}\ \forall\, j,s,k;\ j\neq s \qquad (1)$$

Where: $t(u)$ is a flow delay function, x_{ij}^s is the flow on link ij to destination s, and $D(j,s)$ is the flow originating at node j for destination s.

The method used to solve this problem is the convex combination algorithm, originally suggested by Frank and Wolfe in 1956 as a procedure for solving quadratic programming problems with linear constraints; it is also known as the FW method. LeBlanc (1973) applied the FW method to solve traffic assignment problems.

The non-linear problem of Equation 1 is linearized around a feasible flow vector, $\mathbf{x}^n$. The direction between the current feasible flow vector and the solution of the linearized equation ,y , provides a good direction in which to seek a reduction in the value of the original objective function:

$$\min_{y} \nabla z(\mathbf{x}^n)\mathbf{y} \equiv \min_{y} \sum_{ij} t_{ij} y_{ij} \tag{2}$$

Because of the special structure of this linearized problem, it can easily be solved by the shortest path algorithm.

The direction between $\mathbf{x}^n$ and $\mathbf{y}^n$ is used to minimize the original objective function in order to determine the optimal step size α:

$$\min_{0 \le \alpha \le 1} \sum_{ij} \int_{0}^{x_{ij}+\alpha(y_{ij}^n - x_{ij}^n)} t_{ij}(u)du \tag{3}$$

The line search for the optimal move size can be performed by any interval reduction method.

The equilibrium assignment problem can be summarized as follows:

1 **INITIALIZATION**

Perform an *All or Nothing* assignment based on $t_{ij}=t_{ij}(0)$. This yields flow vector $\mathbf{x}^1$. Set the iteration counter n to 1.

2 **UPDATE TRAVEL TIMES**

Update the link travel times $\left(t_{ij}^n = t_{ij}\left(x_{ij}^n\right) \quad \forall ij\right)$

3 **DIRECTION FINDING**

Perform an *All or Nothing Assignment* with t_{ij}^n. This yields the auxiliary flow vector y_{ij}^n.

4 **LINE SEARCH**

Find α that solves the linear program of Equation 3.

5 **MOVE**

Set: $x_{ij}^n = x_{ij}^n + \alpha\left(y_{ij}^n - x_{ij}^n\right)$

6 **CONVERGENCE TEST**

If the convergence criterion is met, stop; otherwise, go to step number 2.

Formulation of the Assignment Problem based on Simulated Delays

The objective of the present work was to develop an assignment method that can obtain the delays from a simulation program. The problem in this case is that step 4 of the FW algorithm cannot be performed, since the function $t(u)$ is not given analytically but by pointwise sampling. The simulation may be considered a black box capable of producing delay values for given values of flow as shown in Figure 3. The *SDM* functions are assumed to be deterministic (namely without noise); and hence the sampling obtained by *SDM* satisfies the conditions imposed on the functions $t(u)$ (i.e., they are monotone, non-decreasing and positive). The use of *SDM* functions instead of fully specified link-performance functions makes it impossible to evaluate the objective function of the user equilibrium problem. It is possible to approximate $\min z(x) = \sum_{ij} \int_0^{x_{ij}} t_{ij}(u)du$ by numerical integration when $t_{ij}(u)$ is evaluated as $SDM(u)$ at many points, but this is highly time consuming.

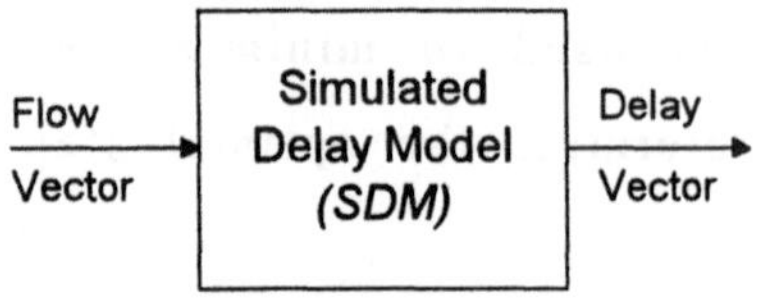

Figure 3: Simulated Flow Delay Model

In contrast to the SATURN and Minneapolis models, the method proposed here does not iterate between full equilibrium assignment and the simulation program. The proposed method utilizes the simulation program whenever travel times have to be evaluated; however, the whole algorithm consists of only one equilibration.

Method of Successive Averages - MSA

The method of successive averages (MSA) is used mainly for stochastic user equilibrium assignment problems. First suggested for use in traffic assignment problems by Sheffi (1985), this method is based on the theory of stochastic approximation (Robbins-Monroe 1954). Its basic assumption is that the minimum of a function can be found when an evaluation of this function involves some random noise. Robbins and Monroe proved that convergence can be accomplished with a predetermined series of step sizes.

The implementation of the MSA algorithm is similar to the implementation of the FW algorithm, except that step 4 is changed to include the predetermined step size sequence α_n, as follows:

$$\alpha_n = \frac{K_1}{K_n + n} \qquad (K_1 > 0,\ K_2 \geq 0) \tag{4}$$

Both the drawbacks and advantages of this algorithm can be attributed to the use of the predetermined step size sequence. The advantage is the simplicity of the algorithm and its insensitivity to noisy simulation results. The main disadvantage is that the algorithm does not use any information about the underlying objective function; thus the convergence is very slow, and appropriate convergence criteria are difficult to define (Powell 1982). Usually this algorithm is terminated after a predetermined number of iterations.

Linear Approximation Method (LAM)

The proposed method is based on a linear approximation of the real underlying flow-delay function. At each iteration of the Frank-Wolfe algorithm, a new flow-delay pair is generated for each network element ij the function $t_{ij}(u)$ is replaced by the secant (straight line), $s_{ij}^{n}(u)$ which goes through the two points $t_{ij}(x_{ij}^{n-1})$ and $t_{ij}(x_{ij}^{n-2})$ of the last two iterations. The FW procedure is then applied to the approximated function $\tilde{z}^{n} = \sum_{ij} \int_{0}^{x_{ij}} s_{ij}^{n}(u)du$; i.e., $\tilde{z}(x)$ is minimized along the straight line chosen as in the FW procedure.

The secant may be written as:

$$s_{ij}^{n} = \psi_{ij}^{n} + \phi_{ij}^{n} u \tag{5}$$

where: ψ_{ij} and ϕ_{ij} are straight line parameters.

If a linear flow-delay relation exists, then the Frank and Wolfe method can easily be implemented. The objective function of the current iteration will be as follows:

$$\min \tilde{z}^{(n)}(x) = \sum_{ij} \int_{0}^{x_{ij}} \left[\psi_{ij}^{(n)} + \phi_{ij}^{(n)} u\right] du \tag{6}$$

The main difficulty in applying the FW method in conjunction with *SDM* functions is that the integral of a pointwise flow-delay model has to be evaluated; however, this step is numerically almost impossible. The proposed method overcomes this inconvenience since the integral is evaluated for the approximated function. The direction to seek the next flow vector is chosen as in the FW procedure. The objective function is linearized around a feasible flow vector $\mathbf{x}^{\mathbf{n}}$ and then $\mathbf{y}$ is defined as the solution of the linear problem:

$$\min_{\mathbf{y}} \nabla\tilde{z}(\mathbf{x}^{n})\mathbf{y} \equiv \min_{\mathbf{y}} \sum_{ij} \tilde{t}_{ij} y_{ij} \tag{7}$$

Where $\tilde{t}$ is obtained by the *SDM* model.

The flow vector of the next iteration, x^{n+1}, is then sought on the interval between x^{n} and y^{n}. As in the case of fully specified functions, this program can be solved by means of the shortest path algorithm.

Moreover, when using the linear approximation method (LAM), the optimal step size can be explicitly calculated, thereby eliminating the need for a line-search procedure. This reduces the computational complexity of each iteration of the algorithm. The step size α can be calculated according to the following expression:

$$\min_{0 \leq \alpha \leq 1} \tilde{z}\left[\mathbf{x}^n + \alpha\left(\mathbf{y}^n - \mathbf{x}^n\right)\right] \qquad (8)$$

If $\mathbf{d}^n$ is defined as the direction between $\mathbf{x}^n$ and $\mathbf{y}^n$, this equation can be expressed as

$$\min_{\alpha} \tilde{z}(\mathbf{x} + \alpha\,\mathbf{d}) = \min_{\alpha} \sum_{ij} \left[\psi_{ij}\left(x_{ij} + \alpha d_{ij}\right) + \frac{\phi_{ij}}{2}\left(x_{ij} + \alpha d_{ij}\right)^2\right] \qquad (9)$$

The solution for α can be found analytically through classical optimization methods, without the need for a numerical optimum-seeking procedure. Elementary algebraic manipulations show that α can be obtained as follows:

$$\alpha' = -\frac{\sum_{ij}\left(\psi_{ij} d_{ij} + \phi_{ij} x_{ij} d_{ij}\right)}{\sum_{ij} \psi_{ij} d_{ij}^2} \qquad (10)$$

then

$$\alpha = \begin{cases} 0 & \alpha' \leq 0 \\ \alpha' & 0 \leq \alpha' \leq 1 \\ 1 & \alpha' \geq 1 \end{cases} \qquad (11)$$

The proposed algorithm can be summarized as follows:

1 INITIALIZATION

(a) Calculate an initial delay vector based on *SDM*.

(b) Perform an *All or Nothing* assignment. This yields the flow vector $\mathbf{x}^1$.

2 **Update Travel Times**

Obtain the delay vector with flow vector . $\mathbf{x}^n$

$$SDM(\mathbf{x}^n) = \mathbf{t}^n$$

3 **Linearization**

Calculate the approximated objective function, $\hat{z}(x)$, based on the last two flow vectors, $\mathbf{x}^{n-1}$ and $\mathbf{x}^n$.

4 **Direction Finding**

Perform an *All or Nothing* assignment based on $\mathbf{t}^n$. This yields the vector $\mathbf{y}^n$.

5 **Next Point**

(a) Calculate the step size according to Equation 11

(b) Set $\mathbf{x}^{n+1} = \mathbf{x}^n + \alpha(\mathbf{y}^n - \mathbf{x}^n)$.

(c) Increase iteration counter . n

6 **Convergence Test**

If the convergence criterion is met,stop;otherwise, go to step 2.

Proof of Convergence for a Two Link Network

The proof of convergence on the simple case of a network consisting of one Origin-Destination (OD) pair connected by two parallel links will be presented. Even for this simple case, the proof is not at all straightforward, and hence we choose to present it here in full.

The solution to the simple case of a single OD pair connected

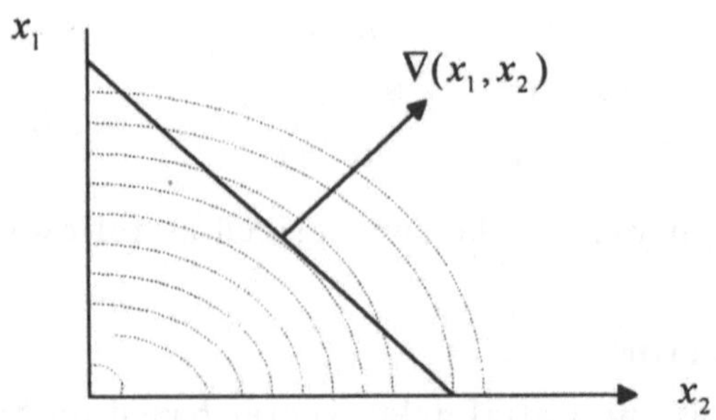

Figure 4. Objective Function and Feasible Region of a Two Link Network

by two parallel links can be represented as in Figure 4. The dotted curves represent the non-linear objective function, and the solid line represents the feasible region. The Frank and Wolfe algorithm would find the solution to the assignment problem, in this case, in only one iteration, by minimizing the non-linear function along the constraint line. The solution to the equilibrium assignment problem for this case is a point on the constraint line at which the gradient is perpendicular to the constraint. The direction of the constraint $x_1 + x_2 = K$ is the vector:

$$\mathbf{g} = \begin{bmatrix} -1 \\ 1 \end{bmatrix} \quad (12)$$

The objective function of the assignment problem is:

$$\begin{aligned} z(x) &= \min \int_0^{x_1} t_1(u)du + \int_0^{x_2} t_2(u)du \\ st{:}\ & x_1 + x_2 = K \end{aligned} \quad (13)$$

The gradient of the objective function is

$$f(x_1, x_2) = \begin{bmatrix} t_1(x_1) \\ t_2(x_2) \end{bmatrix} \quad (14)$$

If the minimum is obtained in the interior of the segment, then the gradient there will be perpendicular to the constraint; this means that the product of the constraint vector and the gradient vector is zero:

$$f(\mathbf{x}) \cdot g(\mathbf{x}) = 0) \quad (15)$$

or,

$$t_1(x_1)(-1) + t_2(x_2) = 0 \quad (16)$$

The solution to this simple example can be represented as in 4. Both delay curves are represented on a single axis. The flow on the second link is expressed as a function of the total demand and the flow on the first link, and the equilibrium flow vector can be determined as the intersection point between the two delay curves.

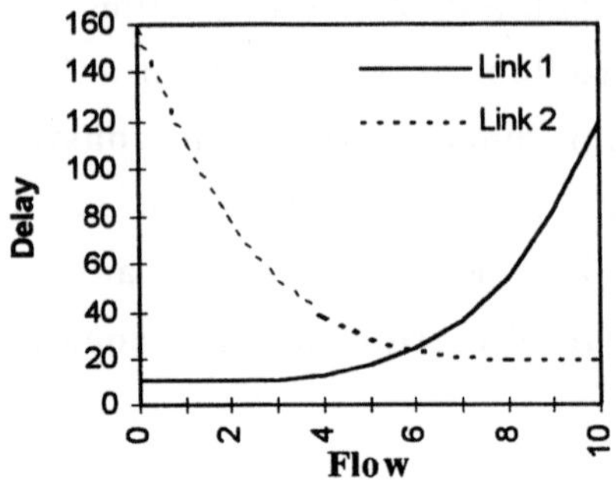

Figure 5: Flow Delay Curves

As mentioned earlier, the algorithm developed in the present work assumes that the underlying flow-delay curves are defined only at discrete points. Therefore in order to find the intersection of the two curves, the delay curves are approximated by straight lines that pass through the values of the delay curves from the last two iterations. In fact, as is easily observed, the algorithm in this case reduces to the solution of the equation:

$$h(u) \equiv t_1(u) - t_2(u) = 0 \tag{17}$$

by the so-called secant method.

Figure 6 shows the convergence characteristic of the secant method applied to find the equilibrium flow pattern for the simple one OD pair network.

As known, the Newton-Raphson's method, as well as the secant method, does not converge for all functions, however, the function of our problem is special: the difference between a convex, monotone, non-decreasing function and a convex, monotone, decreasing function.

We shall prove:

Theorem 1

If $h(x) = f(x) - g(x)$

Where: $f(x)$: convex monotone non-decreasing function

$g(x)$: convex monotone non-increasing function

then the Newton-Raphson method, as well as the secant method, converges to the zero, z, of $h(x)$ regardless of the choice of initial points. Our algorithm uses, in fact, the interval-restricted version of the secant method. Thus, we restrict the possible values of the variable to an interval $[a,b]$; if the iteration formula yields $x_n > b$, it is replaced by $x^n = b$; and if $x^n < a$, it is replaced by $x^n = a$.

Hence, the following version of the theorem will be needed:

Theorem 2

Let h be the same as in Theorem 1. Assume that $h(z) = 0$ for some $z \in [a,b]$, then both, the Newton-Raphson method, and the secant method when restricted to interval, converge to z.

The full proof of the theorems will appear elsewhere. Here we present only an outline of the proof of Theorem 1 for the Newton Raphson (NR) method.

Lemma 1:

Let v be a non-negative function. Then v is the sum of a non-negative, non-decreasing function, u, and a non-negative, non-increasing function, w, if and only if the following holds for all finite sequences $x_1 < x_2 < \ldots\ldots < x_{2n+1}$:

$$v(x_1) - v(x_2) + v(x_3) - \text{K} + v(x_{2n+1}) \geq 0 \qquad \textbf{(18)}$$

The "if" part of the lemma is hard and not necessary for the rest of the discussion; hence, only proof of the "only if" part will be given here.

Proof:

It is enough to prove Equation 18 for u and w. For u collect the terms in Equation 18 as:

$$u(x_1) + (-u(x_2) + u(x_3)) + (-u(x_4) + u(x_5)) + \text{K} + (-u(x_{2n}) + u(x_{2n+1})) \qquad \textbf{(19)}$$

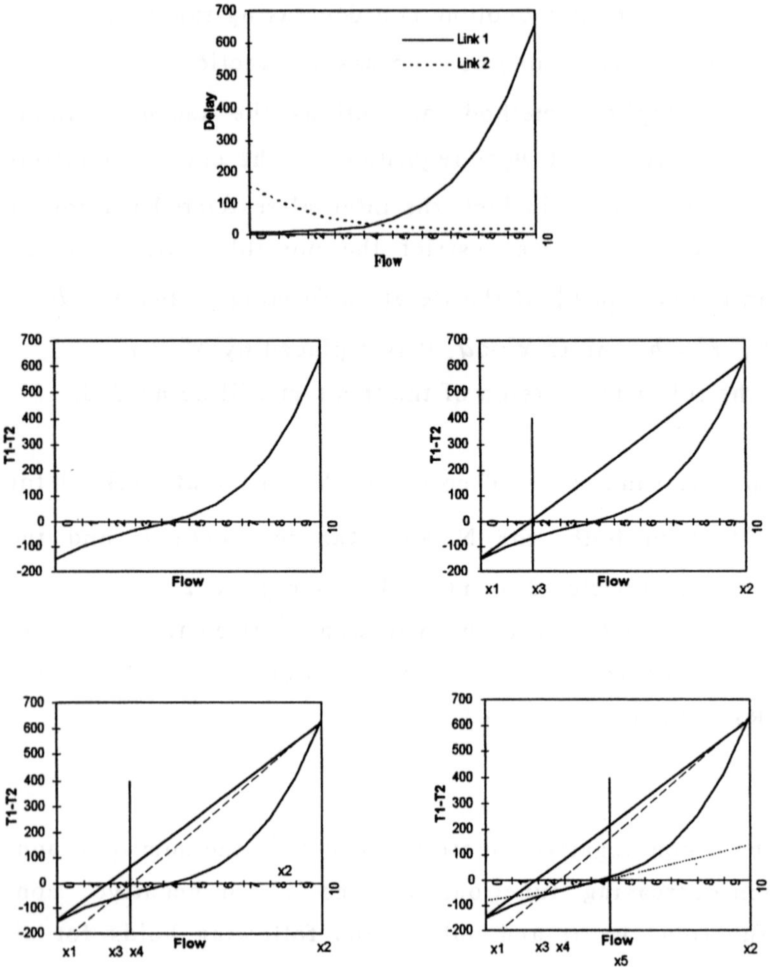

Figure 6. Convergence Characteristic of the Secant Method

By the assumption about u, all terms are non-negative. For w, collect the terms as:

$$(w(x_1)-w(x_2))+(w(x_3)-w(x_4))+\mathrm{K}+(w(x_{2n+1})-w(x_{2n}))+w(x_{2n+1}) \quad \textbf{(20)}$$

By the assumption about u, all terms are non-negative.

In the situation of Theorem 1; $u(x)=\dot{f}(x)$ and $w(x)=-\dot{g}(x)$ are non-negative, u is non-decreasing and w is non-increasing. Hence, as a special case of the lemma, the following is obtained when $x_1 < x_2 < x_3$:

Let x_n be the sequence of iterations in the NR method. For a given n, let r be the minimal value of an $x_i (i < n)$, such that $x^i > z$ if such i exists; otherwise let $r = z$. Let l be the maximum value of an x^i $(i < n)$, such that $x_i < z$ if such i exists; otherwise let $l = z$. Here is the main step in the proof of the theorem

Lemma 2: If , $l < z < r$ then $l \leq x^n \leq r$

Proof: By induction about n. Since $l < z < r$, we have $l = x^j$, $r = x^k$ for some $j,k < n$

Hence, $n > 2$.

By the induction hypothesis:

$$j = \max\{i: i < n, x^i \leq x\} \qquad k = \max\{i: i < n, x^i \geq z\} \qquad (22)$$

Hence, either $j = n-1$ or $k = n-1$ Without loss of generality, assume that $j = n-1$. Then, since h is non-decreasing, $x^n > x^{n-1}$. Assume for contradiction that $x^n > x^k$. Since $k \leq j$, then by the induction hypothesis Equation 22

$$x^{k+1} \leq x^j \qquad (23)$$

(see Figure).

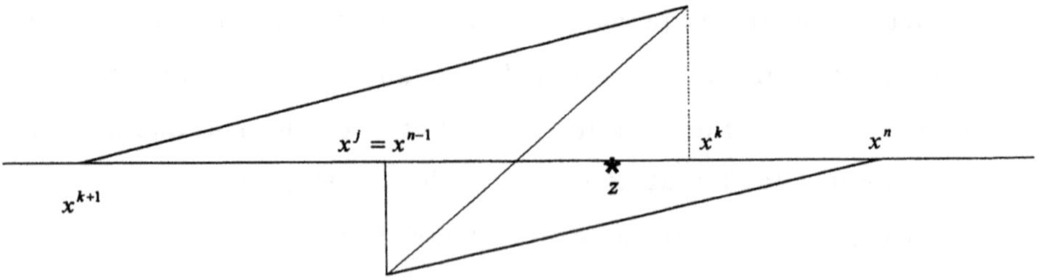

By definition of the NR method,

$$\frac{h(x^k)}{x^k - x^{k+1}} = \dot{h}(x^k)$$
$$\frac{-h(x^j)}{x^n - x^j} = \dot{h}(x^j) \qquad (24)$$

By Lagrange's theorem: a point $x^j < y < x^k$ exists such that

$$\dot{h}(y) = \frac{h(x^k) - h(x^j)}{x^k - x^j} \qquad (25)$$

By Equation 23, $x^k - x^j \le x^k - x^{k+1}$. By the negation assumption ($x^n > x^k$), this obtains: $x^k - x^j < x^n - x^j$. Hence, by Equations 24 and 25

$$\dot{h}(y) > \frac{h(x^k)}{x^k - x^{k+1}} + \frac{-h(x^j)}{x^j - x^{j+1}} = \dot{h}(x^k) + \dot{h}(x^j) \qquad (26)$$

This contradicts Lemma 2 when applied to the triple . $x^j < y < x^k$

The rest of the proof of Theorem 1 for the NR method is relatively easy. From Lemma 2, we know that the sequence $\{x^i \colon x^i > z\}$ is a descending one and hence convergent to a limit R. Similarly, the sequence $\{x^i \colon x^i < z\}$ converges to a limit L. By continuity arguments, it may easily be shown that in fact $R = L = z$, thus proving the theorem.

The proof of the convergence of the secant method is similar to the proof shown here and will be given elsewhere (Aharoni *et al.* 1992).

Examples and Results

The simulated delay values for the LAM method were computed from the BPR function. In principle, the delay values for the LAM method can be obtained from a conventional simulation program like TRANSYT. The present case, since we wished to compare the LAM method to the two other methods (FW, MSA), it was necessary to base the calculations on the same delay functions. If different models were used for each of the methods, there would be no basis for comparison. It should be remembered that different equilibrium flows are obtained for different delay functions with the same assignment algorithm.

In order to determine the ability of the algorithm to produce satisfactory results, the proposed method was tested on three different networks. It was applied by using the known BPR functions as *SDM* models, but on the assumption that the delay values are the result of a pointwise *SDM* model. Various assignments were performed by using different parameters of the delay curves in order to simulate different deterministic simulation programs. The proposed methodology was compared to Sheffi's method of successive averages (MSA). LeBlanc's implementation of the Frank-Wolfe procedure with the explicitly specified BPR functions was used as the basis of comparison to evaluate the accuracy of the two methods. Many iterations of the algorithm were performed to ensure convergence.

The LAM method was applied to the following networks:

- a link network presented by Sheffi (1985)
- the Sioux Falls network presented by LeBlanc (1973). This is a 24 node 76 link network.
- the network of the city of Jerusalem consisting of 639 nodes, 1621 links and 9774 OD pairs.

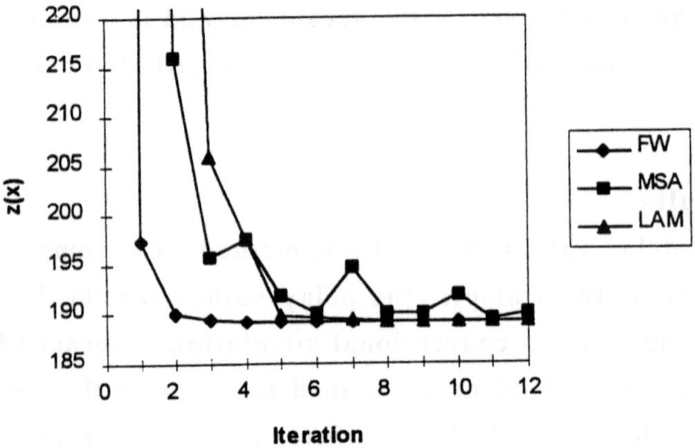

Figure 7. Convergence Pattern for the Three Link Network

The proposed method was initially applied to the very simple three-links network presented by Sheffi (1985). Figure 7 shows the convergence pattern of the three methods. It can be seen that the proposed method converges asymptotically to the exact solution. For this small example, the performance of the proposed methodology is better than that of the MSA method in two aspects. First, it steadily converges to the exact solution, and second, the number of iterations necessary to achieve acceptable results is significantly smaller.

The method was also applied to the classic Sioux Falls network, presented in the original work by LeBlanc (1973). This network consists of 24 nodes and 76 links. Table I presents assignment results for various combinations of the BPR function parameters (α, β).

$$t = t_0 \left[1 + \alpha \left(\frac{q}{C} \right)^{\beta} \right]$$

Where q represents the flow, and C the capacity of the link.

Table 1. Objective Function Values for the Sioux Falls Network.

Parameter		Method			Results	
α	β	FW	MSA	Linear	FW-MSA	FW-Linear
0.15	1	673465.25	673465.06	673465.42	0.00	0.00
0.15	2	658047.13	658092.72	658048.79	0.0069	0.0003
0.15	3	652008.50	652094.23	652008.11	0.0131	-0.0001
0.15	4	649040.33	649380.82	649056.72	0.0525	0.0025
0.15	5	647383.66	647816.21	647397.8	0.0668	0.0022
3.00	1	1251805.93	1262714.46	1251836.6	0.8714	0.0025
3.00	2	920306.65	938585.44	920602.65	1.9862	0.0322
3.00	3	797500.07	814918.08	800769.58	2.1841	0.4100
3.00	4	741256.24	756196.98	744504.26	2.0156	0.4382
3.00	5	711110.66	732911.87	713986.46	3.0658	0.4044
4.50	1	1052317.34	1081560.57	1054371.8	2.7789	0.1952
4.50	2	1052317.34	1081560.57	1054371.8	2.7789	0.1952
4.50	3	868625.09	895160.58	873920.02	3.0549	0.6096
4.50	4	784554.39	812435.01	790197.29	3.5537	0.7192
4.50	5	742024.08	769401.7	748640.45	3.6896	0.8917

As can be seen in Table I, no matter what function was used, the results of the proposed model were closer to the exact solution than those of the MSA method. Furthermore, the convergence characteristics of the proposed method do not deteriorate significantly with an increase in the sensitivity of the network elements to congestion.

The method was also applied to a real-size network, that of the city of Jerusalem. This network consisted of 639 nodes, 1621 links and 9774 OD pairs. As was done with the other cases, the results were compared to Frank-Wolfe and MSA algorithms. The result of this is presented in Figure 8. Here again, the proposed method performed better than did the MSA method (see Table II).

Each iteration of the proposed algorithm is shorter than that of the FW algorithm because of its simpler search algorithm (exact calculation of the step size). Thus the total running time of the LAM procedure was about 14% shorter than the FW and almost equal to

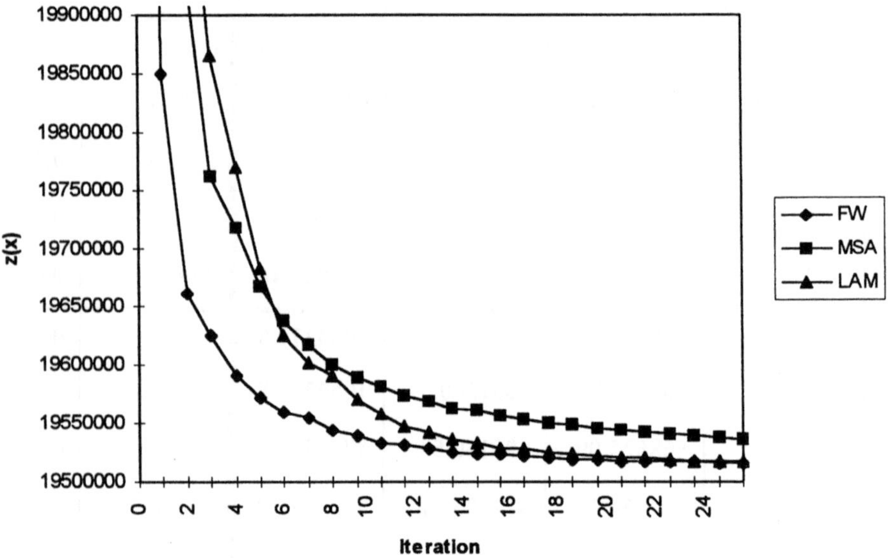

Figure 8. Results of the Jerusalem Network

the MSA running time.

Table II: Objective Function Values for the Jerusalem Network

Parameter		Method			Results	
α	β	FW	MSA	Linear	FW-MSA %	FW-Linear %
0.15	4	19514929	19534992	19516417	0.1028	0.0076
3.5	4	22356849	22406119	22335632	0.2204	-0.0949
4.5	2	27659564	27816213	27663233	0.5663	0.0133
4.5	4	22931286	23018710	23009239	0.3812	0.3399
0.15	2	19611948	19612851	19611994	0.0046	0.0002

Concluding Remarks

The present paper provides a full proof of convergence of the Linear Approximation Method (LAM) for a simple case of a two-link network. Although we do not have the proof for the general case, we have good reason to believe that our LAM method can be used for real size networks. This is supported by two facts:

1. The proof of convergence for the simple case indicates that the method has a strong mathematical basis and there is no evidence that contradicts the convergence in larger networks.
2. The performance of the method was validated on various networks and was found to perform satisfactory on all the networks tested. The networks used for this validation consisted of different topologies and sizes. For all cases, the results of the LAM were compared with the MSA algorithm and to the Frank and Wolfe algorithm as explained previously. For all the cases studied, the value of the objective function of the LAM method was closer to the "real" objective function than that obtained by the MSA algorithm. Moreover, the difference between the resulting flow

vector obtained by the LAM algorithm was closer to the Frank and Wolfe flow vector than that obtained by the MSA algorithm. This results validate the fact that the present method can be used for real size networks whenever the delay values are obtained from a deterministic SDM model.

Procedures which perform stochastic assignment are of great interest. The ability of the LAM method to perform stochastic assignment was not fully investigated. The main problem which may arise when the method is applied to solve stochastic assignment is that the slope of the secant, in some iterations, may be negative. This contradicts the properties necessary for the convergence of the Frank and Wolfe method. There are different ways to overcome this problem (use the slope of the previous iteration, or use a zero slope). Due to the complexity of stochastic assignment the convergence characteristics of the LAM method when using stochastic SDM functions need further investigation.

References

[1] R. AHARONI, A. REGUEROS, J.N PRASHKER and D. MAHALEL. *"Global Convergence of the One-Dimensional Newton-Raphson Method and the Secant Method. Working Paper"*, Technion, Israel Institute of Technology - Haifa, 1992.

[2] Traffic Assignment Manual, Bureau of Public Roads, Urban Planning Division, US Department of Commerce, Washington DC, 1964

[3] M. BECKMAN, C.B McGUIRE, and C.B WINSTEN. *"Studies in the Economics of Transportation"*. Yale University Press, 1956.

[4] M. FRANK and P. WOLFE, *"An Algorithm for Quadratic Programming. Naval Research Logistics Quarterly"*, Vol. 3, pages 95-110, 1956.

[5] A. HALATI and D.E. BOYCE. *"Simulation Evaluation of In-Vehicle Navigation Systems - Methodology and Application"*. Preprint, Paper No. 920568, Transportation Research Board, January 1992.

[6] M.D. HALL, D. VAN VLIET, and L.G. WILLUMSEN. SATURN - *"A Simulation-Assignment Model for the Evaluation of Traffic Management Schemes. Traffic Engineering & Control"*, pages 168--175, 1980.

[7] T. HU, R.W. ROTHERY and H.S. MAHMASSANI. DYNASMART: *"DYnamic Network Assignment-Simulation Model for Advanced Road Telematics"*. Technical Working Paper DTFH61-90-C-0074-TWP1. Center for Transportation Research, The University of Texas at Austin. January 1992.

[8] L.J. LEBLANC. *"Mathematical Programming Algorithms for Large Scale Network Equilibrium and Network Design Problems. Ph.D. thesis"*, Northwestern University, 1973.

[9] W.B. POWELL and Y. SHEFFI. *"The Convergence of Equilibrium Algorithms with Predetermined Step Sizes"*. Transportation Science, 16(1):45--55, 1982.

[10] Y. SHEFFI. *"Urban Transportation Networks: Equilibrium Analysis with Mathematical Programming Methods"*. Prentice-Hall, Inc., 1985.

[11] Y.J. STEPHANEDES, I.Z. ARGIROPOULOS, and P. MICHALOPOULOS. *"On-Line Traffic Assignment for Optimal Freeway Corridor Control. Journal of Transportation Engineering"*, 116(6):744--755, 1990.

[12] D. VAN VLIET. SATURN - *"A Modern Assignment Model. Traffic Engineering & Control"*, pages 578--581, 1982.

[13] D.J. WILDE. *"Optimal Seeking Methods"*. Prentice-Hall, Inc., 1964.

Springer-Verlag and the Environment

MIX
Papier aus verantwortungsvollen Quellen
Paper from responsible sources
FSC® C105338

If you have any concerns about our products,
you can contact us on
ProductSafety@springernature.com

In case Publisher is established outside the EU,
the EU authorized representative is:
Springer Nature Customer Service Center GmbH
Europaplatz 3, 69115 Heidelberg, Germany

Printed by Libri Plureos GmbH
in Hamburg, Germany